Microcomputer Control of Telescopes

Mark Trueblood
Russell Genet

Willmann-Bell, Inc.
P. O. Box 3125
Richmond, Virginia 23235 ☎ (804)
United States of America   320-7016

Serving Astronomers Worldwide

Since 1973

The following publishers have generously given permission to use illustrations from copyrighted works: From **Textbook on Spherical Astronomy,** edited by W.M. Smart and R.M. Green. Copyright 1977 by Cambridge University Press. Figures 17, 29, 71, 92, 93, 95, 96, and 125. Reprinted by permission of Cambridge University Press. From **Physics: Parts I and II,** by D. Halliday and R. Resnick. Copyright 1960, 1962, 1966 by John Wiley & Sons, Inc. Table 12-1. Reprinted by permission of John Wiley & Sons, Inc. From **Classical Dynamics of Particles and Systems** by J.B. Marion. Copyright 1965 by Academic Press, Inc. Figure 13-13(a). Reprinted by permission of Academic Press, Inc. From **Servomechanisms: Devices and Fundamentals,** by R.M. Miller. Copyright 1977 by Richard W. Miller. Figures 4.13, 8-2, 9-2, 9-4, 11-13, 12-5, 12-6, and 12-8. Reprinted by permission of the publisher, Reston Publishing Co., a Prentice-Hall Co.

Library of Congress Cataloging in Publication Data

Trueblood, Mark, 1948-
    Microcomputer control of telescopes.

    Bibliography: p.
    Includes index.
    1. Telescope--Automatic control. 2. Microcomputers
3. Astronomy--Data processing. I. Genet, Russell.
II. Title.
QB88.T78   1985   522'.2   85-631
ISBN 0-943396-05-0

Printed in the United States of America
    86 87 88 89 90 91 10 9 8 7 6 5 4 3 2

DEDICATIONS

I dedicate this book to my wife, Pat, who supported this effort with her love, and who helped prepare the manuscript; to my parents, who encouraged my interest in astronomy and telescope building at an early age; to my good friend Andrew J. Tomer, who rekindled my interest in astronomy and in the problems of building and computerizing telescopes; and to Irvin M. Winer, who, as a professor of Physics at Wesleyan University, took an unusually personal interest in his students, who was not afraid to pursue offbeat or unpopular areas of science if they held promise of leading to Truth, and who had a refreshingly honest approach to coming to terms with himself, those around him, and the Universe at large. It is after Irv, who died in middle age in early 1982, that I have decided to name my mobile telescope facility, which is now under construction.

In recognition of the large amount of material from my Masters thesis that serves as the foundation of this book, I acknowledge the help, encouragement, guidance, and good suggestions on the preparation of my thesis offered by my advisor, Prof. Michael F. A'Hearn, and by my examination committee, Prof. Andrew S. Wilson and Prof. David Zipoy, all from the Astronomy Program of the University of Maryland. Warm thanks are also given to Dr. Joan B. Dunham, who introduced me to the wonders of Kalman filtering and who critiqued the sections on least squares and Kalman filtering, Dr. David W. Dunham, who has cajoled me into going on enough lunar and minor planet observing expeditions that I have become interested in pursuing them further, and who provided information on the process of predicting minor planet occultations, George Kaplan, who suggested how to treat azimuth tilt and field rotation, and who supplied many of the equations in the new epoch J2000.0 form, and to my co-author, Russell M. Genet, who offered many fine suggestions on control theory and mechanical error sources during the preparation of the thesis. Thanks are due also to Farrand Controls, Inc. and Mesur-Matic, Inc. for permitting me to quote extensively from their sales literature when discussing the principles of operation of their products.

Mark Trueblood
Winer Mobile Observatory

I also dedicate this book to my wife, Ann, who is an active member of the Fairborn Observatory, and helped on this book in many ways; to the pioneers of microcomputer telescope control, particularly Louis Boyd, Frank Melsheimer, and Kent Honeycutt; and to the many other pioneers in this field whom I would have liked to have known. Those who do the pioneering are usually too busy to write books about things, yet it is they who solve the hard problems when it is not yet known if solutions exist.

It was a couple of years ago when I decided that a book on microcomputer control of telescopes was really needed. Perry Remaklus of Willmann-Bell, Inc. kindly agreed, after seeing an outline, to publish such a book. About six months later, with no appreciable work yet done on the book, I received a large package from Mark Trueblood. The note said "... contains a number of chapters on my thesis on microcomputer control of telescopes. Do you suppose it might be a suitable basis for a book on this topic?" I always wondered how Darwin felt when he received a package from Wallace. "Dear Darwin ... ." I immediately recognized that it was a goodly part of a book in the rough, and proposed to Mark that we coauthor this book. I am delighted that he accepted, as the final product is much better than I could have done myself, I learned a lot from Mark, and acquired a good friend in the process.

Two amateur astronomers have been particularly influential in my work in the area of microcomputer telescope control. One is Louis Boyd, who not only developed an automatic microcomputer-controlled telescope, but, more importantly, who shared his knowledge freely and with complete candor. Louis's many ideas in software and hardware form the basis for the Automatic Photoelectric Telescope (APT) described in this book, and it has been a particular pleasure to exchange ideas with him over the past four years. The other is Douglass Sauer, who worked directly with me on the control systems at the Fairborn Observatory, and who has discussed with me just about all of the conceivable options in telescope control.

Two professionals have also been quite influential with respect to my work in the area of microcomputer control of telescopes. One is Dr. Kent Honeycutt, of Indiana University, who's microcomputer control system at the Goethe-Link Observatory was among the very first in operation. His many helpful ideas in this area, as well as his suggestions on this book were most helpful. Dr. Frank Melsheimer, of DFM Engineering, has, over the past several years, contributed many ideas in terms of both telescope design (mechanical) and telescope control. He has been a gracious host on my many visits to Boulder, Colorado, and his donation of a prototype mount to the Fairborn Observatory made possible one of the Automatic Photoelectric Telescopes described in this book. His suggestions for improving this book have also been helpful.

Warm thanks are in order to George C. Roberts, who kindly supported the work on the dual-channel photometer, to Dr. Kenneth Kissell, who provided key optics and materials, to Peripheral Technology, for their donation of the 6809 microcomputer system used on the APT, and to the STD Bus manufacturers who donated cards and equipment to the Fairborn Observatory. Without their kind generosity, important aspects of microcomputer control would have gone unexplored.

Finally, I appreciate the generosity of the editors of Byte magazine for allowing the use of some material from an article of mine in an altered form in this book.

Russell Genet
Fairborn Observatory East

FOREWORD

Astronomy and microcomputers were a natural pair from the start. The data gathering routine at the telescope is characterized by many time-robbing interruptions due to the sun, moon, and the clouds. It is therefore important to ensure efficient use of the clear dark hours that remain to the observer by automating the process as far as is practical. The data gathering is also very number-intensive, calling for just the kind of help the computer can best provide. Fatigue and sleepiness are important problems in late night observing for astronomers, but not for computers. One of the major benefits of a computerized telescope and instrument is the error-free data logging it can provide, even at four AM in the morning. As a pure science, astronomy seems always to be underfunded. It is therefore significant that microcomputers are inexpensive, at least compared to earlier generations of computers. Though the advantages of computers to astronomy have been evident for decades, only since the advent of the microcomputer has it been realistic to automate small and moderate size telescopes.

This book is a testimony to the rather amazing progress that has been attained recently in applying microcomputers to small telescopes. Much of this progress has been possible because of the diligence of these authors and others in making available to a wide audience the often hard-won knowledge of how to implement a telescope automation project on a small budget. Practical knowledge turns out to be a very important ingredient in the conduct of astronomy, perhaps otherwise the most impractical of the sciences. The reader who chooses to apply the content of this book will soon appreciate the generous use of examples and construction hints.

The development of small telescope automation, and indeed small telescope astronomy, has benefitted from a delightful union of amateur and professional interests. The work is mutually beneficial to the two groups, and in some cases the merging is so complete that one can scarcely make a distinction. Amateurs in all of the experimental physical sciences except astronomy have practically disappeared because of the enormous expense of modern research equipment. Astronomy was headed down this same sad path until the rapid development of microcomputer automation. The potential of microcomputers in astronomy to preserve the ability of individuals to do useful science "just for the fun of it" should not be underestimated. Indeed this preservation may be one of the most important and pleasing results of the publication of books such as this volume.

R. Kent Honeycutt
Indiana University

CONTENTS

A. OVERVIEW

During the past decade, amateur astronomers have developed the means to build telescopes of a size suitable for serious research in photoelectric photometry and other fields, and have demonstrated the interest and ability to carry out such research. Many of these same people have purchased hobbyist or personal computers, and are looking for a way to blend these two interests together. Genet (1983) describes some aspects of how personal computers can be used in astronomy, and Ghedini (1982) and Meeus (1982) have published many useful astronomical formulae in a form which is convenient for implementation on a personal computer.

This book seeks to fill a gap in the books that are currently available on using personal computers in astronomical applications. Most of the other books stress computing that can be done at the leisure of both the hobbyist and the computer, and that uses only the basic computer and standard peripheral devices (disks, printers, etc.) as they come from the computer store. This book is primarily concerned with how to connect a non-standard computer peripheral device (a telescope) to a computer, and how to program the computer to perform time-critical computations. This is only one example of the more general problem of real-time control, so if viewed in this larger context, this book should find an audience among those interested in **any** real-time control application, such as robotics. One of us (Genet, 1982b) wrote a previous book on real-time control with the TRS-80, but it was primarily devoted to data logging and instrument control in astronomy, not telescope control.

At the same time that the revolutions in optics, telescope construction, and personal computers were affecting amateur astronomy, basic research budgets for professional astronomers did not keep pace with the growth in astronomical knowledge. Answering the growing volume of questions makes increasing research efforts desirable, despite the increasing number of research-quality instruments. Although space-borne telescopes will open up new areas of research, the limited number of such instruments will keep the main burden of astronomical research on ground-based telescopes.

One way to perform more research on the same number of instruments is to increase the research throughput per instrument. This can be accomplished by computerizing the telescope and associated data acquisition instruments to minimize non-observing time. This unproductive time includes activities such as slewing to the next object's coordinates, acquiring the object, adjusting the tracking rate, and logging the data. Several large telescopes have been successfully computerized, but the expenses these projects experienced, both in hardware and software, are beyond the reach of many smaller observatories (see Appendix A). Similar gains can be realized on smaller telescopes using modern hardware and control techniques. Much useful research can be done on instruments of modest size, especially if they are computerized.

1

The use of computerized telescope facilities owned by amateur astronomers can help produce greater quantities of useful research. Cooperative efforts between amateurs and professionals in the area of photoelectric photometry have already proven beneficial to both parties. The International Amateur-Professional Photoelectric Photometry (IAPPP) group has demonstrated this since its inception in 1980, and we expect even greater benefits for basic research in the future. In addition, professionals who are interested in computerizing a small telescope or an observing instrument may find this book useful.

The continuing evolution in microcircuit technology has produced personal computers which are both powerful and inexpensive. Several astronomers, both amateur and professional, have built useful telescope and instrument control systems using personal computers and inexpensive position feedback hardware. Such systems can be incorporated into the manual control systems of existing telescopes without disturbing the functioning of the manual controls. Programs can then be added to provide high accuracy tracking, computer control of data acquisition, real-time data reduction, and even an interactive graphics display of the acquired data. This would allow an observer doing photometry, for example, to correct the data for extinction immediately after the observing session, and to combine his corrected measurements with data gathered on previous evenings to display the developing light curve.

The field of microcomputer control of telescopes is in its barest infancy, and it is not possible at this point in time to know where it is heading. We would, however, like to hazard a few guesses and suggest how this book might relate to future developments in this field. In terms of the number of microcomputer-controlled telescopes, we suggest that:

(1) The greatest number will be commercial systems made by existing telescope manufacturers, such as Celestron®and Meade®Instruments. We would expect that these would be self-contained systems that operate from a fixed program in EPROM. We hope that these systems will be able to communicate, via an RS-232 serial line, with external personal computers, so that the latter can provide high-level supervision and simultaneously log data from instruments, such as photometers.

(2) The next greatest number will be complete research-grade telescope systems using commercial telescope mounts specifically designed for computer control. The DFM Engineering mount, which is described in Chapter 8, is the first of these.

(3) The third greatest number will be built entirely by amateurs, except for the personal computer used to control the system. Enterprising and resourceful amateurs will develop a wide variety of telescope mounts, control systems, and control programs, and their ideas will significantly advance the state-of-the-art.

(4) Various specialty systems will evolve over time. The two systems described in greatest detail in this book are specialty systems. One is an automatic photoelectric telescope (APT). The APT is not made unusual by its use of either an equatorial mount or open-loop steppers, but by closing the position loop with a photometer. This approach is unconventional, but not unprecedented. The other is a mobile (trailer-mounted) alt-azimuth telescope -- an unusual system on both counts.

This book is intended for a wide audience of varying interests. The most common reader of this book is expected to be an amateur astronomer with a personal computer, a telescope under 10 inches in aperture, and the desire to use his computer to point his telescope to an accuracy that consistently places target objects into the field of view of a low-power eyepiece. For this reader, the system developed by Tomer using stepper motors in an open-loop control system is described in detail. For those with more demanding requirements, or for those who are just curious, more advanced concepts, equipment, and systems are presented. In Sections V and VI, we have described in the greatest detail the systems with which we are most familiar. While they are somewhat unusual systems, the principles involved and problems encountered are similar to those that will be faced by developers of any microcomputer telescope control system. In the final analysis, each system is unique.

B. ORGANIZATION OF TOPICS

The topic of this book is the use of a personal computer to enhance the normal functions of telescope pointing and tracking. We wish to emphasize the distinction between full **automation** of all telescope control functions, so that human intervention is not necessary, and merely adding a computer to the control system, or **computerization**. Most of the telescope control systems described in this book are not automatic.

There is no doubt that personal computers possess the power needed to solve many telescope control problems. During the 1970's, a few professional observatories constructed computerized telescope control systems using groups of microprocessors or small computers with about the same speed and, typically, less memory than today's personal computers (see Appendix B). More recently, projects undertaken by both professionals and amateurs have demonstrated that a personal computer can be part of a low cost and practical solution to a wide range of telescope control problems.

This book covers most of the problems which must be solved in designing a telescope control system and proposes solutions to them, then assesses the ability of several currently available system components to handle the demands the performance requirements place on them. Systems developed or being developed by Louis Boyd, Andrew Tomer, DFM Engineering, and by the authors are described to illustrate the process of building your own system.

To design a telescope control system, you should first decide what the system will do. Next, choose a design approach to making the system work in the desired fashion. The design choices and problems in the selected approach should be foreseen and analyzed. A good approach for very complex systems, such as the one described in Section VI, would then be to construct the system theoretically to see what will be required to make it work. At this stage, specific pieces of hardware can be examined to see which ones, if any, are capable of meeting the system and component performance requirements. Finally, the control system is built and integrated with the telescope. It is our experience that the integration phase can be the longest and the most difficult phase if the earlier design work was not done properly.

Section I of this book is concerned with the benefits one can expect from computerization, and explores the reasons for computerizing a telescope. Section II is concerned with elementary aspects of modern control theory, and the components of a control system for telescope pointing and tracking. This section deals with topics that you should think about **before** you start designing your system.

Section III covers astronomical and mechanical corrections that must be computed to achieve high accuracy telescope control when using either open loop or closed loop servo control. Although all major sources of error are described in this section, those who do not require the ultimate in accuracy can perform only those corrections that are most significant.

Section IV describes two different systems using Apple II computers, a third system developed at the Fairborn Observatory East using STD Bus modules, and a fourth system using a single board computer. All of these systems are simple and straightforward, and the single board computer system is quite inexpensive. Those with limited budgets who do not need very high accuracy will be most interested in these control systems. Complete hardware and software details of Tomer's Apple-based system and the single board computer system are given, as an aid to getting your own system up and running quickly with a minimum of fuss.

Sections V and VI describe the approaches taken in the systems used by the authors. Section V describes the development of the Fairborn Observatory systems, problems which were encountered as these systems evolved and the approaches taken to solving them, and future growth paths these systems may take. Included are details of the Fairborn Observatory system, which has demonstrated its ability to make accurate photometric observations automatically much more rapidly than a human could using a manually controlled telescope. Section VI is Trueblood's account of the system he has in the final design stages. It includes his reasoning for the choices of system components that he made.

It is our hope that this combination of some control theory, detailed information about available hardware, examples of working systems, and practical advice (with a little bit of our preaching) will get you started on your own system. The Appendices contain a great deal of valuable information on sources of hardware. We recommend that you become very familiar with what hardware and software are currently available and engineer your system to suit your particular observing program, rather than blindly copy one of the systems presented in this book. That way, you will have a system that works the way you want it to, not the way somebody else thinks it should.

Chapter 2. WHY CONTROL TELESCOPES WITH MICROCOMPUTERS?

There are many reasons why one might want to control a telescope with a microcomputer, but they can be conveniently split into two categories: cost/benefit reasons, where the increased efficiency of scientific data gathering warrants the cost of microcomputer control, and all other reasons. Other reasons will be treated first.

A. VARIOUS REASONS

Given that there are many telescopes, microcomputers, and inventive amateur and professional astronomers, it is only human nature that successful attempts (and a few unsuccessful ones!) will be made to get a microcomputer to control a telescope. The reasons given will be various, but in many--perhaps even most--cases, the real reason will be that they were there and it seemed like a fun thing to do. While some will be content to just use their microcomputers for analysis, projecting star charts, and star war games, many simply will not be able to resist the challenge of computerized telescope control. It is towards these hardy souls, more than any others, that this book is aimed.

There are situations in which computer control of a telescope is helpful or even essential. A good example is the automatic south pole telescope (Giovane, et al, 1983). No one will be there to operate it. Other examples are daytime IR astronomy, which requires accurate automatic pointing, as the stars are generally not visable to the human eye, and the Multiple Mirror Telescope, which simply cannot be controlled quickly enough by a human to keep its mirrors aligned (Stephenson, 1975).

A number of amateur astronomers have noted that the necessity of getting up for work each morning seriously cuts into the clear-night observing time. Being familiar with computers, a few of them have naturally turned to the microcomputer to control their telescopes and related instruments as a way out of this dilemma. The work by Skillman (1981) and by Boyd (Boyd, Genet, and Hall, 1984) is particularly worthy of mention. Skillman's system gathers photometric data from a short-period variable star and nearby comparison star hour after hour, while Boyd solved the problem of automatically locating and measuring many different stars in a single night. It would only be fair to mention that this problem was partially solved in the mid-1960's using a large remote computer in Tucson to control the 50-inch telescope at Kitt Peak (Goldberg, 1983), and using a minicomputer in a pioneering effort by McNall (1968) and others at the University of Wisconsin. A few telescopes at other observatories have had similar capabilities added to them or designed in from the start over the last 15 years.

Microcomputers can be used to control or assist in a wide or narrow range of observing functions. A good example of a truly simple and low-cost application is that of Rafert (1983). At a cost of less than $100 in non-computer parts, two stepper motors were added to the 24-inch telescope at Appalachian State

University, and controlled by a Commodore microcomputer. The objective was not to point the telescope anywhere in the sky automatically, but simply to move back and forth between variable and comparison during photometry. Initial telescope pointing and final centering (each time) was still performed by the astronomer. The astronomer was relieved, however, from moving the telescope between the stars and rough-centering it in the finder. Those readers who are experienced photometrists will appreciate how much effort such a simple telescope rough-pointing system can save.

Besides greater data gathering efficiency, microcomputer control of a telescope and its instrumentation can increase the reliability of the data. Data logged manually towards the end of a late night observing session tend to be prone to errors. An additional benefit of freeing the astronomer from the data logging and telescope pointing functions is that he is free to think about the astronomy while observing. This allows him to make prompt decisions about the wisdom of beginning or continuing an observation, what the next object to be observed should be considering the observing conditions, and so on. In some observing programs, such as long period variable star photometry, even these functions can be assumed by the computer, which makes possible the automatic photoelectric telescope described in Section V. This frees the astronomer from all data gathering functions, and allows him to devote his complete attention to designing the observing program, interpreting the reduced data, and publishing the results.

When a computer is introduced into the control system of a telescope, it should be combined with other elements of the observing program in a manner which is consistent with them, and which enhances the program overall. A'Hearn (1984) points out that some research programs, such as comet photometry, require a great deal of judgement and intervention by a human observer during the process of gathering data. His system, which uses an Apple II computer and photometer control software written in FORTH, is quite flexible, and accepts human judgement as an input into the observing process. The roles of the computer and the observer are carefully defined in this system.

Before proceeding to some cost/benefit considerations, it might be good to point out that there are a number of reasons why one might NOT want to control a telescope with a computer. We credit the best argument in this direction to Douglas S. Hall. He pointed out that a number of perfectly good photometrists who regularly gathered high-quality data of considerable scientific value were totally ruined when exposed to microcomputers. Instead of being out there observing, they were spending all their time trying to control their photometers and telescopes with microcomputers. However, with great insight into dynamic processes, he later pointed out that as the number of microcomputer enthusiasts was far greater than the number of photoelectric photometrists, that the loss of good photometrists to microcomputers would be more than offset by the conversion of microcomputer enthusiasts into good photometrists!

Another reason for not wanting to computerize a telescope is that there is certain merit to the old dictum KISS (keep it simple, stupid!). There is not much to go wrong with setting circles. Equipment failures can be very time consuming and frustrating--even expensive. Computer-controlled telescopes are usually complicated--definitely not KISS! However, any astronomer who has paid good money to buy this book will not be dissuaded by any of these cautions.

B. COST/BENEFIT REASONS

Many computerized systems of various types have been developed in response to the expectation that the expense of system development and maintenance will be outweighed by an increase in productivity or more efficient utilization of resources. The system user has a problem which the proposed system should solve. Often the problem is felt as too much demand for a scarce resource, for example, observers requesting more telescope time than there is available. Whether computerized systems at large observatories are cost effective is still a topic of controversy. Racine (1975) argues that non-observing time (time spent acquiring the object, setting up instruments, etc.) that can be reduced by computerization is a small fraction of total telescope time, so the reduction does not enhance science throughput enough to justify the cost. Boyce (1975) argues that time spent collecting photons can be as low as 7% of total telescope time, and that computer control can improve this low duty cycle. Hill (1975) claims "for about 10% of the price of the observatory we had increased our output rate by about a factor of four and we expect that to increase". He does not indicate how output rate was measured. Four times as much data gathered does not necessarily mean four times as much useful science was done, and it must be the useful science done that is the measure of system benefit.

The proportion of non-observing time depends on the type of observing being done and how the telescope is being used. For example, UBV photometry may have a large non-observing-to-observing time ratio which may exceed 1. Much of this non-observing time is spent setting photometer controls and pointing the telescope, both of which can be done faster and more accurately by computer. On the other hand, the time spent pointing a telescope for a long plate exposure may be a small fraction of the duration of the exposure. Factors such as time spent removing and attaching equipment (which cannot be controlled by a computer) and whether an astronomer can be doing something useful while the computer controls the telescope, as opposed to controlling the telescope himself, vary from one observatory to the next. Before a small observatory, presumably working with a limited budget, decides to computerize its telescope, it would be wise to make a candid assessment of the observing programs supported by the observatory and how much they would benefit from any proposed computerization project.

One might argue the value of computerizing a small telescope. However, Abt (1980) has shown that for telescopes at Kitt Peak ranging in aperture from 0.4 m to 2.1 m, the most cost effective telescopes were the 0.4 m telescopes, in terms of both the number of published papers and the number of times these papers were cited per dollar of annual expense. Although computerization can bring about greater cost savings on larger telescopes per dollar spent on the computerization project, the dramatic price drops in recent years for fast and reliable computer hardware have made computerization of smaller telescopes economically feasible. One should not, therefore, dismiss the concept of computerizing small telescopes because of the history of high costs incurred by projects undertaken at large observatories a decade or more ago.

The reasons for using a microcomputer in a telescope control system are many and varied. We suggest that you develop a very clear idea of what benefits you expect from a computerized control system and what it will cost before proceeding with the construction of your system. An important cost element which is often overlooked is the cost of the software, which can be several times the cost of the hardware. To aid in the assessment of the hardware costs involved, the next section deals with the design of control systems and the hardware used to build them.

A. CONTROL OPTIONS

After the decision has been made to proceed with a telescope control project, and the system performance requirements have been specified, the next step is to select the overall design approach. For telescopes to perform useful work, they must be pointed quite accurately in the sky, and then must track objects over an extended period of time with considerable accuracy. For example, in photoelectric photometry with a typical small telescope, a star might be centered in a 60" diaphragm with an accuracy of at least 15", and must stay centered with this accuracy for about two minutes of time. This sort of accuracy is routinely obtained with Celestron 8-inch telescopes on modest mounts in backyard observatories. Pointing and tracking accuracies that keep objects centered within a couple of arc seconds for hours is not at all uncommon when these same mounts are engaged in astrophotography, with the astronomer making constant drive rate corrections manually. When one recalls that there are 1,296,000" in a circle, this sort of accuracy may be surprising. It is achieved in simple systems by having a sharp-eyed observer at the controls. The observer detects deviations of the system from its desired position (errors), and activates the controls to reduce these errors. In this system, the observer's eye is the error sensor, the observer's brain is the computer that determines the appropriate response, and the observer's fingers implement the response using the manual telescope control system. If this series is thought of as a "loop", then it is the observer that "closes the loop" in this case.

All telescope pointing and tracking systems are closed-loop systems in some sense (when feedback from humans and photometers are counted), as there appears to be no way to achieve the high accuracies demanded by most astronomers other than by detecting and correcting for deviations from desired functioning. There are, however, several ways to close the loop in telescope control systems. In this book we will be concerned with those systems in which, for one reason or another, a microcomputer is involved somewhere in this process. We have chosen to classify the methods by which the loop is closed into two broad groups.

The first type is the "classical" closed-loop servo system. This is the one treated in some detail in this chapter and in Trueblood's example in Section VI. The classical system is one in which there are angle position sensors on each axis of the telescope (e.g., optical shaft angle encoders), and a microcomputer which compares the actual position (given by the encoders) with the desired position (the apparent place of the star) and issues appropriate commands to bring these two closer together. There are two key points here: First, in the classical system, only angular position at the telescope axis (or in many systems, at the final worm, which is one step removed from the axis) is sensed, along with, perhaps, motor velocity. No direct sensing of the star itself is performed. This means that no random or systematic error between the encoder and the image plane, such as tube or mount flexure or atmospheric refraction, is directly sensed. Instead, systematic errors (which are thought to be understood) are modelled in software, and model constants are measured by making a special calibrating observation for each type of error. It is then expected that after such calibration and modelling, the residual errors that remain after the modelled errors have been removed will be small enough to be ignored.

The second key point in the definition of the classical system is that the microcomputer is almost continuously in the loop, updating the control commands and recalculating telescope flexures, atmospheric refraction, and other systematic errors. As the computer must also do other tasks, this means that a multi-tasking interrupt-driven system is almost a necessity, and the computer must take a break regularly, perhaps once per second, to make the calculations contained in the control algorithm and to issue the resulting commands to the motors.

We have chosen to call the second type of telescope control system "non-classical" or "other approaches", where "other" in this case means other than the classical approach. The detailed discussion of these systems is deferred until Genet's example in Section V. However, a brief introduction is appropriate here. Generally, it is the goal of nonclassical systems to get around some of the difficulties of the classical system, although new difficulties are introduced in the process.

The main difficulty with the classical approach is that it senses errors at the telescope axes, not the stars themselves. Since the ultimate goal is to point the telescope at the stars, any unmodelled errors are ignored in the classical approach, or one has to try to correct for them some other way -- a task not easily accomplished, and one that can load down the computer. The non-classical approaches generally try to detect the error at the star itself, and as the error is detected and corrected here, no complex calculations are needed, and the computer is not loaded down. However, star sensors are generally more difficult to implement than shaft angle sensors, so one has essentially traded one set of problems for another set.

There are, however, important exceptions to this. First, if the sensor is the astronomer's eye, then implementation may be rather straightforward and inexpensive. Examples of this are the stepper motor telescope control system developed by Rafert mentioned earlier, and Tomer's system described in Section IV.

Second, if for some other reason, an electronic star sensor of one type or another, such as a photoelectric photometer, must be part of the instrumentation at all times, then one can use this sensor to close the loop directly on the star itself. A number of classical systems have star sensors such that once the star is acquired, the star sensor, not the optical encoders, assumes control of the telescope. A good example of this is the Vienna 1.5-m telescope (Stoll and Jenkner, 1983). In such a system, the job of the classical control system is to point the telescope just accurately enough to permit the star tracker to lock onto the guide star. Lock-on range in some systems can be as large as 1-2'. Some telescopes with permanent photoelectric sensors have purposely extended their acquisition range to the point where optical shaft angle encoders and correction computations of all kinds have been eliminated entirely. The telescope moves from one area of the sky to another totally "open loop" (i.e., without any position feedback), and when it arrives in the area of the sky where it "thinks" the star ought to be, it then searches for it with the photometer, acquires it, and then centers it -- all under microcomputer control. Such a system is described in Section V.

B. THE ROLE OF POSITION FEEDBACK

The purpose of a telescope control system is to point the telescope at the target object and track it with an accuracy dictated by the observing requirements. In the classical control approach, this is done by controlling the angular motion of the telescope about its axes. The system determines where the telescope ought to be at any particular time, then sends commands to the drive motors to move the telescope so that it arrives at the desired position at the proper time. This kind of control system is an example of the general class of servo control systems.

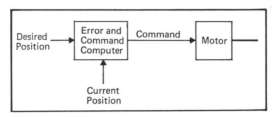

3-1 Typical Open-loop Servo

The simplest form of control system is the open-loop system. As shown in Figure 3-1, the open-loop system finds the difference between the desired position and the current position and computes an error signal which forms a command to the telescope drive motor.

An example of an implementation of the open-loop system using analog electronics is the typical variable frequency drive corrector used to control the RA synchronous motor on a small telescope. The observer sets a switch or adjusts a potentiometer ("pot") to obtain the desired drive rate, which is indicated by a mark near the frequency knob.

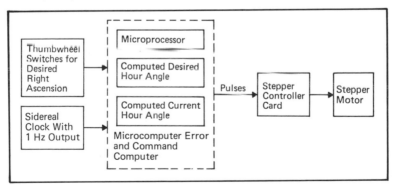

3-2 A Digital Open-loop Right Ascension Drive

An example of an implementation of an open-loop telescope drive with digital electronics is using a microcomputer to control a stepping motor. Figure 3-2 shows a block diagram of a digital open-loop right ascension drive for an equatorially mounted telescope. The microcomputer computes the desired hour angle from the desired right ascension stored in the thumbwheel switches and from the sidereal time in the sidereal clock register. By counting the pulses it sends to the stepper and by using a stored constant representing the gear ratio between the motor and the RA shaft, it can compute the current telescope position, after a previous calibration on one or more stars of known position. The microcomputer then adjusts the rate at which it sends pulses to the stepper controller based on the difference between the computed desired hour angle and the computed current hour angle. After each pulse is sent, the current hour angle register is updated.

This system can be as complex as is needed. The desired hour angle can be corrected by the microcomputer for precession, nutation, aberration, refraction, polar axis misalignment, and other systematic errors down to arc second accuracy, since typical microcomputers available today have adequate CPU power to perform these calculations in real-time. However, achieving arc second accuracy in calculating the desired hour angle does not guarantee arc second accuracy in pointing the telescope.

The problem with all open-loop systems is that once a command is given, there is no way to know if the command was executed accurately. Software models of systematic errors are either not accurate enough, or depend on knowing the positions of all the parts in the tube, mount, and drive train at each instant. These systematic errors include backlash and other gearing errors, flexure, and mechanical misalignments. Such structural element position information is not generally available to the computer. If it were, the algorithms needed to model all systematic errors to very high accuracy would tax the resources of most microcomputers to compute them in real time. Even if these systematic errors were modelled perfectly, random displacements due to wind gusts or other error sources would still introduce significant errors. One can take precautions to size the motors properly and to design a good mount and drive. Such precautions will reduce errors sufficiently for most small telescope applications, but for high-accuracy computer control of large telescopes, it is conventional to close the loop somehow.

There are other reasons for using a closed-loop servo. If the telescope is slewed by hand, all current computed pointing information in an open-loop system is in error, perhaps by as much as 180 degrees. This is also true when a non-incremental motor is used for turning the telescope axes, since the computer has no way of knowing how much the motor has moved the telescope. Thus some situations require a means to sense the telescope position independently of the commands sent to the motors.

Achieving high accuracy pointing in open loop (stepper-based) systems requires special care. The drive system must be highly linear (or the non-linearities must be modelled to high accuracy) and free of backlash. Slewing acceleration and slew speeds must be set so that no steps are lost for any reason. Finally, the telescope structure needs to be quite rigid to avoid excessive flexure if the computation of such corrections is to be avoided (although an open-loop system can compute and correct for flexure). Many of the telescope mounts and drives described in this book were specifically designed to meet these stringent open-loop requirements. Any system that meets these requirements has an unusually good mount, and can just as easily be used in a closed loop system. If a well designed telescope and mount are used, and all the corrections discussed in Section III are computed often enough with adequate precision, one could achieve very good accuracy. How good is an open issue, since to our knowledge, no very accurate completely open loop telescope control system has been attempted. Depending on the amount of work put into tuning such a system, pointing accuracies of 10" might be possible.

In contrast, several professional observatories have built closed loop servo systems for telescope control which regularly achieve 5" overall pointing error or better in all areas of the sky. In the classical closed loop approach, a device is mounted on the telescope axis to feed the true angular position of that shaft back to the current position register. In this case, the microcomputer uses the actual shaft angle position to compute the error signal.

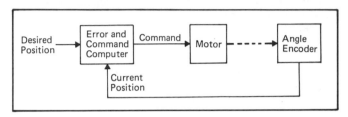

3-3 Typical Closed-loop Servo

As shown in Figure 3-3, the closed-loop system is identical to the open-loop system except that feedback is used to generate the current position information, instead of computing it. In the example given above of a variable frequency drive corrector, if the observer sees the stars slowly drifting through the eyepiece field, he usually corrects the drive rate. In this case, he is "closing the loop" and providing position feedback to the system (in the non-classical sense). Even in systems in which electronic feedback is used for automatic drive rate corrections, multiple feedback loops can exist, with the operator in the outermost loop guiding the telescope with a hand paddle.

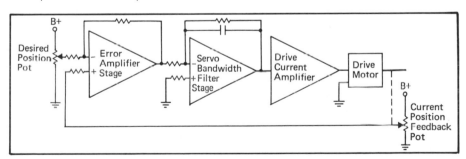

3-4 An Analog Closed-loop Declination Drive

Any closed-loop control system is called a "servomechanism", or "servo". This term is used regardless of whether DC servomotors, steppers, or some other type of actuating device is used. A household furnace controlled by a thermostat is a common example of a servo. An example of an implementation of a closed-loop servo controlling a declination drive using analog electronics is shown in Figure 3-4. The observer dials in the desired declination angle by adjusting a linear pot with a pointer knob and a calibrated dial. The feedback is provided by a linear pot turned by the declination axis, and the error computer is an operational amplifier (op-amp) which tends to produce an output proportional to the difference between the "desired" and "current" voltages generated by the two pots. The output is thus proportional to the error, and can be used to drive a DC motor to point the telescope. Tracking in RA can be accomplished using the same circuit with a modification that adds a constant voltage offset to the error (command) voltage. This approach can give quite acceptable performance for very low cost--assuming high accuracy is not required.

The purpose of the filter in the circuit shown in Figure 3-4 is to limit the bandwidth of the servo and adjust the relative phase between the forward and feedback paths. This is required to help prevent overshoot and oscillation, both of which are the result of high gain in the amplifier and inertia in the motor and telescope. By limiting the maximum rate at which the control voltage can change to that at which the motor can reasonably respond to a step change, the servo performance can be tailored to the mechanical properties of the system. The filter helps to damp the system to permit it to find the desired position in the least time.

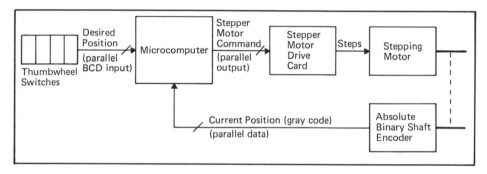

3-5 A Digital Closed-loop Declination Drive

An example of an implementation of a closed-loop servo controlling a declination drive using digital electronics is shown in Figure 3-5. The microcomputer acts as the error and command computer using the desired position input from the thumbwheel switches as the reference input and the binary shaft encoder output as the current position feedback. The microcomputer converts the shaft encoder output in the form of a binary code to an angle in degrees (declination) or hours (right ascension) and compares it to the desired position stored in the thumbwheel switches. The difference is the error, which the microcomputer converts to the number of stepping motor steps required to reduce the error to zero. The microcomputer then computes a step rate profile designed to ramp the stepping motor up to some speed, sustain that speed, then ramp down so that the motor has moved the required number of steps to reduce the error to zero in the shortest possible time. Actually, the error is not reduced to zero, but to half of the angle the controlled shaft moves through in one motor step, or the resolution of the shaft encoder, whichever is greater.

In computing the number of steps required to minimize the error, the microcomputer can correct the desired position input for precession, nutation, aberration, and refraction to obtain true apparent position in the sky from the mean catalog position entered in the thumbwheel switches, and it can correct the current position feedback number to reflect flexure, non-perpendicularity of the rotation axes, misalignment of the polar axis, non-linearity of the encoder, misalignment of the optical and mechanical axes, and other sources of systematic error discussed in Chapter 7. The corrected desired and current positions then are used to determine the number of steps required to minimize the error. In this way, systematic errors can be modelled and subtracted from the error signal. The overall system accuracy is then determined by the accuracy (and expense) of the encoders used for the position feedback, as well as the accuracy of the systematic error models.

Note that there is no servo bandwidth filter explicitly shown in Figure 3-5. The microcomputer fills this function by sampling the desired position and current position inputs periodically, rather than continuously as in the analog circuit. The rate at which these inputs are sampled, and the rate at which commands are sent to the stepper motor drive card determine the servo bandwidth. These rates are under direct control of the software, so the system can be tuned without changing the hardware.

C. EQUATION OF MOTION

Any servo is essentially a damped harmonic oscillator, an example of which is an automobile body suspended on its chassis with springs and shock absorbers. If your car has old, worn out shock absorbers, and if you push down on a fender suddenly, the car will bounce up and down on its springs several times. This continuous bouncing that dies out slowly is underdamped motion. If the shocks are new, when the car is pushed down, it comes all the way back up quickly, then stops. This is critically damped motion. If the shocks are too stiff, the body rises slowly from being pushed down, and eventually comes all the way back up. This is overdamped motion.

The entire ensemble of the telescope and its control system behaves as a damped harmonic oscillator. In addition, the mechanical telescope structure and the electronics in the control system, taken separately, act as damped harmonic oscillators. All damped harmonic oscillators satisfy the general equation

$$ J \frac{d^2\Theta_o}{dt^2} + F \frac{d\Theta_o}{dt} + K\Theta_o = K\Theta_i \qquad [3.1] $$

where Θ_i is the commanded input angle

Θ_o is the resulting output angle of the telescope axis

K is the system torque constant

J is the total moment of inertia of the driven component

F is the system damping.

Note that if one merely substitutes electric charge for angle, inductance for inertia, resistance for damping, and capacitance for torque constant, we have the equation for an electronic oscillator.

This equation gives the relationship between an input command angle Θ_i, and how the servo reacts to that command, given by the output angle Θ_o. J includes the motor rotor moment of inertia, J_m, and the load moment, J_l. If a gear train of reduction N:1 is placed between the motor and the load, the load moment is reduced by the factor $1/N^2$. F includes the motor damping, $F_m = T_s/S_o$, where T_s is the motor stall torque and S_o is the motor's unloaded free (maximum) speed. Figure 3-6 shows the speed versus torque curve for a typical servo motor.

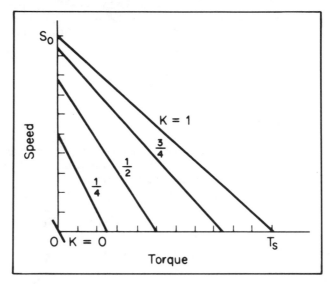

3-6 Speed versus Torque Characteristics for Different Values of Control Voltages
Courtesy Reston Publishing Co.

The system damping, F, contains sources of damping besides the motor damping, including the effect of viscous damping cups attached to the motor, or velocity damping obtained by attaching a tachometer-generator to the motor and feeding back its signal to the error generator. The latter technique reduces the error signal magnitude by an amount proportional to the motor velocity. K is given by $K_s K_a K_v N$ where K_s is the gain or transfer function of the device generating the commanded input angle signal (such as the volts/radian generated by a pot turned by the observer to command the servo to point the telescope), K_a is the gain of the servo amplifier, K_v is the motor torque constant, and N is the gear ratio between the motor shaft and the load.

To understand the behavior of the servo, consider the homogeneous quadratic equation

$$J \frac{d^2}{dt^2} + F \frac{d}{dt} + K = 0 \qquad\qquad [3.2]$$

which has roots

$$-\frac{F}{2J} \pm \sqrt{\frac{F^2}{4J^2} - \frac{K}{J}}$$

[3.3]

Three separate physically meaningful cases result:

Case 1. Negative, real, and unequal roots.

$$\frac{F^2}{4J^2} > \frac{K}{J}$$

Define the damping ratio DR = $F/2\sqrt{KJ}$, >1 in this case. This is an overdamped system.

Case 2. Negative, real, and equal roots.

$$\frac{F^2}{4J^2} = \frac{K}{J} \quad \text{or DR = 1}$$

This is a critically damped system.

Case 3. Conjugate and complex roots with negative real parts.

$$\frac{F^2}{4J^2} < \frac{K}{J} \quad \text{or DR < 1}$$

This is an underdamped system.

The damping ratio DR is the ratio F/F_c of actual damping to critical damping. Thus for critical damping, the damping constant F_c should equal $2\sqrt{KJ}$.

When a new position is entered into the servo, in the underdamped case, Θ_o will attain the desired position rapidly, then overshoot. The error signal will reverse sign, and drive the motor in the opposite direction. This oscillation continues for a time determined by the damping ratio. When F = 0, the servo oscillates forever at a natural frequency of $\omega_n = \sqrt{K/J}$. For the critically damped and overdamped systems, Θ_o approaches the desired value monotonically and exponentially. Of the three cases being considered, a system which is very slightly underdamped settles to within some defined tolerance of the desired position and stays there sooner than the other two systems. Therefore, the system should be designed to be slightly underdamped to obtain the most rapid response, with an overshoot equal to the error tolerance. If the time to achieve the desired position is not of utmost importance, and oscillations must be tightly controlled, the control system should be overdamped. Most telescope control systems are overdamped, in part to compensate for the typical underdamping of the mechanical structure, and to

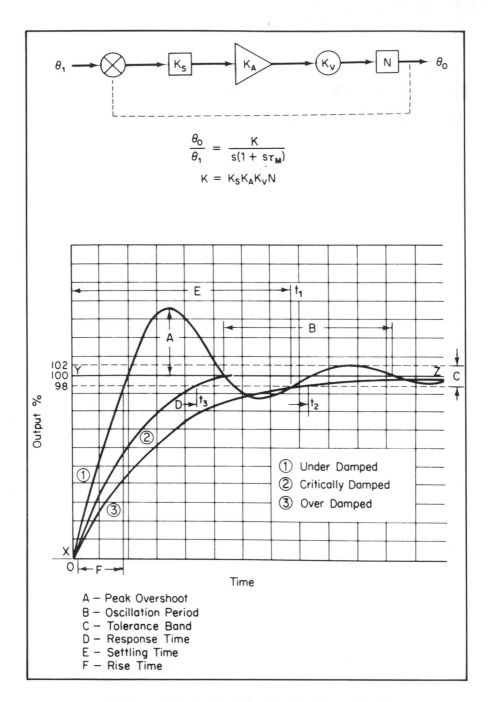

3-7 System Response Curves for Three Conditions of Damping,
Courtesy Reston Publishing Co.

protect the drive from damage. Figure 3-7 shows the system responses for the three cases.

D. THE ROLE OF VELOCITY FEEDBACK

In an underdamped system, the damping ratio is less than 1, which indicates the gain, K, is too large for a given system damping F and for a motor matched to the inertia J. It is desirable to have a large system gain K to keep the system dead band low. Dead band is the range of values over which the input signal Θ_i may vary without causing the output Θ_o to respond. For a DC servo motor, it is the threshold voltage at which the motor just starts to turn as it overcomes internal friction. The dead band is given by $2V_c/(K_s K_a)$, where V_c is the maximum control voltage applied to the motor. By using a large value of K_a, the dead band is reduced. A telescope pointing system must have a very small deadband, since to achieve, say, tens of arc seconds pointing accuracy or better in Θ_o, any change in Θ_i that gives no response in Θ_o produces an error in Θ_o. The large gain K_a needed to reduce the dead band also tends to produce an underdamped system, so the total system damping F must be increased to compensate. Note that the dead band may not be uniquely defined in terms of absolute positions if the system has substantial hysteresis.

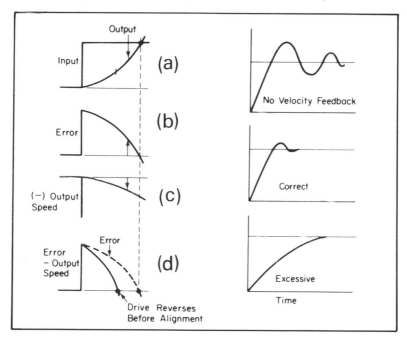

3-8 Effect of Velocity Feedback
Courtesy Reston Publishing Co.

To demonstrate the method of increasing F with velocity feedback, one begins with a servo with no damping, which is the extreme case of underdamping. With a step input, the high gain forces the output to rise rapidly, as shown in Figure 3-8(a). At the moment of the step input, the error is greatest, but decreases as the output moves to the desired position, as shown in Figure 3-8(b). When the output reaches the desired position, the motor speed is highest, so that the motor inertia, along with the load inertia, carries the output through the desired point into the characteristic underdamped oscillations.

The velocity feedback needed to achieve proper damping can be provided for a DC servo motor using a tachometer-generator. This is a small DC electric generator which, when coupled to the motor shaft, produces a voltage proportional to the motor speed. This motor speed signal is first inverted to provide negative feedback, then combined with the position error signal and the input control signal to produce a modified error signal, as shown in Figure 3-9.

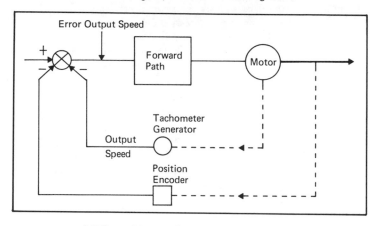

3-9 Servo Diagram Showing Velocity Feedback

Figure 3-8(c) shows the velocity feedback signal produced by the tachometer-generator, and Figure 3-8(d) shows the modified error signal that is fed to the forward path. If the gains of the position and velocity feedbacks are properly adjusted, the servo will not overshoot the desired position and oscillate, but rather will be properly damped. A tachometer-generator is used in analog systems, but in digital systems using a digital processor as the error computer, the processor can compute shaft speed easily and rapidly using the system clock and only position feedback. To avoid excessive noise from the differencing operation, the position feedback device must have high resolution.

The phase of the feedback signal with respect to the forward path signal is important in determining the stability of the servo. Figure 3-10 shows that oscillations are enhanced, rather than attenuated, if the feedback is out of phase with the forward path. Electronic servo circuits that oscillate can be improved by placing phase-compensating filters in the feedback path. This can also be accomplished in software in a digital servo by adjusting the software execution timing.

Despite the availability of sophisticated structural analysis programs, it is usually not worth the trouble to attempt to predict the behavior of a telescope control system in advance. Instead, experience has shown that a better approach is to use good design in all aspects of the telescope and the control system, and to tune the system after it is built. Being aware of the nature of the problems that might be encountered, the telescope designer can then diagnose and solve these problems as they manifest themselves in the completed telescope. One example of

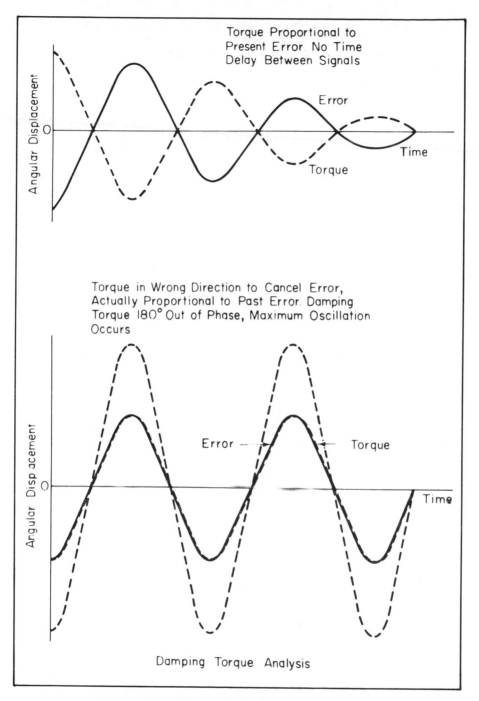

3-10 Analysis of Damping Techniques and an Error Signal
Courtesy Reston Publishing Co.

the tuning that is required is adjusting the servo bandwidth. Because of the disparity between the tracking and slewing rates, two different filters may be required in the control system, each adjusted to give the best performance for its mode of operation.

Another example of a system parameter to be tuned after the system is built is the forward loop integral. The system described by the equation of motion given by equation [3.1] has a servo lag phase error which is proportional to the size of a new step function input and inversely proportional to the bandwidth of the servo. That is, the servo filter that limits the servo bandwidth also limits how rapidly the servo can respond to a new input command. Since the servo cannot respond immediately to the new command, a phase delay is introduced between the input and the output. One can compensate for this delay by measuring the characteristic time constant for the servo and adding a correction term to the error signal. This can be done with additional circuitry in an analog servo, or with software in a digital servo, and has the effect of adding a forward path integral term to the equation of motion. This permits the use of a lower system gain without increasing the dead band.

These are only two examples of the kind of adjustments that must be made to a control system after it is built to achieve the desired level of performance. The process of tuning is made easier when the telescope, mount, drive, and control system all work together with each other, rather than against each other.

E. MATCHING THE TELESCOPE AND THE CONTROL SYSTEM

Several projects were undertaken during the last 15 years to retrofit a computerized control system to an existing telescope. Many of these projects were quite successful in meeting their design goals and functional requirements, but a few projects failed because the telescope and control system were not well matched. Often, the approach taken in those projects that failed was to rely on the computer software to correct mechanical design deficiencies in the telescope tube or truss, mount, or drive train. Recent projects in which both the telescope and the control system were designed for each other tended to enjoy much greater success. Based on this history, equal attention must be paid to the design of the telescope tube or truss and mount, the drive train, and the computerized control system. Each of these segments of the total system will be treated briefly in the following paragraphs.

First, consider the mechanical parts of the telescope, taken as a whole. If the rotor of the motor which drives the telescope is locked, and then the tube or truss is struck with a quick tap, the telescope will oscillate with a characteristic natural frequency. This mechanical behavior is described by the same equation (3.1) used to characterize the electronic and electro-mechanical behavior of a servo circuit. Mechanically, telescopes have very little damping. The oscillations die out exponentially after some number of periods of the natural resonant frequency of the mechanical structure. Since the locked rotor frequencies of telescopes of conventional design tend to be on the order of 1 Hz, it can take a relatively long time for oscillations to die out. This causes a problem for any control system, since momentary gusts of wind, or short, frequent pointing commands cause the telescope to vibrate in a manner that can damage the drive mechanism, smear photographic images, and lengthen the time after a pointing command is given before reliable data can be gathered.

This situation can be remedied either by (1) increasing the amount of damping, or (2) raising the natural frequency of the structure. In the first approach, the number of oscillations the structure makes is reduced, which shortens the time it takes them to damp out. In the second approach, the period of each oscillation is reduced, which also reduces the total damping time by reducing the time it takes for the structure to make a given number of oscillations. Returning to the automobile example again, increasing the damping is analogous to using stiffer shock absorbers. Although this works well for automobile suspensions, it is just not practical to place shock absorbers (or other means of increasing the damping) throughout a telescope's tube, mount, and drive structures. This leaves the second approach--raising the natural frequency of the structure.

The natural frequency of a mechanical structure is determined by the stiffness of its elements and the moment of inertia about the axis being rotated by the drive. A structure can be made stiffer by adding more mass to strengthen key structural elements. This by itself would raise the natural frequency, but it also tends to increase the moment of inertia, which lowers the natural frequency. There is some tradeoff point where the natural frequency is maximized by balancing added stiffness with added inertia. It is this tradeoff point that should be sought in the telescope design, not just minimum moment of inertia, or maximum stiffness. With proper design, a locked rotor natural frequency as high as 10 Hz can be achieved on large telescopes, and even higher frequencies on smaller telescopes. The DFM Engineering small telescope mount described in Chapter 8 was designed with a high natural frequency to aid computer control of the mount.

The moment of inertia about any axis is minimized when all mass elements attached to that axis are as close to the axis as possible. Thus symmetrical mounts, such as fork and yoke types, will tend to have lower (more desirable) moments of inertia, while the moments of inertia of other mount types, such as the German equatorial and cross axis mounts, tend to be significantly higher.

Next, consider the telescope drive system. Most telescopes of conventional design use a worm gear drive. The problem with this kind of drive is that it forces the control servo to provide considerable damping in the overall system, and tends to limit the slewing speed. To see why this is so, consider what happens when the telescope is at maximum slew speed, and the slew motor is suddenly stopped. The energy of the mass of the telescope and its mount are dissipated by driving the worm gear into the worm, most likely damaging both gears. The use of non-back-drivable gears forces the control system to provide a very gradual deceleration from slew rate to track rate.

The fact that longer ramp times are required for decelerations when worm gears are used often means the peak slew rate of the drive is never reached before it is time to ramp down to the tracking rate. For some observing programs, this may not be important, but for others, it can be critical. For example, the research program of an automatic photoelectric telescope (described further in Section V) depends on minimizing the time to acquire stars. This is achieved using frequent short motions to execute spiral search patterns.

A better approach to drive design is to eliminate gears altogether. Since the drive contributes to the locked rotor natural frequency of the telescope, it should be very stiff. Although band and chain drives have several advantages over a worm gear drive, and have been used successfully on small computerized telescopes, they are not adequately stiff for larger telescopes. The chain drive also suffers from considerable periodic error. The band drive does not have this drawback, but on larger telescopes, the band would have to be undesirably wide before it would have the required stiffness (Melsheimer, 1984). Some telescopes use helical gears instead of worm gears. Helical gears are back-drivable, and good helical gears have very small errors, since they tend to be averaged out by having several teeth in contact at once. However, the best approach is to use a disk and roller in friction contact. The friction drive is both back-drivable and completely free of tooth-to-tooth gear errors. For these reasons, it is the drive train approach taken on the DFM Engineering telescope mount described in Section IV, and the Trueblood portable telescope described in Section VI.

The natural frequency of interest when moving the telescope is not really that with a locked rotor, but that of the system with the motor powered up and moving the telescope. Suppose the tube and mount were both perfectly stiff, and gusts of wind hit the telescope tube at regular intervals. If the motor and (back-drivable) drive are not sufficiently stiff and can not develop sufficient torque to overcome such load variations, and if the wind gusts come at the right frequency, the motor would lose ground when a gust comes along, then make it up again when the gust subsides, then lose ground again when the next gust arrives. The result is telescope oscillation, rather than proper tracking. This means that drive trains should be designed with high inherent stiffness, and the motor should have adequate torque to overcome varying torque loads.

The traditional drive design, especially on smaller telescopes, has been to drive the tube assembly using a long, thin (2 inches in diameter or smaller) axle which is driven at the end opposite to that attached to the telescope tube. This approach is to be avoided because it is not very stiff. As shown in Figure 3-11, motor torque applied at one end of the axle tends to wind the axle up like a spring, with the telescope tube oscillating on the opposite end in response to varying wind and friction loads. A much better approach is to attach a drive disk directly to the bottom of a fork mount in RA, and to the tube assembly in Dec. The result will be a telescope mount with a high natural frequency that is a pleasure to use and easy to control by computer.

The final element to be considered is the control system. Depending on the telescope design and the requirements imposed by the observing program, position feedback, velocity feedback, both, or no feedback can be used in the servo. Despite the tendency to think of tracking as a time ordered sequence of pointing commands, in reality, pointing (slewing) and tracking are quite different operations. One may want to use some form of feedback for one operation, and go open-loop in the other. Appendix D summarizes the servo approaches used in successful projects a decade ago. In his systems, which are available commercially to professional observatories, Melsheimer (1983, 1984) uses a hybrid system of a stepper for tracking and a DC servo motor for slewing. When the control system is commanded to slew the telescope, the computer issues a speed command to the DC servo motor through its control amplifier. A tachometer on the motor provides velocity feedback to the servo amplifier, which uses this feedback to adjust the voltage on the motor to ensure it is turning at the commanded speed. Separate angle position encoders on the telescope axes are used to provide accurate position feedback. Therefore, both

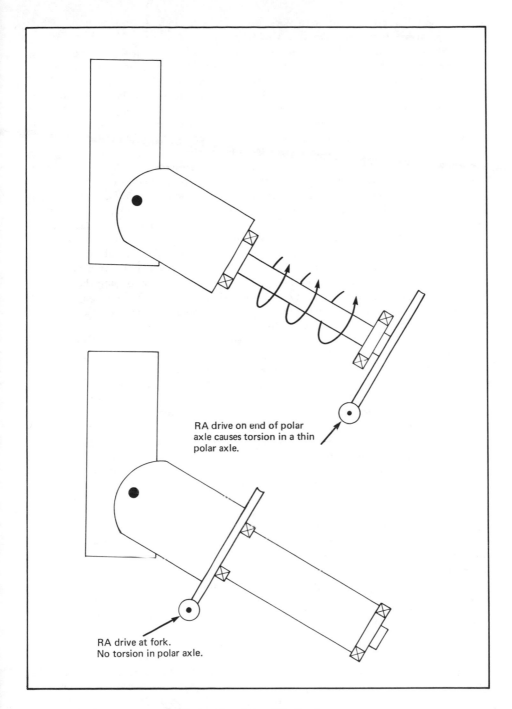

RA drive on end of polar
axle causes torsion in a thin
polar axle.

RA drive at fork.
No torsion in polar axle.

3-11 Avoiding Polar Axle Torsion

position and velocity servo loops are used for pointing. An open loop system employing stepper motors controls the tracking.

In his APT, Boyd performs open loop slews, then closes the pointing loop by using the photometer to lock onto the target star. He then uses simple open loop tracking, since the amount of time he spends on any one object is relatively short. Boyd regularly achieves better than 12' pointing accuracy using stepper motors and a chain drive. Both the short-term and long-term tracking accuracy are adequate for UBV photometry using a 10-second integration period, and a 1' diaphragm. Only precession corrections are computed in this system.

If modelling of drive train errors is to be avoided in systems requiring high accuracy, or if DC servo motors are used instead of stepper motors, some form of position feedback is usually employed. (In a well designed drive, drive train errors capable of being modelled are usually small enough to be ignored.) For reasons that are explained in Chapter 5, incremental optical shaft angle encoders are often the most cost-effective means of providing both position and velocity feedback. These devices send out a given number of "ticks" or pulses per revolution of the encoder shaft. If an incremental encoder shaft is attached to the axis of a telescope, and if that axis is driven at a constant rate, then the encoder will put out a stream of pulses at a constant rate. Position feedback can be obtained by feeding these pulses (and rotation direction information also provided by the encoder) to a counter which counts up when the telescope axis is turning in one direction, and counts down when the axis is turning in the opposite direction.

The pointing accuracy that is required determines the resolution of the encoder. For example, if 5" pointing accuracy is required in slewing, then the encoder resolution and its gearing to the telescope axis can be chosen so that the encoder sends out at least one pulse for every 5" rotation of the telescope axis. Close attention should be paid to the absolute accuracy specification of the encoder, and the method of gear reduction used (if any). Note that to achieve better than 5" accuracy, expensive encoders with 20 or more bits of resolution and accuracy are usually attached directly to the telescope axis being controlled. A mechanically geared-up encoder of lower resolution (and cost) could, conceivably, achieve these accuracies, provided great care is taken in the gearing.

Most small- to medium-sized telescopes with encoders use them for closed-loop pointing, but still track objects open loop. The reason for this is that closing the tracking rate (velocity) loop requires encoders of much higher resolution (and expense) than is required for the position servo. Also, it has been found that the accuracy of open loop tracking in well-designed telescopes is more than adequate for most purposes. Even though the tracking is open loop, tracking rate corrections (such as for refraction) can be included in the control software.

The encoder resolution requirements for a velocity servo can be made apparent with an example. Suppose that the research program to be pursued requires the telescope to drift from the object due to tracking errors no more than 1" in ten minutes. With 600 seconds in ten minutes, and a nominal tracking rate of about 15" per second in RA, an accuracy of 1" in 9000", or roughly 0.01%, is required. Now consider what would happen if an encoder with 5" resolution is used to provide position feedback by sending pulses to a counter read by the control computer. The first problem is that a deviation in position of 1" cannot be detected by an encoder having only 5" resolution. The second problem is that sufficient tracking rate information cannot accumulate in the required amount of time. At the nominal 15" per second tracking rate, the encoder sends out pulses at the rate of three per second, so the computer need not read the pulse counter any more frequently. The computer could execute a loop every one-third second in which the actual telescope position (determined from the counter) is compared to the desired telescope position (computed using the star's RA and sidereal time), and a new motor command is generated to compensate for the detected error. The problem with this approach is that the changes in motor speed commanded by the computer would often be greater than 0.01%, so the speed of the RA axis would change every one-third second by more than this amount to compensate for errors that accumulated over the last one-third second.

The effect of these relatively large and frequent speed variations on a perfectly stiff telescope and mount would be to jiggle the image at the program loop rate (3 times per second). Since the telescope and mount are not perfectly stiff, additional oscillations could result. If the program loop execution rate is the same as the natural frequency of the telescope, these oscillations would be reinforced, and the image would suffer perceptibly. Therefore, the frequency of speed changes in a closed-loop system should be much higher than (or much lower than) the natural frequency of the telescope. Furthermore, the resolution of the velocity feedback encoder should be high enough to keep the amplitude of speed changes within the allowed tracking rate error. Short term tracking rate requirements for many telescopes imply an encoder resolution of about 0".3 to 0".05 with very high tick-to-tick accuracy. Such high resolution encoders are very expensive.

It is clear, then, that the various elements of a telescope, both mechanical and electronic, should be directed toward the particular observing program, and matched to make an integrated system. The characteristics of two of these elements, motors and position encoders, are discussed in greater detail in the following two chapters. Those readers who wish to pursue control theory beyond the level presented here are directed to Miller (1977) for a good introduction to the subject that is more intuitive than mathematical, and to Goldberg (1964) for a more complete and theoretical discussion.

A. A STANDARD TELESCOPE PROBLEM

In this chapter, various kinds of motors and motor controllers that can be interfaced to microcomputers are described. Although this kind of technical information is interesting and useful, it is not the only point of interest. New motors are continually being introduced with improvements over earlier designs, and specifications of current offerings are changing as the technology advances. Appendix E lists the names and addresses of several manufacturers of motors and controllers appropriate to telescope control. (We recommend that the manufacturers listed there be contacted, and specific applications be discussed with their engineers. One long distance telephone call may save hundreds of dollars and many hours of grief.)

Another aim of this chapter is to demonstrate how to evaluate various motor options. Once the decision has been made to build a particular type of control system, a particular motor, gear, and motor controller combination can be selected that meets the system performance requirements. To help compare the performance of three motor and gear combinations, we have devised a hypothetical model of a telescope. The moment of inertia of this model is computed first. This defines the load that each motor and gear combination must move. In succeeding sections, each type of motor is described, then its applicability to the hypothetical telescope is assessed. In this way, a motor and gear combination can be selected. An example of this process for an actual telescope is presented in Chapter 15.

When retro-fitting a telescope with a computerized control system, one might consider using the existing drive motors and controls, by simply connecting these controls to the computer. When this is too difficult or is otherwise inappropriate, or when designing a new telescope, motors and controllers should be selected with a view to how easily they can be integrated into the computerized system. This is as important as their ability to provide the needed torque and slew, set, guide, and track rates. To do this, one should compare different servo design proposals with numerical calculations using actual motor and load characteristic data.

A hypothetical alt-az telescope is depicted in Figure 4-1, with the numbers in the figure identifying the mass elements described below. The mass elements define the standard load on the azimuth motor, using the moment of inertia equations in Figure 4-2 to compute the moment about the azimuth axis. English units (pounds, feet, etc.) are used. Consequently, since the pound is a unit of force, the slug (one pound/foot/sec/sec) is the unit of mass used in the table. In those cases in which two moment arms are involved, both are given.

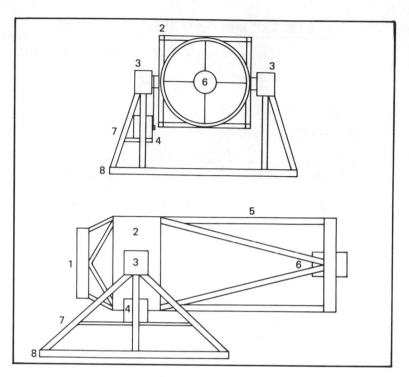

4-1 Hypothetical Telescope for Comparing Motor Drive Designs

Mass Element	Weight (pounds)	Moment Arm(s) (feet)	Moment of Inertia (slug-ft^2)
1. Mirror cell and supports	250	1.5	17.6
2. Altitude axis support box	100	1.0, 1.46	4.8
3. Altitude axis bearing housings	100	1.67	8.6
4. Altitude axis drive	100	1.67	8.6
5. Tube truss	160	3.0, 1.25	22.8
6. Secondary mirror and motor	75	6.0	84.2
7. Altitude axis support beams	120	2.0	15.0
8. Azimuth bearing ring	100	2.0	12.5

Total weight = 1005 lbs.
Total moment of inertia about the azimuth axis = 174.1 slug-ft^2

The total weight of 1005 lbs. is distributed in such a way as to yield a moment of inertia through the vertical azimuth axis of 174.1 lb-ft-(sec^2) or 33,427 oz-in-(sec^2).

The following paragraphs describe various types of motors, how they would be integrated into a computerized control system, and how their characteristics affect telescope control performance, based on this hypothetical model.

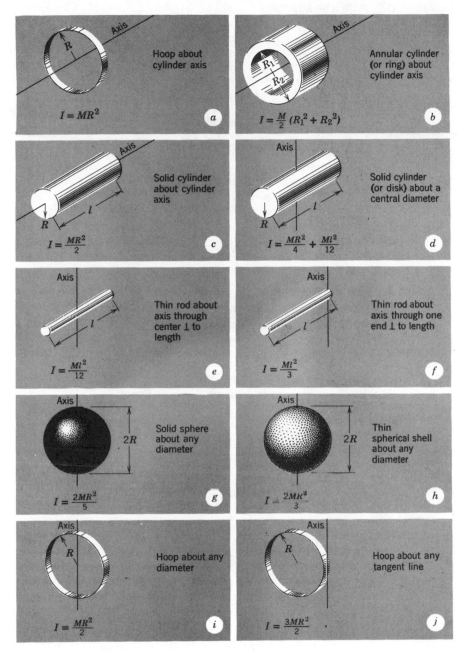

4-2 Moments of Inertia for Various Body Shapes
Courtesy of John Wiley & Sons, Inc.

B. LARGE DC TORQUE MOTORS

The first motor configuration is a large DC torque motor mounted directly on the driven axis. This configuration is used on a few modern small telescopes of 0.1 to 1.0 meter in aperture designed for special applications (low Earth orbit satellite tracking) that require very high speed, optimum servo performance. Most telescopes used for ordinary astronomical research do not use this approach, for reasons that are discussed below.

Torque motors are DC servo motors that use a permanent magnet stator. They are often frameless, which means that the usually large stator housing is bolted to the telescope mount, and the rotor has a hole through it so it can be press-fit mounted onto the telescope axis shaft. The telescope shaft bearings maintain correct rotor alignment inside the stator. These motors are intended for applications requiring high torque to inertia ratios. Although pulse width modulation control can be used with torque motors, linear proportional control is used in the high performance telescope control applications requiring DC torquers. Figures 4-3 and 4-4 show typical torque motors. An example of the largest practical motor is the Inland Model T-12008, which is the motor used in the calculations below.

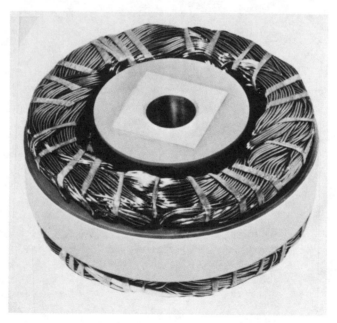

4-3 A Typical Brushless DC Torque Motor
Courtesy Inland Motor, Specialty Products Division

One advantage of a direct drive torque motor is that gearing errors are eliminated, since the telescope axis is driven directly by the motor. However, when DC torque motors are used in a simple analog servo system, they can suffer from an even larger error--the motor dead band, which is the error signal voltage (corresponding to an error angle) to which the motor just fails to respond. For example, to achieve a 1" dead band, assuming a peak voltage V_s of 49.8 volts and a servo amplifier gain K_a of 20 volts/volt, the control gain K_s, which is $2V_s/K_a$, must be 4.98 volts per arc second. To avoid exceeding the maximum applied motor voltage of 49.8 volts, the error signal cannot represent more than 10 arc seconds. This now poses problems in generating the error signal, given that the error can be as much as 180°. Conversely, if a servo gain K_s of 4.98 volts per 180° were used so as never to exceed V_s, and assuming a servo amplifier deadband (input bias current, converted to a voltage through the input resistor) of 15 uV (typical of high power servo amplifiers), the motor deadband is (15 uV/4.98 V) x 180°, or 0.000542 degrees (703", or 11'.7). At the 15 uV error signal level, the current is only 20 V/V x 15 uV/0.970 ohms (winding resistance), or 309 uA, which produces only 0.0012 lb-ft of torque. This is well below the 1.0 lb-ft motor friction, which is due mainly to the brushes. To produce 1.0 lb-ft to just overcome the friction, the error signal must be 0.0124 V. This produces a dead band of 0.45°, which is too large for most applications.

4-4 Open Frame DC Torque Motor
Courtesy Inland Motor, Specialty Products Division

The deadband is a problem only in simple analog servos, where a potentiometer or a similar device is used to generate the reference voltage representing the desired position. In more complex analog servos, or in digital servos, the control gain K_s is easily adjusted according to the size of the error. In a digital servo, the computer can apply the maximum voltage to the motor until it senses the error is less than 10" (the error signal value which has acceptable dead band). At this point, the program can revert to the linear mode until the desired position is reached. This would not be practical, since with a maximum error of 10" in the linear mode, the inertia of the telescope would cause significant overshoot at slew speeds. A more practical control approach is to vary K_s in fixed ranges, according to the current value of the error signal. This adjustment of the control gain is another example of the use of a forward path integral.

Aside from the higher cost of the large torque motor ($6000) and servo amplifier ($1500 or more), the power requirements for DC torque motors can be high. The Inland Model T 12008 develops 68 lb-ft of torque, but requires a 5200 watt servo amplifier (assuming 50% efficiency). Dissipating 5200 watts near the telescope optical path can cause air currents that degrade the image. There is usually no need to supply this much power when gears can be used to increase the effective motor torque, which permits the use of a smaller motor. This can reduce motor power consumption to 100 watts or less. Furthermore, most telescope mounts could not bear the weight of the T 12008 (190 lbs) on each axis without sacrificing some instrument weight capacity.

Despite these problems, the acceleration of the torque motor is excellent. Even assuming the motor only delivers its continuous torque (which is lower than its peak torque) throughout a slew, the torque of 68 lb-ft (13,056 oz-in) accelerating a mass moment of 33,427 oz-in-(sec^2) gives an angular acceleration of 0.39 rad/(sec^2). That is, it can move the telescope through 180° in less than 6 seconds (counting ramp up and ramp down times)! Although this is impressive performance, that much mass moving that quickly could rip tangled cables to shreds or severely injure an unwary bystander before the observer could intervene. Furthermore, the telescope would be difficult to point accurately without a tachometer generator to provide velocity feedback to damp the system.

Thus from the standpoint of cost, power, weight, and ease of control, direct coupled DC torque motors are inappropriate to the typical telescope control problem (although they are appropriate in those few instances in which extremely rapid slewing or tracking performance is required). A DC torque motor which is smaller than the T-12008 would have less weight, power consumption, and cost. However, its dead band would be about the same, and small variations in the load due to friction or wind gusts would cause large over- or undershoots when the driven shaft is near its desired position, since a smaller motor would not have adequate torque to compensate for these variations.

C. SMALL DC SERVO MOTORS

Torque motors are but one member of the general class of DC servo motors. Other examples include wound-field, permanent magnet (PM), moving coil, and printed circuit types. Figure 4-5 shows a typical servo motor. Figure 4-6 shows a sample of the wide range of available servo motors.

The wound-field types use coils in both the armature (rotor) and the field (stator). By controlling the current in either the armature or the field, the speed of the motor is controlled. It is better to use armature control in telescope control applications, since the requirement for larger driving signals is usually offset by greater ease of control.

PM motors use a permanent magnet in the stator instead of field coils. They usually have two wires instead of the four found in a wound-field type. Advantages include higher starting torque, a linear speed-torque relationship, smaller size, and lighter weight. PM motors usually use brushes for commutation, but brushless types are now available. This is the type of servo motor used most frequently in modern telescopes.

4-5 A Typical AC Servo Motor
Courtesy Vernitron Corporation

4-6 Some of the Many Available Servo Motors
Courtesy Vernitron Corporation

The moving coil motor is a PM motor with a different type of construction. Rather than using a conventional cylindrical armature, they use either a flat disk for the armature, or a shell armature that has no iron; epoxy or fiberglass holds the copper conductors in the armature. A variation is the printed circuit motor, which uses copper foil on printed circuit board material to form a flat disk armature. Although moving coil motors have low inertia, high acceleration, and other good qualities, they are more expensive than the conventional PM motors, and therefore are rarely used in telescope control applications.

The second motor configuration to be tested against the standard telescope model is a small PM instrument DC servo motor geared to each telescope axis using precision gears. A typical motor to use in this case is the Inland Motor Model T-1814, which has 85 oz-in of stall torque.

Assuming that a 359:1 reduction is used to gear down the motor, the load moment of inertia is reduced by $1/(359)^2$ to 0.26 oz-in-(sec)^2. When combined with the rotor inertia, this yields a total inertia of only 0.28 oz-in-(sec)^2. Therefore, if the continuous stall torque were applied when the load is at rest, the acceleration would be 85 oz-in/0.28 oz-in-(sec)^2, or 304 rad/(sec)^2, which, when divided by the gear ratio, accelerates the telescope from rest at a rate of 0.85 rad/(sec)^2. Although this acceleration figure is higher than that for the DC torquer, the top speed of the torquer is significantly higher, since there are no gears between the torquer and the telescope axis to reduce the speed.

To compute the overall gain K of the servo, assume the control gain, K_s, is 4.98 V/π radians, or 1.585 V/radian; the servo amplifier gain, K_a, is 20 V/V; the torque sensitivity, K_v, is 36.8 oz-in/A; and the gear reduction, N, is 359. Therefore, K is 1.585 V/radian x 20 V/V x 36.8 oz-in/A x 359 (/4.07 ohms), or 102,910 oz-in/rad, which gives a damping constant for critical damping, F_c, of 368.6 oz-in/rad/s. This makes the servo heavily overdamped, which eliminates the possibility of overshoot, but increases the time required to settle to the new position. This is not a disadvantage, since most telescope control systems need to be overdamped to the point where it may take a half minute to slew 180 degrees. Assuming the same servo amplifier characteristics as in the previous example, the dead band is reduced by the amount of the gear reduction from 703" in the previous example to about 2".

The performance of the geared servo motor makes it entirely acceptable as a telescope drive, with the only disadvantage being that it is more difficult to interface to most computers than a stepper motor.

D. STEPPER MOTORS

The third motor configuration to be considered is a stepper motor geared to each telescope axis using precision gears. A stepper is typically a DC motor with a permanent magnet armature consisting of a dozen or more magnet faces and four or more field windings. An example of a small angle stepper motor is shown in Figure 4-7.

4-7 A Typical PM DC Stepper Motor
Courtesy The Singer Company

When a current is generated in one of the field windings, the armature is forced to move through an angle determined by the number of armature magnet faces and field windings, then it comes to rest. This action is a "step". By turning successive field windings on and off in a set pattern, the motor can be made to step repeatedly until the armature shaft rotates through 360 degrees, as shown in Figure 4-8. Steppers are available in sizes from a fraction of an oz-inch of torque (for quartz "analog dial" wristwatches) up to 3000 oz-inches or more, and from 2 steps per revolution up to 50,000 steps per revolution. Typically, a special drive card is needed to convert a train of digital TTL pulses into the correct sequence of field winding currents, although Cybernetic Micro Systems now makes integrated circuits that interface easily to a microcomputer bus and convert commands sent in standard ASCII character codes to stepper motor coil currents. A typical motor for telescope control is the Superior Electric M092 Slo-Syn stepper with 200 steps per revolution.

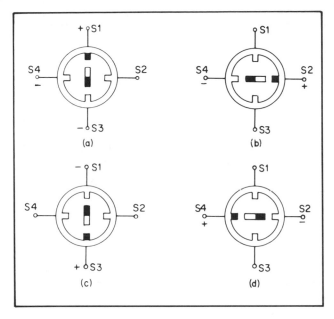

4-8 Stepper Motor Operation
Courtesy Reston Publishing Co.

The maximum speed of the motor is determined by the maximum number of pulses per second it can translate into steps. The servo and stepper motors share the property that their torques decrease rapidly with increasing angular velocity. However, the two motors differ, in that the servo motor torque versus speed curve is essentially linear, while the stepper motor torque curve is quite non-linear. The characteristics of the circuit used to drive the stepper motor are as influential as any other factor in determining a stepper's torque versus speed curve. Circuits that compensate for the dynamic impedances of the motor coils enable the motor to deliver higher torque at high speeds, but still do not make the torque versus speed curve linear. A typical stepper motor torque curve is shown in Figure 4-9.

Because of the divergence of the running torque and start-up torque versus speed curves at even moderate step rates (with start-up torque being lower), the controller should ramp a stepper up to speed rather than accelerating instantaneously from a standstill. Even if ramped, steppers may suddenly seize at some point in the ramp when the load is such that the rate of stepping causes the rotor to oscillate at a harmonic of the mechanical natural frequency. This can be avoided by using the design practices discussed in the previous chapter. Mechanical rotational dashpots are sometimes added to steppers to increase damping, which decreases the per-step settling time. Ramping a stepper permits it to achieve higher step rates, and is necessary in worm gear drives to prevent gear damage, and to reduce telescope oscillations when a slew is completed. The automatic photoelectric telescope described in Section V initially ran without ramping, but was modified to include this feature. In that system, a step size of 1".6 is used with a tracking rate of 11.5 steps/second and a slewing speed of about 5000 steps/second.

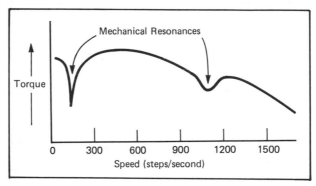

4-9 Typical Stepper Motor Speed vs. Torque Curve

In evaluating stepper performance, note that the stepper dead band is essentially its step size. For motors with 200 steps/revolution, each step is 1.8 degrees. Gear reductions should be chosen to make each step an appropriate size. In systems with separate slew and track motors in which the track motor is a stepper, the step size should be considerably smaller than the seeing disk, and small enough to avoid any noticeable vibration, say 1/2 to 1/20 arc second. In systems where there is only one stepper motor per axis, then the step size and slew rate must be balanced against each other. Having high maximum step rates eases this tradeoff, but it is possible that one may have to live with a large step size or a slower slew speed than one might have otherwise wanted. In some applications, small step size is only important for tracking in RA. In these cases, the step size can be large (1" or more) in Dec, and in RA while slewing. A higher gear reduction ratio is usually needed for a stepper motor than that typically used with a servo motor.

Despite the need for care in design, ramping, and additional gearing, steppers can give performance rivalling servo motors, and at reasonable cost. Servo motor and stepper motor costs are roughly equal when the motor, controller, power supply, computer interface, and gearing are all taken into account. All of the items needed to build a telescope axis drive using either a servo motor or a stepper motor can be purchased for about $1000 per axis--considerably less than for a large DC torque motor. If you have a sharp eye at the surplus electronics stores, the patience to wait for a bargain, and the willingness to build some of the electronics yourself, you can cut the price to $300 per axis, or even less.

Stepper motors have several significant advantages over other types of motors. First, they can be used without position or velocity feedback in simple open-loop systems, such as those described in Section IV. This eliminates the cost of shaft angle encoders and their computer interfaces, if they are not needed for other reasons. The computer can determine the number of pulses to be sent to the stepper motor, and using the (geared down) step size, an estimate of the telescope's position can be made in software. This is not possible using either the DC torquer or the smaller DC servo motor.

The second advantage of the stepper motor is that it is easier to interface to a computer. All that is needed is a source of pulses. Examples of simple electronic circuits to provide pulses at a rate under software control are given in Sections IV through VI. Pulse output interface cards to drive stepper motors are available for several different computer buses. Although DC servo motor controllers are available for the STD Bus, Q-Bus, Multibus, and a few other standard computer buses, in general they are quite expensive ($1000 and up).

For these reasons, steppers are often the best choice for a low cost computerized telescope control system.

One disadvantage of both steppers and small servo motors is the need for gear reduction. Aside from introducing some errors, which can be reduced to a reasonable level using precision gears, the use of gear reduction requires that these motors be capable of a high maximum speed, which is often difficult to achieve. To solve this problem, and because motors are not capable of full torque at high speeds, a separate slew motor is often used, especially on telescopes larger than one meter in aperture. Although it requires an additional computer interface per axis and the slew motor must have higher torque than the tracking motor, small telescope slew motors often can be operated at a single speed, so it may be connected to line voltage with a simple on/off switch under computer control (e.g., an Opto-22 module). Most slew motors on large telescopes are ramped, however, to prevent gear damage.

One form of mechanical reduction which is quite useful is precision belting, such as that available from Winfred M. Berg, Inc. A good toothed belt cushions stepper pulses slightly, which helps to prevent stalling due to mechanical resonances. Toothed belts have negligible backlash, and the slight amount of elasticity in the belt tends to average out tooth-to-tooth belt errors. Belts are most appropriate as the first stage in the drive train right after the stepper, so their errors can be reduced by later gear stages.

The use of two motors per axis is expensive. To avoid this expense, a high maximum step rate is needed so a single stepper per axis may be used. One method of obtaining this high step rate is to employ a higher than normal voltage to drive the stepper when it is slewing the telescope. The field windings have the electrical characteristics of a pure resistor in series with a pure inductor. When the normal operating voltage of the stepper motor, say 6 volts, is applied to the field winding, the resistor limits the total current flow, and the inductor uses the inrushing current to build up a magnetic field about its windings. It takes a relatively small, but finite amount of time to build up the magnetic field in the windings. This time is only small when compared to the length of time between steps at low step rates; at high step rates, it can be a significant fraction of the time between steps. This reduces the torque for high step rates, which effectively limits the maximum step rate for most steppers to about 500 to 1000 steps per second.

By using an overvoltage of a few times the normal operating voltage to slew the motor, say 24 volts for a 6 volt motor, the time it takes to build up the field is reduced, since the initial current is larger. This produces full torque at higher step rates. By shortening the drive pulse "on" time, the energy stored in the field windings is delivered at a faster rate, so the power (at any given step rate) is increased. This technique can be used only for high step rates, since when the motor is stopped or stepping slowly, the amount of time the drive voltage is switched to any given winding is long enough for large currents to build up and overheat the winding. A variation of this approach, which can be used for all motor speeds, is to apply the higher voltage only for a short time at the beginning of a step, then to reduce the drive voltage to the rated voltage. Circuits to implement this "bi-level" approach can be found in the literature of some of the stepper motor manufacturers. The concept is illustrated in Figure 4-10.

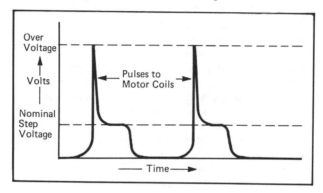

4-10 Bi-Level Stepper Driver Characteristics

Another method of improving the speed range of a stepper motor is to drive the motor in half steps instead of full steps. Half steps are achieved by driving two field windings simultaneously, rather than just one at a time. This produces 400 steps in a 200 step per revolution motor, which allows more freedom in choosing mechanical reduction ratios. It also produces smoother steps, which means the motor is affected less by mechanical resonances. Superior Electric, Mesur-Matic, and other manufacturers make stepper motor controllers capable of half step operation. The use of half steps (or smaller) is recommended over the use of full steps, which disturb both the motor and the entire drive train with large "kicks" of torque that tend to induce vibrations. Some of the drive systems described in Section IV make use of half stepping to obtain smoother operation and a wider speed range.

The half step concept has been extended by both Compumotor and Mesur-Matic to microstepping. Mesur-Matic achieves this by placing several small current steps per full motor step on pairs of field windings, while Compumotor uses a pulse width modulation technique. The net effect of both methods is to divide the current flow between the pairs of windings. The Compumotor Model 106, shown in Figure 4-11, can be purchased with either 25,000 steps per revolution (standard), or 50,000 steps per revolution (optional). The technique used by Mesur-Matic to divide each full step of a 200 step per revolution motor into 4 microsteps is illustrated in Figure 4-12. Mesur-Matic offers step division factors from 2 to 64, for up to 12,800 steps per revolution. Step rates up to 500,000 steps per second or more are possible with microstepped motors. At the other end of the speed range, microstepping allows motors to be operated at about one-tenth the speed achievable using conventional stepper drivers without introducing noticeable vibration. This provides the wide speed range required for drives using one motor per axis, and for alt-az mounts. One skilled in electronics could build a stepper driver that can be switched between two or more step sizes to help solve the step size/slew rate tradeoff problem mentioned earlier.

4-11 Compumotor Very Small Angle Stepper and Drive
Courtesy Compumotor Corporation

The speed range of any motor is necessarily limited. Generally, DC servo motors have a wider speed range than stepper motors. Among the steppers, those that employ microstepping have the widest speed range. If separate slew motors are used, then speed range will not be a problem. However, if one is willing to make the necessary compromises, then the conventional stepper is the least expensive single-motor-per-axis approach. If better performance is desired, then a microstepper or DC servo motor approach is preferable.

One characteristic of steppers which should be taken into account is the motor's absolute accuracy specification. When a torque load (as opposed to a pure friction load) is placed on a stepper at rest and powered up, the holding torque of the stepper is offset by the load, which produces a shaft position error. This effect is depicted in Figure 4-13. The following table gives a rough idea of shaft error (deviation from the no load position) as a function of load torque (expressed as a percentage of the motor's holding torque) for a typical 200 step per revolution motor.

Load Torque (% of holding torque)	Error (degrees)
25	0.3
50	0.6
100	1.7

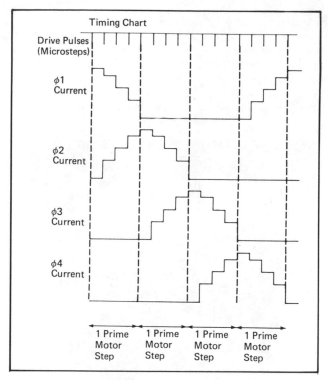

4-12 Microstepping Timing Chart
Courtesy Mesur-Matic Electronics Corporation

In a dynamic situation with no load on the motor initially, when the motor starts to turn at some constant rate, a constant load torque will introduce a "phase error" (shaft position error) of an amount given roughly by the table above. This error will remain constant as long as the torque load causing the error remains constant, and will act similarly to servo lag position error. However, as the torque load changes with changing telescope pointing angles, the motor shaft position error will vary by amounts similar to those given in the table for the static case. For the dynamic case, the motor torque of importance is the pull-in torque, not the holding torque. Gearing should be chosen so that the change in torque load on the motor for all telescope pointing angles is a small fraction of the motor's pull-in torque. Gear reduction not only amplifies the motor's torque, it also reduces the effect of a given motor shaft position error on the driven telescope axis.

As an example of how a manufacturer specifies a stepper's absolute accuracy, Compumotor specifies the unloaded absolute accuracy of its Series M106 microstepping motors as ±5'. The loaded error varies with the load, and is specified as being approximately equal to the unloaded error plus a factor which is proportional to the residual friction loading when stopped. This additional error is given as 1' for a friction load of 1% of the motor torque. This is roughly linear, so a 50% load would produce a dynamic shaft position error of 50'. When added to the unloaded error of 5', this would produce a total motor shaft error of 55', or almost one degree (with no gear reduction). Each motor step in a motor with 25,000 steps per revolution is only 52", so the total shaft error can be as large as 60 times the

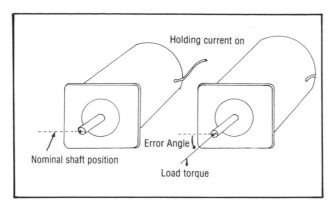

4-13 Stepper Error Induced by Load Torque

step size if the motor's torque is much too small to handle load variations.

Another effect, unique to microstepped motors, is torque ripple. The motor is microstepped by applying different currents to two windings simultaneously. When one of the motor's permanent magnet poles is nearly aligned with a field winding, almost all of the current is in the one field winding nearest the magnet pole. This produces a slight dip in the torque at this rotor position. With a constant load, this reduction in torque shortens the step size slightly. Conversely, when the magnet pole is in the middle between two field windings, both windings are driven by roughly equal currents. This raises the torque slightly, which lengthens the step size. In the Compumotor M106 with 25,000 steps per revolution, this effect causes the step size to vary sinusoidally between 50" and 70". The error does not accumulate over a complete 360 degree rotation of the rotor, but it does accumulate temporarily between the motor poles to a maximum value of ±5', as shown in Figure 4-14. Since the individual step error (as much as 20") is a large fraction of the nominal step size (52"), and the maximum error between motor poles (5') is so large, the motor should be geared to the telescope axis in a manner that reduces this effect to a tolerable level.

When dealing with such small step sizes, step-to-step errors due to small variations in the spacing and shape of the rotor magnet poles and the placement of field windings (both of which vary from motor to motor and from step to step within a particular motor) become a significant effect. Compumotor corrects for these effects by measuring each unloaded step size at the factory, and storing correction current values in a PROM that is placed in the motor controller. This feature customizes the field winding currents for each microstep to correct for motor geometry variations, but it cannot eliminate the torque ripple errors.

Another error source in microstepped motors is hysteresis. If the motor is stepped n steps in one direction, then the same number of steps in the opposite direction, the shaft does not end up at the original position. As much as 3' of the 5' total unloaded shaft error in the Compumotor M106 can be due to hysteresis (White, 1984).

The effects of dynamic load torque and torque ripple errors can be reduced through gear reductions between the motor and the driven shaft. By choosing the gear reduction and motor torque carefully, a microstepper can be a very effective telescope drive motor. A design example using microsteppers is discussed in Chapter 15.

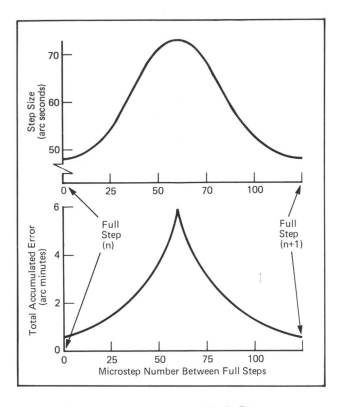

4-14 Microstepper Torque Ripple Error

Steppers are available with many options. Figure 4-15 shows a stepper with an integral gearhead reducer, and Figure 4-16 shows a stepper with an integral lead screw, which would be useful for moving a Cassegrain secondary for focus adjustment.

4-15 A Stepper with Integral Gearhead Reducer
Courtesy Vernitron Corporation

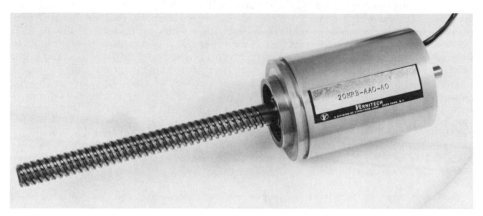

4-16 A Stepper with Integral Lead Screw
Courtesy Vernitron Corporation

E. SERVO MOTOR CONTROLLERS AND COMPUTER INTERFACES

Servo motors require a linear power amplifier to drive them. For a DC motor, the amplifier must accept a DC voltage from the computer and drive the motor at a speed proportional to this input (control) voltage. Usually, a tachometer attached to the motor generates a feedback voltage proportional to the motor's speed. This serves as a second input to the servo amplifier, which adjusts the current in the motor coils to keep the motor speed proportional to the control voltage.

Desirable servo amplifier characteristics are low input bias voltage and current (the voltage or current that just barely produces no output), and very linear output voltage or current (without velocity feedback) or linear motor speed (with velocity feedback) response to the control voltage. Good servo performance is expensive. For example, the Torque Systems C0401 servo controller, a popular servo amplifier used in telescope control systems by DFM Engineering, Inc., costs about $460. With a Torque Systems "snapper" servo motor and tachometer combination unit, which sells for about $220, the motor and controller total price is about $700.

A variation from the traditional linear servo amplifier which is less expensive and easier to interface to a computer is the technique of pulse width modulation (PWM). Silicon General markets the SG1731 DC motor PWM chip that converts a DC control voltage to a series of pulses with widths that are proportional to the control voltage. This chip can drive small servo motors directly, or large DC torquers with external power transistors. It can also accept tachometer inputs for speed control. If this approach looks promising for your application, a discussion of your motor performance requirements with the manufacturer of the telescope drive motor will aid in your assessment of the ability of PWM to meet these requirements. For a wide variety of uses, PWM can provide adequate control of DC servo motors inexpensively. In those applications requiring better performance with larger motors, linear proportional control can give more precise control, but at a higher price.

There are two basic ways to attach DC servo motors to computers. The first is to use an interface card with the servo amplifiers built-in. This is exemplified by the Buckminster Corporation C-1600 card for the LSI-11 (they offer a similar card for the STD bus) which can control up to four servo motors (two on the STD Bus) directly without the need for additional amplifiers. The cost is about $1300, or $325 per motor. The other approach is to use a card employing a digital-to-analog (D/A) converter to apply a control voltage to a separate servo amplifier. The computer loads a number into a bus register which the card converts to a corresponding voltage. Such cards are available for a large number of computers, including the Apple, IBM PC, LSI-11, and STD bus. The ADAC 1412DA four-channel 12-bit D/A interface card for the LSI-11 gives 4096 separate control voltages (motor speeds) for each channel, and sells for about $545. Note that the resulting tracking resolution is one part in 4096, or roughly 0.02%, which may not be enough.

Motion Science offers a servo motor/PWM amplifier combination using either an STD Bus or RS-232 interface. The continuous torque of their motors ranges from 200 to 650 oz-in., with speeds to 3000 rpm. The system is expensive, up to $2,000 per axis.

A low cost closed loop DC servo motor control system can be built very inexpensively. Such a system would not be very accurate, but it would place the object within the field of the finder. This is all that many applications require. The system can be built with power operational amplifiers (about $10 for 100 watts) driving common DC motors. Conductive plastic potentiometers can provide the angle position feedback. The computer interface would consist of a 14-bit Burr-Brown D/A converter to drive the op amp, and a 14-bit Burr-Brown A/D converter to read the voltage on the wiper of the potentiometer. This drive system would be used for slewing only, then a normal synchronous motor and drive corrector would be used for tracking. The AC drive corrector (with a set speed) would also be used to center the object in the field of the main telescope. An estimate of parts costs is given below.

Item	Cost (new)
Conductive plastic potentiometers (2)	$ 60
DC motors (2)	20
Inexpensive gears (2 sets)	60
DC power op amps (2)	20
A/D's and D/A's	80
Miscellaneous interface chips, etc.	80
Total	$320

Costs could be reduced further by using parts bought at radio amateur "ham fests" and surplus electronics stores. Using these sources, the total cost for both axes might be $100 or less. Using this approach, the computer would read the potentiometers, correct them for non-linearities using a calibration curve determined from star measurements, and adjust the motor control voltage based on the potentiometer readings. The computer program would slow down the motors as the desired positions are approached, to avoid overshoot. The only correction such a system has to compute is precession. A refraction correction could be added if observations are to be made near the horizon. If higher accuracy in pointing and tracking are required, the system will cost more, due primarily to the higher costs of more accurate shaft angle encoders and their computer interfaces.

F. STEPPER MOTOR CONTROLLERS AND COMPUTER INTERFACES

Unlike servo motors, steppers require a controller that turns the winding currents on and off in a precisely defined sequence. This can be done directly by the computer at low cost. Eight single bit outputs (four per motor) can be toggled by the computer in the proper drive sequence. These outputs drive power transistors, which feed coil currents to the stepper. This requires a sizeable number of the available CPU cycles of an 8-bit microcomputer, but if other computations do not consume very much of the CPU, this is probably the least expensive way to drive a telescope with a computer, especially since steppers are readily available at many surplus electronics stores.

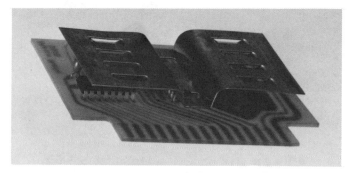

4-17 A Mesur-Matic Model 3616 Stepper Driver/Translator
Courtesy Mesur-Matic Electronics Corporation

The next step up in price is to use a commercial stepper controller. This approach reduces the load on the CPU, at the cost of adding more hardware to the system. Figure 4-17 shows an inexpensive driver card capable of either full or half steps. It accepts TTL-level or CMOS inputs on two separate input lines. These can be in the form of either pulse rate and direction, or up-pulses and down-pulses (pulse one line for CW rotation, pulse the other line for CCW rotation). Although some inexpensive stepper controllers, such as the one shown in Figure 4-17, can accept pulse rates up to 1 MHz, they typically do not compensate for motor characteristics which change with speed. This usually limits the top speed of the motors driven by this driver to under 1000 full steps per second, since the motor torque falls off rapidly after about 500 full steps per second. The computer could be used to toggle a single output bit at the desired step rate. This output would serve as the signal to the driver card. A second bit could be used to indicate motor direction to the controller card. This is the approach taken in the APT system described in Section IV. The APT stepper driver is capable (with ramping) of up to 5000 half steps per second with ordinary Slo-Syn steppers. Most of the parts for this driver can be obtained at Radio Shack for about $30.

More sophisticated driver cards are offered by Superior Electric (manufacturer of the Slo-Syn steppers) and others. Figure 4-18 shows the Mesur-Matic Model 3620 stepper driver, which is able to supply almost three times the current as the Model 3616 shown in Figure 4-17. In addition, high step rate performance is improved. Mesur-Matic also offers the Model 3640A Microstepper Driver/Translator for producing very small step angles and very high step rates.

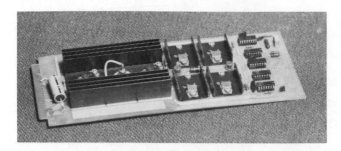

4-18 A Mesur-Matic Model 3620 Stepper Driver/Translator
Courtesy Mesur-Matic Electronics Corporation

If the microcomputer generates the pulse train, a good deal of the CPU can be consumed in this activity. In the APT, the CPU has little else to do, so here CPU loading is not an issue. However, in those applications where the CPU must perform other tasks, an interface card can be used to generate the pulses that are translated by a stepper controller into motor steps. Pulse output cards designed to drive steppers are available for many different computers. One example is the ADAC 1604/POC for the LSI-11. This card has four output channels for driving up to four motors. Each channel can send different pulse rates from 100 to 50,000 steps per second in either speed/direction or CW pulse/CCW pulse mode. The computer determines the pulse output rate by loading a number into a bus register on the card. From that point on, the card sends out pulses at the commanded rate without further reliance on the computer until the computer loads a different number into the register. Besides the pulse output lines, this card has two separate input lines per channel for limit switch sensing.

Another approach is the Superior Electric stepper driver that uses the General Purpose Interface Bus (GPIB), otherwise known as the Hewlett-Packard Interface Bus (HPIB) or the IEEE-488 standard. This is shown in Figure 4-19. Since many computers are available with GPIB interfaces, this driver or one using a similar approach could become popular.

Rather than offering board-level products designed for a particular computer or employing an industry standard interface, some manufacturers offer integrated circuits designed to perform both computer interface and driver functions. Three such products from Cybernetic Micro Systems accept commands over a parallel TTL input (such as an 8-bit microcomputer bus) in a standard ASCII character shorthand. These commands can be stored in the chip and executed when needed. The command repertoire is very robust, and includes commands for setting an acceleration ramp slope, total number of steps to send, and destination position. Programming commands such as start, jump, loop, and quit program mode are also included. These very intelligent chips can drive low-power steppers directly, or more powerful motors with external driver transistors. The CY500 ($95) can handle full or half steps at rates up to 3300 steps per second, the CY512 ($145) can handle rates up to 8000 steps per second, and the CY525 ($195) can handle rates up to 10,000 steps per second. The CY525 has an added feature that permits continuous program looping to send out an uninterrupted stream of step pulses at the specified rate. It also permits the stepping rates and slopes to be changed while the program is executing.

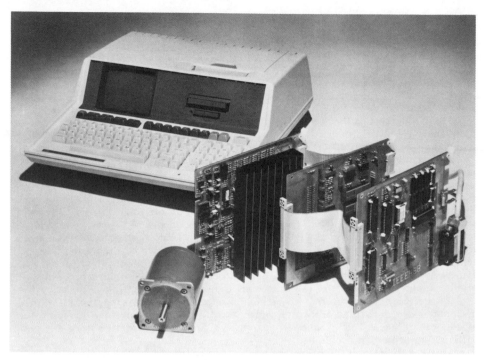

4-19 A Stepper Driver Using the IEEE-488 Interface Standard
Courtesy The Superior Electric Company

Like the chips from Cybernetic Micro Systems, the PPMC-101B/201A from Sil-Walker, Inc. is also designed to connect directly to an 8-bit microcomputer bus, but instead of driving the stepper directly, it is designed to send the desired number of pulses per second to a separate driver card. This chip offers intelligent control of acceleration rates and total number of steps to be sent, at rates up to 5000 pulses per second for the PPMC-101B, and 10,000 pulses per second for the   PPMC-201A. Unlike the CY5xx chip family, it does not store programs.

Sigma Instruments makes a very useful hybrid analog/digital device that also has a ramp capability, but does not interface to a computer bus. Their device consists of two chips and a few discrete components, and is relatively inexpensive (about $100). To operate the Sigma ramper, a steady stream of pulses is provided with the exact number of pulses desired in a move at the steady-state rate desired. The Sigma ramper works by "borrowing" pulses during the ramp up, and then adding these pulses back in, after the input pulse train has stopped, to provide the ramp back down. This allows ramping to be done easily in systems which do not contain a computer.

The last approach to computer control of steppers is an intelligent controller card that combines the "smarts" of the CY5xx chips with a standard bus interface. Rogers Labs uses the CY512 in its stepper controller for the Apple II microcomputer. The A6 T/D card plugs directly into the Apple II backplane, and serves as the interface to the stepper driver card. The R2 D23 card handles two Size 23 steppers rated at 35 oz-in. The complete package sells for $365 and includes both cards, two steppers, and software, and provides an inexpensive totally integrated system for Apple II owners. The R2 D34 card and two Size 34 motors rated at 220 oz-in may be substituted, for a total package price of $595. The Size 23 motor power supply is optional at $145, or the user can furnish his own. A similar approach is used by Whedco in its STD Bus intelligent stepper interface card, which allows the CPU to load motor drive parameters (ramp characteristics, speed, number of steps to send, etc.) and then handle other computing chores while the intelligent stepper controller card controls the motor. This card is described in greater detail in Section IV.

The main point to these intelligent chips and boards is that once the computer loads the slew profile and motor control parameters into the chip or board, the computer can go off and do other things. This permits sophisticated control of steppers without requiring equally sophisticated real-time programming techniques or most of the available CPU cycles.

One final point to bear in mind when designing the drive system is how the motor/gear combination affects the stability of the control servo. For example, direct-drive systems using a large DC torquer can be difficult to fine-tune, and usually require both position and velocity servo loops and some even have acceleration loops (Klim and Ziebell, 1985). This is because structural resonances are reflected directly into the velocity loop, and the inertia of the motor is small compared to the inertia of the load. This usually requires a separate compensation (filter) network for each servo loop, and can easily lead to stability problems. A smaller motor geared to the telescope has a much larger inertia as "seen" by the load (larger by the factor n^2, where n is the gear reduction), so the structural resonances are decoupled from the motor for frequencies above the locked rotor resonance frequency. The bandwith of the velocity loop can therefore be high without any stability problems.

This is only a small sample of the wide range of motors and controllers that are available. Having discussed in some detail how these components can be used in telescope control systems, we now turn our attention to methods of providing position feedback from the telescope axes to a computer, so the control system can "sense" where the telescope is actually pointing.

Chapter 5. **ANGLE POSITION SENSING DEVICES**

In a classical closed-loop telescope control system, one or more position feedback sensors may be placed anywhere between the motor shaft and the telescope focal plane. Once a position feedback sensor is calibrated so that its output can be converted to apparent position in the sky, the sensor will automatically correct, through its calibration, for any systematic errors (e.g., gear errors) from the motor shaft up to the position feedback sensor. The position sensor cannot be used to correct for errors between it and the focal plane. These errors (mount flexure, tube flexure, collimation errors, etc.) may be modelled by the error and command computer, which then determines a new position feedback sensor reading corrected for the modelled errors. By placing the position feedback sensor as close to the focal plane as possible, one can use the sensor calibration to remove systematic errors directly, rather than increasing the computational burden of the error and command computer in computing these errors.

Many methods are available for providing shaft angle feedback for closed-loop servo telescope control, with a wide range of accuracies and prices. Most telescopes using control systems with position feedback employ optical shaft angle encoders, so particular attention should be paid to this portion of the chapter. Inductosyns and synchros are also used in larger telescope applications. We would expect most modest-sized telescopes controlled by microcomputers to be open-loop systems, a few to use precision potentiometers for low cost position feedback, and most of the rest to use incremental optical encoders of moderate accuracy. Other types of sensors are covered to stimulate interest and ideas for improving the accuracy of low-cost position feedback devices. Appendix F lists the names and addresses of several manufacturers of position angle feedback devices.

A. PRECISION POTENTIOMETERS

One frequently-used form of shaft angle encoder is the precision potentiometer. Two examples are shown in Figure 5-1. This device is typically used with a well-regulated DC power supply to provide a DC voltage proportional to the potentiometer's shaft angle. In analog telescope control servos, it can be used for both position control input and for actual position feedback. Precision op-amps can be used to provide the error signal which is then amplified to drive a DC servo motor. This very simple system was described in the previous chapter.

The characteristic that determines a potentiometer's suitability for position encoding is its conformity to the desired function of resistance versus shaft angle. The function which is usually desired in a telescope control system is a linear relationship between resistance and shaft angle. Any deviation from linearity is a characteristic of the potentiometer, and cannot be offset by adjusting other system parameters. Wirewound potentiometers are quite common, and can be obtained on the surplus market with a linearity of 0.1% (0.36° if used directly on the measured axis) for roughly $15 - $25. The accuracy of these potentiometers is limited

5-1 Precision Potentiometers
Courtesy Vernitron Corporation

primarily by the potentiometer's resolution. For direct coupling to the telescope axis, the resolution is 360° divided by the number of turns of wire used to make the winding. Since 1000 turns or more of wire is typical for these potentiometers, their resolution roughly equals their linearity, or 0.1%.

The other parameter that influences the accuracy of wirewound potentiometers is temperature stability, which may be on the order of 0.05% per degree C. This limits the overall accuracy to about one-half degree of arc for typical observing conditions. This may be adequate for those who only wish to find Messier objects. A way around the temperature stability problem is to use the ratio of resistances (or voltages) of the two legs of the pot instead of the absolute resistance (or voltage) of just one leg. This can be done in an analog circuit by placing one leg between the inverting input of an op-amp and ground, and the other leg between the inverting input and the output of the op-amp. In a microcomputer-based system, both legs of the pot can be connected to the computer using A/D converters, and the ratio of voltages can be computed directly. However, power supply voltage ripple and random noise on the supply voltage to the pot will still introduce errors into the readings of the voltages on the two legs, and limit the overall accuracy.

Several suppliers manufacture so-called "infinite resolution" potentiometers, which use a thick film of conductive plastic as the resistive element. The resolution of such potentiometers is not limited by the pot itself. In analog systems, the resolution of "infinite resolution" pots is determined by the bias voltage and drift of the DC amplifier used in the servo. In digital systems, the sample-and-hold amplifier of the A/D converter and the resolution of the A/D converter itself determine the resolution. These factors, plus the voltage regulation of the DC supply feeding current to the pot limit the accuracy of these pots to about 0.1 degree of arc. Conductive plastic potentiometers may be obtained on the surplus market for about $10 - $30.

5-2 Multi-Turn Conductive Plastic Potentiometer
Courtesy Vernitron Corporation

Potentiometers are packaged in many different shapes, sizes, and configurations. Both wire-wound and conductive plastic pots are available in multi-turn versions which can be geared to each axis to take advantage of the increased resolution and accuracy, since many of the pot errors are decreased by the gear ratio. This arrangement does introduce gear errors, but if high precision anti-backlash gears are used, the gear errors on a multi-turn pot should be less than the intrinsic errors of a single-turn pot. Figure 5-2 shows a 10-turn conductive plastic pot. Figure 5-3 shows two linear potentiometers which could be used for determining the position of a Cassegrain secondary when its separation from the primary is used to determine the location of the focus.

When using a pot in a control system, it is important to ensure that the voltage response of each leg of the pot is linear. This means the impedance of the computer interface that senses the voltage must be orders of magnitude greater than that of the pot. If one makes the total resistance of the pot too low, the larger resulting currents produce self-heating of the pot, which can cause hot spots in the resistive element of the pot. Taking the ratio of the voltages of the two legs of the pot does not compensate for these hot spots. This means that the pot should be of high enough resistance to avoid significant self-heating, and the input impedance of the circuit that senses the voltage should be very high (10 to 100 megohms).

There are two common methods for measuring a voltage with a microcomputer. The first is to use an analog-to-digital (A/D) converter, which is a module or chip designed to convert a voltage input directly to a digital value. Typical A/D converters come in resolutions from 6 to 16 bits, with 8, 10, and 12 bit converters being the most popular. The resolution of the A/D converter determines how many different voltage values can be sensed. For example, a 12-bit A/D converter with an input range of 0 to 10 volts DC divides that 10 volt range into 2^{12}, or 4096, different possible output values.

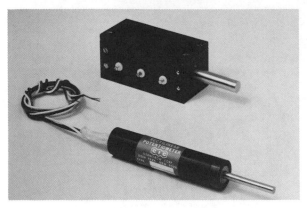

5-3 Linear Potentiometers
Courtesy Vernitron Corporation

The second way for a microcomputer to sense a voltage is to apply the voltage to the input of a voltage-to-frequency (V/F) converter, then count the pulses from the V/F converter for a fixed period of time (e.g., one second). V/F converters are available with a frequency range of 0 to 1 MHz over a voltage range of 0 to 10 volts DC. This gives much greater resolution than most ordinary A/D converters, and V/F converters are typically at least as linear as A/D converters. A simple counter circuit, which is very easy to interface to a microcomputer, can count the pulses from the V/F converter. Several companies make both A/D converter cards and pulse counting cards for popular standard computer buses.

These methods of sensing voltages with a microcomputer are not limited to potentiometers, but can also be used for many of the other sensor types discussed later.

B. VARIABLE RELUCTANCE TRANSFORMERS

The rotary variable differential transformer (RVDT) is another type of angle position sensor. As the shaft of an RVDT is turned, an iron core or secondary coil is moved through the field of the primary coil, which changes the magnetic reluctance of the transformer. A regulated AC voltage is applied to the primary coil, and the output is an AC voltage on the secondary coil that is linearly dependent on the shaft angle. The coils can be wound to provide the AC output voltage in phase for all shaft positions (a V response) or 180° out of phase for shaft angle positions less than a predefined zero (straight line response). RVDTs usually use a ferrite core with a cardioid shape to linearize the response. While most RVDTs allow a full 360° of rotation, the output is linear over only about 80-120° of rotation. Later designs have extended this to 340° of rotation, but this could still present some difficulty in calibration when a full 360° of rotation is needed, such as for the azimuth axis on an alt-az mount.

In an analog servo, two RVDTs may be connected in a master/slave circuit with a servo amplifier and motor, so that the motor shaft angle always agrees with that of the master RVDT, whose shaft is turned by the observer to indicate desired position. This is analogous to the simple servo using pots as both input and feedback devices. In a digital servo, the RVDT output would be converted to a DC voltage using a precision true RMS filter, then fed to an A/D converter connected to the control computer.

RVDTs are more accurate than precision potentiometers, with overall accuracy of a few arc minutes possible with precision units. The temperature stability is better than potentiometers, but they are expensive ($200 on up) and are rarely seen on the surplus electronics market. They typically are not used for telescope position feedback because of their cost, the expense of the required support circuitry, and the difficulty of building an interface to connect them to a computer.

Microsyns are also variable reluctance transformers, but are typically constructed with a cylindrical ferrite armature that rotates inside a stator with four field poles, two each for the input and output coils. Microsyns are used when the total angular motion is only a few degrees, and thus are unsuited for telescope axis angle feedback.

C. RESOLVERS

A resolver is a transformer with two or more windings, with at least one on the stator and one on the rotor, that are coupled so that the output voltage is proportional to the sine or cosine of the rotary shaft angular position. In this way, it differs from an RVDT, which has a linear output. Figure 5-4 shows some examples of resolvers. Most resolvers available today use two stator windings and two rotor windings to enable the calculation of complex trigonometric functions in an analog system. Figure 5-5 shows the internal construction of a typical resolver.

5-4 Three High Accuracy Resolvers
Courtesy The Singer Company, Kearfott Division

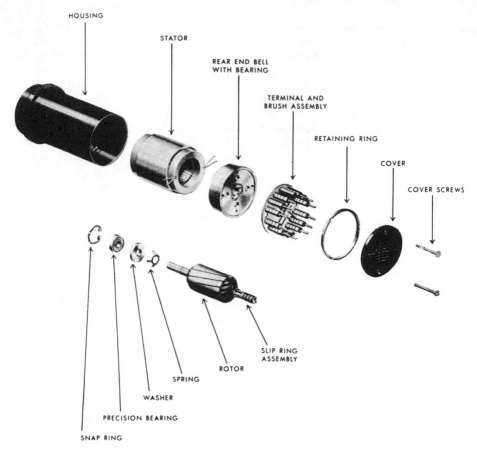

HOUSING

STATOR

REAR END BELL
WITH BEARING

TERMINAL AND
BRUSH ASSEMBLY

RETAINING RING

COVER

COVER SCREWS

SLIP RING
ASSEMBLY

ROTOR

SPRING

WASHER

PRECISION BEARING

SNAP RING

5-5 Exploded View of a 4-Coil Resolver
Courtesy Reston Publishing Co.

Some resolvers are capable of very high resolution and accuracy--18 bits (262,144 counts per revolution) or greater. They are also very expensive. The Northern Precision Laboratories Model 801397-2 with 18 bit resolution costs $11,500. This is somewhat less than a new optical shaft angle encoder of the same resolution and accuracy, but such high resolution resolvers are often a bit larger and heavier, require more power, and are more complex than optical shaft angle encoders.

Very high resolution and accuracy can be obtained using two resolvers, one tied directly to the axis of the telescope, and the other geared to the axis such that it rotates several times with one turn of the main shaft. This type of resolver is called "two-speed", and uses two lower resolution resolvers, each turning at a different speed. Figure 5-6 is a block diagram of such a two-speed resolver configuration, and Figure 5-7 shows the kind of electronics board used to combine the outputs of, for example, a 14-bit and an 11-bit unit to obtain 22-bit (roughly 0".3) resolution. The 256:1 gearing gives three overlapping bits between the two units, which the electronics board uses to sort out gearing errors and intrinsic resolver errors. The

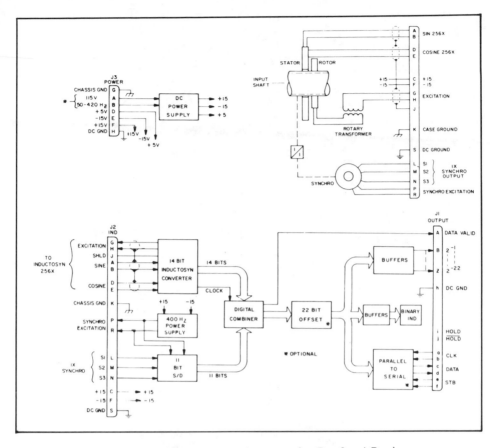

5-6 Electrical and Mechanical Layout of a Two-Speed Resolver
Courtesy Northern Precision Laboratories, Inc.

result is a unit with an RMS error of about 1". Figure 5-8 shows a two-speed resolver with very high resolution and accuracy.

Northern Precision Laboratories sells circuit boards which provide the interface between their resolvers and computers using the Multibus. Adapter boards are available to connect a Multibus to a Q-Bus or an STD-Bus, as well as to a few of the more popular personal computers.

If one is able to obtain a high resolution resolver inexpensively on the surplus market, it would be worth considering as an angle position feedback sensor, but surplus optical shaft encoders of higher resolution are usually more readily available for less money. Therefore, resolvers are not expected to see much use in telescope control applications, except, perhaps, at larger professional observatories. Analog Devices offers a series of modules which makes it simple and inexpensive to build a computer interface for common types of resolvers.

5-7 Digital Circuit Board with 800 Hz Oscillator and VCO
Courtesy Northern Precision Laboratories, Inc.

5-8 Two-Speed, Multi-Turn, Absolute Resolver
Courtesy Northern Precision Laboratories, Inc.

D. SYNCHROS

Synchros are transformers that are used in pairs to transmit shaft angle position over large distances without a direct mechanical linkage. The synchros are connected together in an electrical circuit such that when one shaft is turned, the other shaft also turns to the same angle. The name comes from the fact that synchros are 'self-synchronous'. Other names are selsyn, autosyn, or teletorque. Synchros are also used for torque transmission or for voltage indication. Synchros could be used as a pair to form an open-loop analog control system, in which one synchro is the input device for giving desired position, and the other is directly attached to the corresponding telescope axis, and drives the telescope to the desired position. Such a scheme, though possible, would be impractical, since the second synchro would need the characteristics of a large DC torque motor.

A variation of the synchro that would be useful as an angle position feedback sensor is the synchro/resolver. This is a single synchro whose output, rather than being wired to another synchro, is electronically decoded into a useful form. A function module is available from Computer Conversions Corp. for about $500 that converts the three-phase synchro outputs to linear DC with ±6 arc minute accuracy. A similar unit from Vernitron is shown in Figure 5-12. A Scott T transformer is available from Magnetics, Inc. for $180 that converts the three-phase synchro outputs to sine and cosine voltages with 10 arc second accuracy. These could be fed to A/D converters, which are then read by the control computer and converted to an angle.

Analog Devices offers 14-bit synchro-to-digital converter modules costing about $200 that utilize a Scott T transformer and a quadrant selector, sine and cosine multipliers, and a subtractor to compute a binary number representing the angle. Two synchros can be linked together with gears and two-speed logic in a manner analogous to the two-speed resolvers. The two-speed logic is available for about $1500.

Synchros are available for about $25-$50 each on the surplus electronics market, and the synchro-to-digital converter modules are available at reasonable prices. One could use just the Scott T transformer and an A/D converter, and do the sine/cosine calculations in the microcomputer, but that increases the computation load on the computer. Analog Devices offers all the modules needed for about $300-400 per axis that yield roughly 10 arc minute accuracy. They have also written a 191-page book on the subject of synchros, resolvers, and Inductosyns which is an excellent introduction to these subjects.

E. INDUCTOSYNS

A variation of resolvers which is expected to be seen in telescope control applications is the Inductosyn (registered trademark of Farrand Controls, Inc.), a kind of resolver using optical encoder technology. One factor limiting the accuracy of a resolver is that wires are wound in a coil. The placement of the wire turns in the coil varies slightly from turn to turn, introducing nonlinearities which are difficult to model. The Inductosyn is made using a photoresist technique to deposit a metal pattern on a stable dielectric substrate.

The Inductosyns made by Farrand consist of two aluminum or hot rolled steel disks, each containing a pattern of loops in metal foil on a dielectric substrate. The disks are mounted coaxially with the sides containing the foil pattern facing each other. One disk is the stator (fixed to the case) and the other is the rotor (rotates with the axis whose position is being sensed). The rotor and its foil pattern are shown in Figure 5-9, while the stator is shown in Figure 5-10. These patterns are deposited with an accuracy of 1" or better.

The device works by passing an AC signal of 200 Hz to 200 kHz (nominally 10 kHz) through the foil pattern of the rotor, then detecting the AC signal that is generated on the stator by inductive coupling across a small air gap between 0.003 and 0.008 inches wide. This is shown in Figure 5-11(a). The induced voltage is a maximum when the loops are facing. As shown in Figure 5-11(b), the induced AC voltage amplitude is zero when the loops are oriented midway between each other, then goes to a minimum when aligned with turns opposite to each other. With a foil pattern pitch of P and a rotor input voltage amplitude V, the relation between the

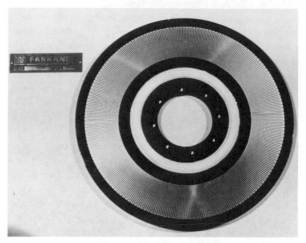

5-9 Inductosyn Rotor
Courtesy Farrand Controls, Inc.

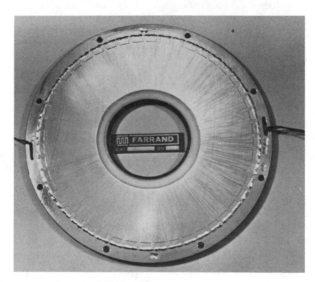

5-10 Inductosyn Stator
Courtesy Farrand Controls, Inc.

induced voltage V_a and displacement x from perfect alignment is given by

$$V_a = k \ V \cos \frac{2 \pi x}{P}$$

To provide a reference for comparing the voltage amplitudes, the stator has a second pattern laid down 90 degrees out of (spatial) phase, or P/4, with the first. This produces a signal

$$V_b = k \ V \sin \frac{2 \pi x}{P}$$

where k is a coupling ratio which depends on the size of the air gap and other factors. Both signals have a constant phase offset from the excitation AC signal in this mode of operation. However, by exciting the stator quadrature coils with signals phase shifted in time, a phase shifting operating mode can be achieved, which undergoes a shift of 360° for each displacement of one cycle length.

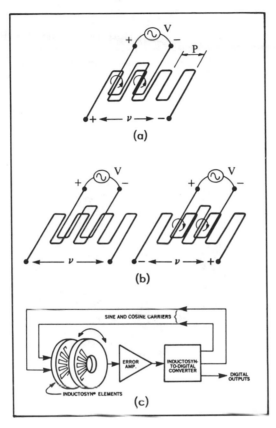

5-11 Principles of Inductosyn Operation
Courtesy Farrand Controls, Inc.

The rotor can have from 18 to 2048 poles (the units in Figures 5-9 and 5-10 have 360 poles). When used in normal (non-phase shifting) mode, the stator output voltages act like a resolver of 18 to 2048 speed. The two outputs can be subdivided electronically to provide very high resolution. Electronics which can be purchased from Farrand convert the quadrature signal into digital pulses to make the unit act like an incremental optical encoder. This package divides the sine (or cosine) wave from each pole into 1000 or 2000 parts. Other divisors are available as options. A standard 12-inch Inductosyn with 2000 poles and electronics using a divisor of 2000 has 4,000,000 counts (pulses) per 360° shaft rotation, or 0".324 resolution. The accuracy is quite high, since the signal is averaged around all the poles of the unit, so small random pattern errors tend to be averaged out. Production accuracies are typically 2" - 8", but units with an absolute accuracy of 1" can be specially selected. The most common Inductosyn configuration is shown in Figure 5-11(c).

Because the pattern is much simpler to generate and replicate than that of an optical encoder, for very high resolution, (18 bits or more), the Inductosyn can cost less than an optical encoder of equivalent resolution and accuracy. For example, an assembled 20-bit Inductosyn from Northern Precision Laboratories costs about $5,000 versus $20,000 for an equivalent optical encoder. The Inductosyn plates can be purchased new from Farrand Controls for about $1800 a pair, with an extra $1400 for the excitation and conversion electronics, and a $575 charge for selecting a pair of plates with 1" accuracy. Electronics from Analog Devices are less expensive and offer even higher resolution.

Farrand also makes the Inductosyn pattern on straight substrates for linear positioning applications. Such a device would be useful for encoding the position of a Cassegrain secondary mirror. Farrand can select units with 0.000080-inch accuracy, or supply standard units with 0.0002-inch accuracy. This level of accuracy is not needed for encoding focusing positions, since temperature changes from night to night will change the absolute position of the focus by that amount or more. However, that level of repeatability is convenient, so that throughout the course of a single evening, one can shift among several foci and return predictably to the desired focus.

5-12 A Synchro to DC Volts Converter
Courtesy Vernitron Corporation

F. ROTARY DIFFERENTIAL CAPACITORS

An interesting angular displacement transducer manufactured by Trans-Tek, Inc. is the Series 600 variable differential capacitor, which is shown in Figure 5-13. The unit contains a precision variable capacitor which is in an LRC oscillator circuit. As the shaft is rotated, a semi-circular plate attached to the rotor moves with respect to a similar stationary plate, which changes the capacitance of the unit, and, therefore, the frequency of the oscillator. The oscillator output is demodulated and converted to a DC signal of 100 mV per degree of rotation. Electronics inside the unit use 12-16 volts DC external power to generate the AC oscillator carrier and perform all signal conversion. The unit has 0.05% of full scale linearity, and theoretically has infinite resolution. With a full scale of 160°, the linearity is 4.8'. Practically, the unit has 3.6 arc second resolution, which corresponds to a 100 nV signal, close to the limit one would want to try to detect reliably in an environment of EMI/RFI caused by motors and computers.

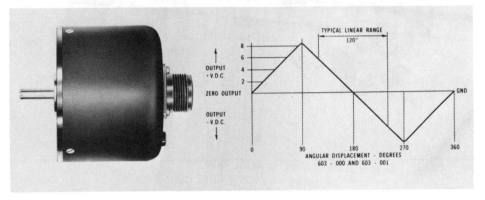

5-13 Trans-Tek Series 600 Rotary Differential Capacitor
Courtesy Trans-Tek, Inc.

The unit has one drawback--its linear response region is limited to at most 160° of rotation. Thus it cannot be used on some types of telescope mounts that require a full 360°, or even 180°, of rotation. If the telescope is a German equatorial and has a bad horizon, or will always be used at elevations of at least 10° above the horizon, the Trans-Tek unit is a viable choice for an angle position sensor. It would also be useful in an alt-az mount if motion in one axis is restricted. The unit itself costs about $350 new (the authors have not seen them on the surplus electronics market), and support electronics, including a power supply and a DC electrometer amplifier, could be constructed for roughly $150 - $200 in parts, which places it within the budget of a small system user. To obtain coverage over a full 360°, three units can be used, but this approach is usually more expensive than using some types of optical shaft angle encoders. We are not aware of any telescope control systems which use this device.

G. OPTICAL SHAFT ANGLE ENCODERS

Optical shaft encoders are devices that consist of a disk, usually glass, on which is deposited by photographic means one or more patterns of alternating transparent and opaque slots or lines which, when rotated through the path of a light source and photodetector, generate a series of pulses. The number of transparent and opaque line pairs around the disk determines the number of pulses or counts per turn of the disk, which is the resolution of the encoder. Examples of two absolute encoder patterns in common use are shown in Figure 5-14. Figure 5-15 shows a few of the wide variety of optical encoders that are available.

Encoders may give an absolute position using a unique output code for each

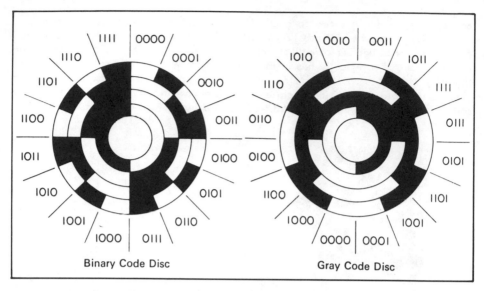

5-14 Two Absolute Optical Encoder Disk Patterns
Courtesy Reston Publishing, Co.

5-15 Some Optical Shaft Angle Encoders
Courtesy BEI Electronics, Inc.

resolved shaft angle (absolute encoder), or they may send out a pulse for each incremental resolved shaft angle step (incremental encoder). An absolute encoder typically has a resolution of 2^N, in which there are N concentric rings of transparent and opaque line pairs, with each successive ring having twice the number of line pairs as the ring inside it, beginning with the innermost ring, which is half transparent and half opaque. A light source and photodetector placed on opposite sides of the disk for each of the N rings give N bits of a binary number representing the shaft angle. To prevent small pattern misalignments from causing false readings taken when several bits are changing state from one count to the next, a Gray code is often used instead of straight binary, since in a Gray code, only one bit changes state from one count to the next. This is shown in the following table (from Spaulding, 1965, p.10):

Decimal	Binary Code	Gray Code
0	00000	00000
1	00001	00001
2	00010	00011
3	00011	00010
4	00100	00110
5	00101	00111
6	00110	00101
7	00111	00100
8	01000	01100
9	01001	01101
10	01010	01111
11	01011	01110
12	01100	01010
13	01101	01011
14	01110	01001
15	01111	01000
16	10000	11000
etc.		

The Gray code is easily converted into a binary number with either hardware or software, using the following rules (from Spaulding, 1965, p.10):

1. The most significant natural binary bit is identical to the most significant Gray bit.
2. If the nth natural binary bit is 0, the (n-1)th natural binary bit is identical to the (n-1)th Gray bit.
3. If the nth natural binary bit is 1, the (n-1)th natural binary bit is the opposite of the (n-1)th Gray bit.

This is expressed in the formula

$$B(n-1) = (G(n-1) \text{ .AND. } (.NOT. \ B(n))) \text{ .OR. } ((.NOT. \ G(n-1)) \text{ .AND. } B(n))$$

where $B(n)$ is the nth natural binary bit, and

$G(n)$ is the nth Gray code bit.

Other anti-ambiguity techniques are in common use, such as window codes, cyclic decimal, U-scan, V-scan, and 2-out-of-5 (a kind of parity code).

While an absolute encoder uses N rings for N bits of resolution, an incremental encoder with the same resolution uses two concentric rings of 2^N line pairs. The two ring patterns are in spatial quadrature with each other. Logic chips can be used to count and store the current shaft position, and the direction of the counter can be determined from the relative phases of the pulses received from the two rings. Thus the shaft angle is determined electronically rather than optically. There is often a third ring with a single mark indicating the zero position. A pulse from this track is used to zero the counting logic.

Either absolute or incremental encoders can be used for telescope control. For resolutions up to about 16 bits, incremental encoders are far less expensive than absolute types, by a factor of as much as 10. A 16-bit absolute encoder has 2^{16} (65,536) counts per revolution, which gives it a resolution of 1,296,000" per circle divided by 65,536, or about 20" per count. An absolute encoder with this resolution costs about $2,000, while an incremental encoder with almost the same resolution sells for $450. Resolutions up to 24 bits and custom mechanical designs are available on special order. For example, a stock 20-bit absolute encoder costs about $10,000, while a custom 21-bit encoder with a 6-inch hollow shaft hole for direct mounting on a telescope axis costs about $20,000 on special order. Multi-turn encoders of either the incremental or absolute type are available which use a lower-resolution (and therefore less expensive) encoder disk which is geared to the input shaft of the encoder assembly. These units can offer the high resolution (but not necessarily the accuracy) of single-turn units at significantly lower cost.

Optical shaft encoders are available on the surplus electronics market if one pursues them and is patient. For example, Trueblood was given a 13-bit absolute encoder in 1971, bought two 15-bit absolute encoders in 1977 for $125 each, and bought two 16-bit absolute encoders in 1981 for $25 each. Ads for optical shaft encoders appear in **Sky and Telescope** magazine at odd intervals. Such encoders can be geared to each telescope axis to yield high resolution (1" or better). Measurements can be made, and algorithms can be developed and implemented in the control software to minimize the errors introduced by the gears to a few arc seconds if a great deal of time is spent taking measurements, reducing the data, and studying the gear error sources. One should be wary of shaft angle encoders obtained on the surplus market, since often the electronics are bad, and occasionally the encoder disk is damaged.

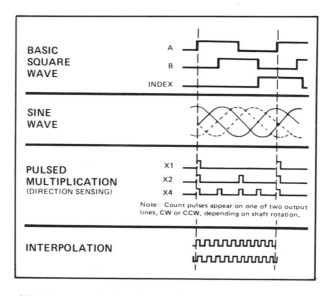

5-16 Characteristic Waveforms of Incremental Encoder Options
Courtesy BEI Electronics, Inc.

The BEI H-25 incremental encoder is a good choice for those who wish to avoid the problems that often accompany surplus optical encoders. The basic unit is small, inexpensive compared to other types of optical encoders (about $450), and has 2540 counts per turn. As shown in Figure 5-16, interpolation logic multiplies the pulses per turn on each output phase by a factor of 5, up to 12,700. The pulses from the quadrature outputs can then be put through optional "steering" circuitry, which produces a series of 50,800 pulses per revolution. These pulses are sent out on one of two lines, depending on whether the encoder shaft is being turned CW or CCW. These two lines can be connected to the "up" and "down" inputs of a series of 74LS193 counters to obtain an integrated position count. Both interpolation and pulse steering circuitry can be placed inside the encoder by BEI at very little additional cost (about $60 in 1983).

Instead of including the pulse steering logic in the encoder, it can be part of the computer interface. An example is the Buckminster Corp. model C-1500 encoder interface cards, made for the LSI-11 Q-Bus and the STD bus. These cards already have the steering circuitry, so the H-25 can be ordered only with the interpolation logic when using this board. (Talk to Buckminster first before ordering either the card or the encoders. They have good advice about TTL output and pullup resistor options for the H-25 that will increase the noise immunity--a must when there are motors and computers radiating interference. This is crucial with incremental encoders, since stray pulses on the line will give inaccurate position readings that accumulate throughout the night.)

Another low cost incremental encoder to consider is the Litton Model 81 1000 count-per-turn (roughly 10-bit) encoder, which sells new for under $200. Similar units for even less may be available by now. For those interested in experimenting with making their own low resolution optical encoder, Opal (1980) describes an encoder yielding 50, 100, or 200 counts per turn. He includes a circuit for converting the two quadrature outputs to pulse and direction outputs.

5-17 Two-Speed Mechanical Encoder
Courtesy Northern Precision Laboratories, Inc.

H. OTHER ENCODER TYPES

A variation on the optical encoder is the mechanical encoder, which is made by the same process used to make etched printed circuit cards. A pattern is photographed onto photoresist over a layer of copper foil on glass epoxy material. The resist is developed, the pattern is etched, then a set of brushes is used in place of lights and photodiodes to make and break one circuit per bit (on an absolute encoder), with the foil and brush acting as a switch. Single-turn mechanical encoders are usually limited to 10 bits or less, since the brush contact width needed to ensure good electrical connection becomes larger than the line widths at resolutions of about 10 bits. Multi-turn mechanical encoders, which use gears to speed up the rotation of the 10-bit disk with respect to the encoder shaft, are available inexpensively with up to 21-bit resolution. Examples of mechanical encoders are shown in Figures 5-17 and 5-18.

5-18 A Mechanical Encoder and Encoder Disks
Courtesy Vernitron Corporation

Another encoder package is the motor/encoder combination, which typically uses a DC torquer and a very high resolution encoder to form a single compact unit. An example is shown in Figure 5-19.

Although there are many methods available for informing your computer of where your telescope is pointing, available technology and market factors tend to favor the potentiometer for moderate accuracy requirements, and the incremental optical encoder for applications requiring higher accuracy. Major observatories will continue to use very high resolution (and cost) absolute encoders mounted directly on each driven axis, but it is expected that potentiometers and incremental encoders will see increasing use in smaller systems. As long as incremental encoders are available with 22' resolution and accuracy for roughly $100 and with 25" resolution and accuracy for $450, they will be used in telescope control systems.

5-19 A Single-Piece Motor/Encoder Unit
Courtesy BEI Electronics, Inc.

Chapter 6. **SOURCES OF SYSTEMATIC ERROR I
—ASTRONOMICAL CORRECTIONS**

When an astronomer plans an evening's observations, he usually obtains an object's right ascension and declination from a catalog which uses the coordinate system of some standard epoch. This is the mean position of the object. This standard epoch is generally different than the one at the time the observation is to be made. In addition, the object's catalog position is not corrected for effects which depend on the time and location of the observation. Thus the catalog position is usually not where the observer would find the desired object. This latter position is the apparent position, and must be computed once or several times a night, depending on the location of the object in the sky, the desired pointing accuracy, and the length of the observation. Since the control computer in a servo must use an object's apparent coordinates to control the telescope, but the astronomer usually enters the mean position taken from the catalog, the mean coordinates must be converted to apparent coordinates before commands can be sent to the telescope. The corrections to be performed vary according to the ephemeris or catalog used as the source of an object's position.

The corrections that must be applied to the catalog mean position to reduce it to apparent position are listed in the following table, in order of decreasing size.

CORRECTION	MAXIMUM MAGNITUDE
1. Precession	40' (roughly 50 years x 50" per year)
2. Refraction	30' (2' at a zenith distance of 60°)
3. Annual Aberration	20"
4. Nutation	17"
5. Solar Parallax	9"/distance in A.U.
6. Stellar Parallax	1"/distance in parsec
7. Orbital Motion	varies (order of 1")
8. Proper Motion	varies (a few arc seconds if you use an old catalog)
9. Diurnal Aberration	0".3
10. Polar Motion	0".1 (random)

Not all of the corrections listed above need be computed. Only those corrections larger than the required pointing accuracy are needed. For many systems, only precession and refraction corrections are made.

For a control system of moderate accuracy, which is only expected to bring an object within the finder on a typical small telescope, no corrections of any sort are needed if the field of view is 2 degrees or more. To center an object in the finder and to bring it well within the field of view of the main telescope, only precession and refraction corrections are necessary. In most cases, the observing is done well above the horizon, and as refraction errors are slight (less than 2' at a zenith distance of 60°, 3' at 70°, and 6' at 80°), refraction corrections are usually not required.

In a simple control system which corrects only for precession, one could calculate the coordinates at the current epoch for all of one's favorite objects just once. Such coordinates would be good for several years before they would require updating. This might be simpler to do than make calculations each time the object is observed.

If refraction calculations are made and one is not taking long exposure photographs near the horizon, then the calculation can be made once just before moving to the object, and need not be repeated.

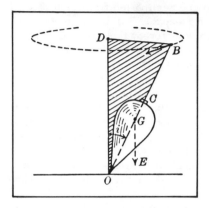

6-1 Precession of a Rotating Body in a Gravitational Field
Courtesy of Cambridge University Press

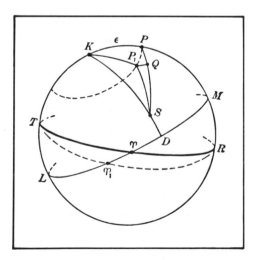

6-2 Equinox Motion Due to Luni-Solar Precession
Courtesy of Cambridge University Press

In any event, it makes sense to correct for only those errors which are significant with respect to the requirements and capabilities of your system. To minimize processor loading, the corrections that are made should be computed only as often as is really required, unless doing it more often just happens to be simpler. Tomer's system (described in Section IV) only corrects for precession and refraction. It is an excellent example of matching the requirements, mechanical assembly, computer, and software to create a balanced system at reasonable cost.

For each correction, the dynamic or physical principle is discussed, the equations that embody the physical principle are presented, the range of values of the correction is computed to assess whether the correction need be done at all to achieve a desired level of accuracy, and algorithms for computing the correction are given which we have chosen for their speed and accuracy when used on a personal computer. In Chapter 14, Trueblood demonstrates how to use these algorithms to assess the CPU loading of the control computer.

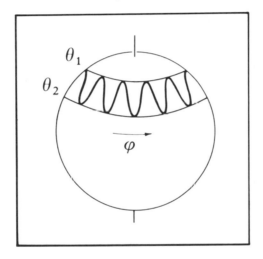

6-3 Nutation of the Pole of a Precessing Rotating Body
Courtesy of Academic Press

A. PRECESSION

1. The Physical Basis of Precession. The rotating Earth in the gravitational field of the Sun and Moon is subject to torques which cause it to precess with a period of roughly 26,000 years. This long-period motion of the mean pole of the Earth about the pole of the ecliptic is called **luni-solar precession.** This is analogous to the precession of a top in a gravitational field, as shown in Figure 6-1. Luni-solar precession produces the motion of the equinox shown in Figure 6-2. In addition to the motion of the mean pole, there is a short period motion of the true pole about the mean pole called nutation, which is discussed below, and shown in Figure 6-3. The planets also perturb the orbit of the Earth, causing the plane of the ecliptic to tilt such that the equinox precesses roughly 12" per century and the obliquity of the ecliptic decreases at roughly 47" per century (Gurnette and Woolley, 1961). This latter effect is called **planetary precession,** and its effect on the obliquity is shown in Figure 6-4. The general precession that results from both of these motions is depicted in Figure 6-5. The motion of the Earth's pole due to both precession and nutation is analogous to that shown in Figure 6-6.

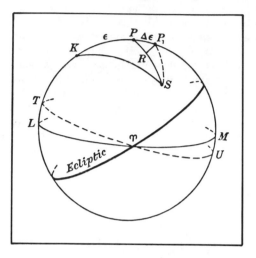

6-4 Obliquity Change Due to Planetary Precession
Courtesy of Cambridge University Press

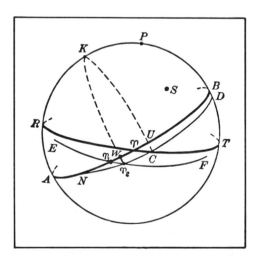

6-5 General Precession
Courtesy of Cambridge University Press

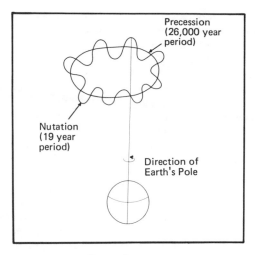

6-6 Combined Effects of Precession and Nutation

2. Magnitude of the Precession Corrections. Smart (1979) combines both luni-solar and planetary precession to obtain an expression for the general precession that is actually observed to occur per year in the right ascension (α) and declination (δ) of an object. He gives the following expressions for the annual changes in these quantities ($\Delta\alpha$ and $\Delta\delta$):

$$\Delta\alpha = (\psi \cos \epsilon - \lambda') + \psi \sin \epsilon \sin \alpha \tan \delta \qquad\qquad [6.1a]$$

$$\Delta\delta = \psi \sin \epsilon \cos \alpha \qquad\qquad [6.1b]$$

where $\Delta\alpha$ = the annual change in right ascension

$\Delta\delta$ = the annual change in declination

ψ = the annual luni-solar precession
  = 50".2564 + 0".000222 t   [6.1c]

ϵ = the angle of the obliquity
  = 23° 27' 8".26 - 0".4685 t   [6.1d]

λ' = the annual planetary precession
  = 0".12 per year   [6.1e]

t = the number of years since 1900.   [6.1f]

Equations [6.1] are given to indicate the magnitude of the precession corrections, which is determined solely by the elapsed time between the two epochs involved. Thus there is virtually no limit to the magnitude of the correction, as expressed by these equations, which account only for the secular effect of precession. In reality, the Earth's pole describes a circle every 26,000 years as it precesses, which defines a cone described by the Earth's rotational axis whose apex is centered at the Earth's center. The apex angle of the cone is the maximum angle of precession. Since this angle is several degrees, if the time between epochs is sufficiently large, precession can dominate all other sources of systematic error.

In actual practice, standard star catalogs tend to be updated every 50 years or so. Using the value of ψ as an order of magnitude estimate for the variation of an object's right ascension and declination over 50 years, ψ (= 50"/year) x 50 years yields about 42 arc minutes. This zeroth order approximation overstates the typical effect of precession, since most accurately computed precession corrections for a 50-year period are on the order of a few arc minutes. Most astronomers want to set their telescopes with an accuracy considerably better than a few arc minutes, so a correction for precession is almost always necessary.

3. Methods of Computing General Precession. There are two methods of computing precession corrections in general use: (1) closed algebraic evaluation of equations for the changes in right ascension and declination, and (2) rotation of the equatorial coordinate system using matrices to represent the rotations. Many computerized systems developed to date employ the mathematically elegant matrix rotation method. However, as is shown below, the evaluation of closed algebraic expressions imposes a lighter burden on the processor when computing either precession alone, or both precession and nutation. These two methods are explained and compared for computational efficiency in the following paragraphs.

Expressions using the first method are presented in Gurnette and Woolley (1961, p.38):

$$\alpha - \alpha' = M + N \sin \frac{\alpha + \alpha'}{2} \tan \frac{\delta + \delta'}{2} \qquad [6.2a]$$

$$\delta - \delta' = N \cos \frac{\alpha + \alpha'}{2} \qquad [6.2b]$$

where α = right ascension using the mean equinox at epoch t

α' = right ascension using the mean equinox at epoch t'

δ = declination using the mean equinox at epoch t

δ' = declination using the mean equinox at epoch t'

M = general precession in right ascension
= $M_m(t - t')$, t and t' in years

M_m = value of m midway between t and t', where

m = 3s.07234 + 0s.00186 T (T measured in tropical centuries from 1900.0)

N = general precession in declination = $N_m(t - t')$, t and t' in years

N_m = value of n midway between t and t', where

n = 20".0468 - 0".0085 T (T measured in tropical centuries from 1900.0)

These equations, which are familiar to most readers, are iterated until convergence to the desired accuracy is achieved.

Another example of this method is provided by Woolard and Clemence (1966, p.262), which expresses $\Delta\alpha$ and $\Delta\delta$ in terms of a simple power series in time. This method uses no iteration and no trigonometric functions, and thus is computationally more efficient.

Equations [6.2] should be used wherever extremely high accuracy is not required, since they are the simplest to program, and take less computer time than other approaches. However, these and other similar approaches suffer from the disadvantage of using coordinate systems in which the various precession motions are not orthogonal. This means that the constants in these equations are valid only over a relatively   short period of time.

Kaplan (1981) cites a more accurate algorithm, which yields an accuracy of at least 0.001 seconds in right ascension and 0".01 in declination. It uses epoch J2000.0, which is the new epoch for the Astronomical Almanac starting in 1984. This new epoch uses a new system of time called Barycentric Dynamical Time (TDB). For computing precession, assume J_s is the TDB Julian date of the starting epoch, and J_e is the TDB Julian date of the ending epoch.

$$\zeta_o = (2306".2181 + 1".39656T - 0".000139\ T^2)\ t \\ + (0".30188 - 0".000344\ T)\ t^2 + 0".017998\ t^3 \qquad [6.3a]$$

$$z = (2306".2181 + 1".39656\ T - 0".000139\ T^2)\ t \\ + (1".09468 + 0".000066\ T)\ t^2 + 0".018203\ t^3 \qquad [6.3b]$$

$$\Theta = (2004".3109 - 0".85330\ T - 0".000217\ T^2)\ t \\ - (0".42665 + 0".000217\ T)t^2 - 0".041833\ t^3 \qquad [6.3c]$$

$$\alpha' - \alpha = \zeta_o + z + \arctan \frac{q\ \sin(\alpha + \zeta_o)}{1 - q\ \cos(\alpha + \zeta_o)} \qquad [6.3d]$$

$$\delta' - \delta = \Theta\ [\cos(\alpha + \zeta_o) - \sin(\alpha + \zeta_o)\ \tan(\alpha' - \alpha - \zeta_o - z)] \qquad [6.3e]$$

$$q = \sin\Theta\ (\tan\delta + \tan\frac{\Theta}{2}\ \cos(\alpha + \zeta_o) \qquad [6.3f]$$

where $(90° - \zeta_o)$ = the right ascension of the ascending node of the equator at epoch J_e on the equator of J_s reckoned from the equinox of J_s

$(90° + z)$ = the right ascension of the node reckoned from the equinox of J_e

Θ = the inclination of the equator of J_e to the equator of J_s

T = the time interval between J2000.0 and J_s
  = $(J_s - 2451545.0)/36525$ in Julian centuries   [6.3g]
  (This definition of T is used throughout this chapter.)

t = the time interval between the catalog and current epochs
  = $(J_e - J_s)/36525$ in Julian centuries   [6.3h]

TDB is computed from the following:

TDB = TDT + 0s.001658 sin(g + 0.0167 sin g)   [6.3i]
    + lunar and planetary terms of order 0.00001 seconds
    + daily terms of order 0.000001 seconds

$$g = (357°.528 + 35999°.050 \ T) \ \frac{2\pi}{360°}$$ [6.3j]

TDT = Terrestrial Dynamical Time
    = TAI + 32s.184   [6.3k]

TAI = International Atomic Time
    = UTC + ΔAT   [6.3l]

UTC = Universal Coordinated Time (the time that is broadcast over WWV and
      WWVB)

Values of ΔAT are published in the Astronomical Almanac, p.B5, in Circular D of
the BIH, and in the USNO Time Service Announcements Series 7. Note that the
definitions of T and t are such that UTC may be used in place of TDB without loss of
noticeable accuracy in telescope pointing. In the above equations, α is the
unprecessed value of right ascension, and α' is the precessed value.
    The second method of computing precession is to rotate the geocentric
rectangular equatorial coordinates from one epoch to another. This can be done in
the following sequence:

1. Transform from right ascension and declination to geocentric
   rectangular equatorial coordinates.
2. Rotate the geocentric rectangular equatorial coordinates.
3. Transform the precessed geocentric rectangular equatorial
   coordinates back to precessed right ascension and declination.

Step 1 can be performed using the following equations (Woolard and Clemence,
1966, p.280):

x = cos α cos δ
y = sin α cos δ [6.4]
z = sin δ

Step 2 is to perform the matrix operation.

$$\begin{bmatrix} x' \\ y' \\ z' \end{bmatrix} = \begin{bmatrix} x \\ y \\ z \end{bmatrix} \times \begin{bmatrix} \text{rotation} \\ \\ \text{matrix} \end{bmatrix}$$

The rotation matrix consists of the direction cosines of the apparent angle produced by precession. Let A, B, and C be the angles in the rectangular coordinate system (x,y,z) between the axes and a radius from the origin to the point in the sky being measured, and A', B', and C' be the angles between the axes of the same coordinate system and a radius to a different position in the sky. This simulates the apparent motion of an object within the coordinate system due to precession. The object appears to have moved through an angle D given by

cos D = cos A cos A' + cos B cos B' + cos C cos C'

The rotation matrix consists of the cosines between the various angles:

$$\begin{bmatrix} x' \\ y' \\ z' \end{bmatrix} = \begin{bmatrix} x \\ y \\ z \end{bmatrix} \times \begin{bmatrix} \cos(x',x) & \cos(y',x) & \cos(z',x) \\ \cos(x',y) & \cos(y',y) & \cos(z',y) \\ \cos(x',z) & \cos(y',z) & \cos(z',z) \end{bmatrix} \qquad [6.5]$$

When the matrix multiplication is performed, the results are as follows:

$$x' = x - 2(\sin^2 (\zeta_o + z) + \cos \zeta_o \cos z \sin^2 \frac{\Theta}{2}) x$$

$$+ 2(\sin \zeta_o \cos z \sin^2 \frac{\Theta}{2} - \sin(\zeta_o + z)) y + (\cos z \sin \Theta) z \qquad [6.6a]$$

$$y' = y + (\sin (\zeta_o + z) - 2\cos \zeta_o \sin z \sin^2 \frac{\Theta}{2}) x$$

$$- 2(\sin^2 \frac{\zeta_o + z}{2} - \sin \zeta_o \sin z \sin^2 \frac{\Theta}{2}) y - (\sin z \sin \Theta) z \qquad [6.6b]$$

$$z' = z + (\cos \zeta_o \sin \Theta) x - (\sin \zeta_o \sin \Theta) y - 2 \sin^2 \frac{\Theta}{2} z \qquad [6.6c]$$

where ζ_o, z, and Θ are the same as in equations [6.3]. Step 3 is to transform the resulting (x',y',z') to (α',δ').

As seen from the above discussion, there are basically two methods of deriving accurate precessed coordinates from catalog coordinates. The first method uses the spherical trigonometric equations [6.3]. The second method uses the matrix equations [6.6]. The table below summarizes the number of trigonometric and arithmetic operations that are required for each method. Since sine, cosine, and tangent functions and their respective inverse functions all require roughly the same amount of time to compute, they are grouped together under the "Trig" heading. Numbers raised to an integer power n are counted as n multiplies. The results are as follows:

Item	Equation	Trig	+	−	×	/
T	[6.3g]	0	0	1	0	1
t	[6.3h]	0	0	1	0	1
ζ_o	[6.3a]	0	3	2	10	0
z	[6.3b]	0	4	1	10	0
Θ	[6.3c]	0	1	4	10	0
α'	[6.3d]	3	4	1	3	1
δ'	[6.3e]	1	2	4	2	0
q	[6.3f]	3	2	0	2	1
Apply	[6.3]	0	2	0	0	0
x,y,z	[6.4]	4	0	0	2	0
x'	[6.6a]	7	5	2	13	1
y'	[6.6b]	1	3	4	13	1
z'	[6.6c]	0	1	2	7	0
α',δ'	[6.4] inverse	4	0	0	2	0

If a value is computed in a prior equation, it is not counted again when it appears later on in other equations used in the same method. For example, sin $(\alpha+\zeta_o)$ was counted for equation [6.3d], so it is not counted again in equation [6.3e]. Method 1 uses [6.3g] and [6.3h],[6.3a]-[6.3c], and [6.3d]-[6.3f], then the resulting corrections are applied to the original (α,δ). Method 2 uses [6.3g] and [6.3h],[6.3a]-[6.3c], and equations like [6.4] to convert (x',y',z') to (α',δ'). This is summarized as follows:

	Trig	+	−	×	/
Method 1	7	18	14	37	4
Method 2	16	17	17	67	4

The relative times for performing various calculations depends on the computer hardware and the algorithms used to perform trigonometric function calculations. To provide a basis for comparing different computing methods, the author has assumed that "trig" functions are evaluated to eight-decimal accuracy using Chebyshev polynomials. These functions would probably be evaluated using eight adds and seven multiplies per function. To allow one to compare different types of arithmetic calculations, the author has defined an add and subtract to be one computation unit, a multiply to be two computation units, and a divide to be three computation units. These values are based on the instruction times for 32-bit floating point operations given in Appendix G. Using these assumptions, the first method requires 272 computation units, and the second requires 532 units. It seems clear that the second method requires considerably more time than the first, but one should not infer that it would take about twice as long, since various kinds of overhead have not been included in this comparison.

B. NUTATION

1. The Physical Basis of Nutation. The action of the Sun and the Moon on the Earth produces the motion of the mean pole of the Earth described above as luni-solar precession. These bodies also produce a motion of the true pole of the Earth about the mean pole, which is nutation. As shown in Figure 6-3, this elliptical motion has a dominant period of roughly 19 years with an amplitude of about 17" in longitude and 9" in obliquity. The principal terms of the series which describe nutation depend on the regression of the node of the Moon's orbit. Other terms depend on the mean longitudes   and mean anomalies of the Sun and the Moon, and on their combinations with the longitude of the Moon's node.

The motion of the true pole about the mean pole is described in terms of corrections to the longitude, $\Delta\psi$, and to the mean obliquity, $\Delta\varepsilon$. These quantities are expressed in two series with 106 terms each, and include all terms of at least 0".0001 in magnitude (Kaplan, 1981, pp.A4-A6). Those terms which depend on the Moon's longitude have periods of less than about 60 days, and are considered to be the short period terms. The remaining terms are longer period, and typically are interpolated in 10-day intervals in tables of apparent places of fundamental stars. Expressions for $\Delta\psi$ and $\Delta\varepsilon$ given below contain all terms, both long and short period.

2. Magnitude of the Nutation Corrections. As stated above, the amplitude of the orbit of the true pole about the mean pole is about a dozen arc-seconds. Computerized systems not requiring this level of accuracy may ignore the nutation correction.

3. Methods of Computing Nutation. The methods of computing nutation parallel those for computing precession. All methods of computing nutation depend on first computing $\Delta\psi$ and $\Delta\varepsilon$.

The tables of terms in Kaplan (1981) give several different coefficients for sine or cosine terms of the same argument. These have been grouped together to yield the following equations, which are accurate to 0".0001:

$$\Delta\psi = - (17".2289 + 0".01745\ T)\ \sin\ \Omega - (1".4006 + 0".00008\ T)\ \sin\ 2\Omega$$
$$+ (0".0242 + 0".00003\ T)\ \sin\ l + 0".0031\ \sin\ 2l + (0".0707 - 0".00017\ T)\ \sin\ l'$$
$$+ 0".0005\ \sin\ 2l' - (1".6341 + 0".00013\ T)\ \sin\ 2F + 0".0002\ \sin\ 4F$$
$$+ 0".0008\ \sin\ D + (1".3346 + 0".00007\ T)\ \sin\ 2D$$
$$- 0".0002\ \sin\ 4D \qquad\qquad [6.7a]$$

$$\Delta\varepsilon = + (9".2179 + 0".00089\ T)\ \cos\ \Omega$$
$$+ (0".6076 - 0".00035\ T)\ \cos\ 2\Omega + (0".0105 - 0".00001\ T)\ \cos\ l$$
$$- 0".0010\ \cos\ 2l + 0".0001\ \cos\ 3l + (0".0200 - 0".00004\ T)\ \cos\ l'$$
$$+ 0".0008\ \cos\ 2l' + (0".7104 - 0".00040\ T)\ \cos\ 2F$$
$$- 0".0001\ \cos\ D + (0".5852 - 0".00034\ T)\ \cos\ 2D + 0".0002\ \cos\ 4D\ [6.7b]$$

where the fundamental arguments at J_s are

Ω = the longitude of the ascending node of the Moon's mean orbit on the ecliptic,
measured from the mean equinox of date
= 450160".280 - (5r + 482890".539) T + 7".455 T^2 + 0".008 T^3 [6.8a]

l = the mean anomaly of the Moon
= 485866".733 + (1325r + 715922".633) T + 31".310 T^2 + 0".064 T^3 [6.8b]

l'= the mean anomaly of the Sun (Earth)
= 1287099".804 + (99r + 1292581".224) T - 0".577 T^2 - 0".012 T^3 [6.8c]

F = the difference L - Ω, where L is the mean longitude of the Moon
= 335778".877 + (1342r + 295263".137) T - 13".257 T^2 + 0".011 T^3 [6.8d]

D = the mean elongation of the Moon from the Sun
= 1072261".307 + (1236r + 1105601".328) T - 6".891 T^2 + 0".019 T^3 [6.8e]

r = one revolution = 360° = 1296000"

Throughout the current decade, values for T will be on the order of 0.8.

     Equations [6.7] and [6.8] yield an accuracy far higher (0".0001) than that
needed for telescope control, so terms in these equations that are not needed should
be dropped. To build a system accurate to one arc-second with roughly a dozen
sources of error, each source of error should be computed with about 0".1 accuracy.
If there are ten terms in the equation to be dropped, then terms less than 0".01 can
be dropped. Actually, since random errors propagate in quadrature, each of 12
terms need only be accurate to 1"/$\sqrt{12}$, or about 0".3. However, the equations that
result in dropping terms less than 0".01 are so improved in efficiency over those
containing all terms that little additional gain in efficiency can be made by taking
this effect into account here.
     Using T = 1, when terms less than 0".01 are dropped, equations [6.7] and
[6.8] become:

$\Delta\psi$ = - (17".2289 + 0".01745 T) sin Ω - 1".4006 sin 2Ω + 0".0242 sin l
     + 0".0707 sin l' - 1".6341 sin 2F + 1".3346 sin 2D   [6.9a]

$\Delta\epsilon$ = + 9".2179 cos Ω + 0".6076 cos 2Ω + 0".0105 cos l + 0".0200 cos l'
     + 0".7104 cos 2F + 0".5852 cos 2D   [6.9b]

Note that by eliminating the superfluous terms, eight trigonometric function
evaluations are eliminated, at a savings of 22 computation units per function.
     Ghedini (1982, p.9) uses a similar set of equations to compute $\Delta\psi$, but gives
none to compute $\Delta\epsilon$. He claims an accuracy of 0".075 for his equations. Meeus
(1982, p.70) also employs an analogous set of equations, but uses the Sun's mean
longitude and the Moon's mean longitude instead of F and D, so his equations
contain different constants.

Once $\Delta\psi$ and $\Delta\varepsilon$ have been computed, $\Delta\alpha$ and $\Delta\delta$ can be found. To evaluate the different methods for doing this in terms of computation speed, the number of arithmetic operations of each type must be counted, using the same method that was used for precession. The table below lists these counts:

Equation	Trig	+	-	x	/	Computation Units
[6.8]	0	14	5	30	0	79
[6.7]	22	21	12	48	0	613
[6.8] + [6.7]	22	35	17	78	0	692
[6.9]	12	9	3	19	0	314
[6.8] + [6.9]	12	23	8	49	0	393

It has been assumed that T was computed at the start of the evening. Equations [6.8] must be evaluated, then used in either the highest accuracy equations ([6.7]) or the lower accuracy equations ([6.9]).

Using the same conversion to computation units that was used previously, equations [6.8] require 79 units, [6.7] require 613 units, and [6.9] require 314 units. Thus the high accuracy equations require a total of 692 units, while the equations with lower, but acceptable, accuracy require 393 units. To compare methods of obtaining (α, δ) from $\Delta\psi$ and $\Delta\varepsilon$, equations [6.8] and [6.9] will be used.

Once $\Delta\psi$ and $\Delta\varepsilon$ have been computed, simple transformations can be computed to convert these deltas of longitude and obliquity into deltas of right ascension and declination.

The first method of computing general precession that was discussed in the previous section used equations in closed form to evaluate the changes in right ascension and declination. The analogous equations for computing these changes that are due to nutation are (to first order):

$$\Delta\alpha = (\cos \varepsilon + \sin \varepsilon \sin \alpha \tan \delta) \Delta\psi - \cos \alpha \tan \delta \, \Delta\varepsilon \qquad [6.10a]$$

$$\Delta\delta = \sin \varepsilon \cos \alpha \, \Delta\psi + \sin \alpha \, \Delta\varepsilon \qquad [6.10b]$$

The second method of computing general precession used matrix multiplications to rotate rectangular equatorial coordinates. Before transforming back to the precessed (α', δ'), the nutation correction can be applied by adding the following changes to the rotated coordinates:

$$\Delta x = - (y \cos \varepsilon + z \sin \varepsilon) \Delta\psi \qquad [6.11a]$$

$$\Delta y = + x \cos \varepsilon \, \Delta\psi - z \, \Delta\varepsilon \qquad [6.11b]$$

$$\Delta z = + x \sin \varepsilon \, \Delta\psi + y \, \Delta\varepsilon \qquad [6.11c]$$

Second order terms are neglected, as they can be at most one unit in the eighth figure.

The nutation correction can also be applied by multiplying the precession rotation matrix in [6.5] by the following matrix before rotating the coordinates:

$$
\begin{bmatrix}
1 & -\Delta\psi\cos\varepsilon & -\Delta\psi\sin\varepsilon \\
+\Delta\psi\cos\varepsilon & 1 & -\Delta\varepsilon \\
+\Delta\psi\sin\varepsilon & +\Delta\varepsilon & 1
\end{bmatrix}
\qquad [6.12]
$$

The two rotation matrices must be multiplied to handle significant second order terms properly. Thus merely adding the two matrices element by element may not produce results of the desired accuracy.

Since all these methods use the same method of computing $\Delta\psi$ and $\Delta\varepsilon$, and they all yield roughly the same accuracy, the criterion for choosing one as the best method is speed of execution. The number of operations of each type are tabulated below:

Equation	Trig	+	-	x	/	Computation Units
[6.10]	5	2	1	8	0	129
[6.11]	2	2	2	9	0	66
[6.12]	2	0	3	2	0	51

These figures assume that once a term involving a trig quantity is evaluated, it can be used elsewhere without the need to evaluate it again.

Using the same method of converting these operations into computation units that was used in the precession algorithm evaluation, the method of equations [6.10] requires 129 computation units, [6.11] requires 66 units, and [6.12] requires 51 units. This is misleading, however, as [6.11] and [6.12] are useful only when combined with the 'second method' of computing general precession. Results obtained using [6.10] are simply added to (α', δ') of the 'first method', while the results from [6.11] are added to the results of the rotation in the 'second method', and results from [6.12] are matrix multiplied with the rotation matrix. The matrix multiplication requires three multiplies and three adds per matrix element, for a total of 81 units.

The total number of computation units for computing precession, computing nutation (without counting the units for computing $\Delta\psi$ and $\Delta\varepsilon$), computing $\Delta\psi$ and $\Delta\varepsilon$, and combining the results are:

Equation	Computation Units
[6.3],[6.8],[6.9],[6.10]	272 + 393 + 129 + 2 = 796
[6.4] - [6.6],[6.8],[6.9],[6.11]	532 + 393 + 66 + 3 = 994
[6.4] - [6.6],[6.8],[6.9],[6.12]	532 + 393 + 51 + 81 = 1057

This clearly shows that when both precession and nutation are computed and the results are combined, the direct evaluation method of using equations in closed form for both precession and nutation is the best. This is an interesting result, in light of the fact that many systems in current use favor the more mathematically elegant, but more computationally intensive, matrix approach (C.f., Huguenin and McCord, 1975).

C. POLAR MOTION

Minor periodic perturbations not taken into account in the theories of precession and nutation, and random realignments of the distribution of the Earth's mass caused by tectonic plate drift and earthquakes, cause the true pole of the Earth to change position by a random amount in a random direction at random times, although the motion tends to make a spiral pattern of increasing radius when plotted over several decades. This motion is depicted in Figure 6-7, and usually amounts to no more than 50 meters per year, which, at the polar radius of the Earth, is usually less than 0".1 in a year. Amounts this small are difficult to measure, and the motion cannot be predicted in advance, only evaluated as an effect on the coordinate system after the motion has occurred. Therefore, there is no theory of polar motion in closed mathematical form.

Considering the magnitude of the correction involved, there is no need to attempt to correct for polar motion in a telescope control system.

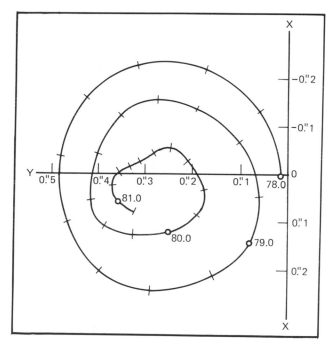

6-7 Motion of the Earth's Rotational Pole Over Several Decades
Annual Report 1980, Federation of Astronomical and Geophysical Services

D. ABERRATION

1. **The Physical Basis of Aberration.** When an observed object moves with respect to a fixed observer, the light travelling from the moving object to the fixed observer takes a finite amount of time to reach the observer. During this time, the object moves a certain amount. The observer receives the light from the direction of the

object when it emitted its light, not the direction of the object at the time the observation is made. The angle between the two directions is a form of aberration called the correction for light time.

This form of aberration is, in general, a constant for each star but is unknown, so it is ignored. The concern of one who is designing a telescope control system is not where the star is at the time of the observation, but where to point the telescope to see the star's light. Light time correction is important only in the solar system, in which the distances that light travels from various objects to the Earth are of the same order of magnitude as the distances between various observing positions, whether caused by observing an object at different places in the Earth's orbit about the Sun, or by sending deep space probes to various parts of the solar system. This makes the light time correction for solar system objects essential. Lack of light time corrections in the ephemerides for Comet Kohoutek caused large pointing errors when radio telescopes were used to observe the comet. The IAU standard for ephemerides of solar system objects after 1979 is to include the light time corrections in the ephemeris position.

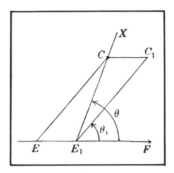

6-8 Geometric Explanation of Aberration
Courtesy of Cambridge University Press

The case where a fixed object (e.g., a star) transmits light to a moving observer is called stellar aberration. In this case, the aberration is due to making an observation in a moving coordinate system. As seen in Figure 6-8, if an observer at the eyepiece E is observing a star whose light enters the telescope at C, the motion of the observer (due to the Earth's rotation and motion in its orbit about the sun) causes motion of the eyepiece to E_1 during the time it takes light to travel the length of the telescope tube. Although the light travels from C to E_1, it appears to the observer to have followed a path from C_1 to E_1. This stellar aberration has three components, corresponding to the three main motions of the Earth: diurnal aberration, caused by the rotation of the Earth, annual aberration, caused by the motion of the Earth in its orbit about the Sun, and secular aberration, caused by the motion of the solar system through space. The stars and the solar system may each be considered to be moving in uniform rectilinear motion, so one cannot distinguish between the correction for light time and secular aberration. Therefore, secular aberration can be ignored.

Planetary aberration is the sum of stellar aberration and the correction for light time, and is applied only to members of the solar system.

2. Magnitude of Stellar Aberration Corrections. If one is observing an object at some elevation angle Θ, then stellar aberration will cause a displacement $\Delta\Theta$ of the object toward the apex of the motion of the observer. Gurnette and Woolley (1961, p.47) give the magnitude of the correction as

$$\sin (\Delta\Theta) = \frac{V}{c} \sin (\Theta - \Delta\Theta) \qquad\qquad [6.13a]$$

or, expanding in powers of V/c,

$$\Delta\Theta = \frac{V}{c} \sin \Theta - \frac{1}{2} \left(\frac{V}{c}\right)^2 \sin 2\Theta + ... \qquad\qquad [6.13b]$$

where V = the velocity of the observer, and c = the velocity of light.

Since the average velocity of the Earth about the Sun is about 18 miles per second, V/c is about 0.0001, which corresponds to about 20". The second-order term in [6.13b] is about 0".001, and may be neglected.

3. Methods of Computing Annual Aberration. Smart (1979, p.184) computes annual aberration in the following manner (expressions for k, L_t, and ε have been updated to J2000.0):

$$\Delta\alpha = Cc + Dd \qquad\qquad [6.14a]$$

$$\Delta\delta = Cc' + Dd' \qquad\qquad [6.14b]$$

where $C = - k \cos \varepsilon \cos L_t$ $\qquad\qquad [6.14c]$

$D = - k \sin L_t$ $\qquad\qquad [6.14d]$

$$c = \frac{1}{15} \cos \alpha \sec \delta \text{ (α in time seconds)} \qquad\qquad [6.14e]$$

$$d = \frac{1}{15} \sin \alpha \sec \delta \qquad\qquad [6.14f]$$

$c' = \tan \varepsilon \cos \delta - \sin \alpha \sin \delta$ $\qquad\qquad [6.14g]$

$d' = \cos \alpha \sin \delta$ $\qquad\qquad [6.14h]$

k = the constant of aberration = 20".49552 $\qquad\qquad [6.14i]$

L_t = the Sun's true geocentric longitude $\qquad\qquad [6.14j]$

ε = the mean obliquity of the ecliptic

= 84381".448 - 46".8150 T - 0".00059 T^2 + 0".001813 T^3

[Kaplan (1981, p.A3)]   [6.14k]

Meeus (1982, p.83) computes the Sun's true longitude L_t from its geometric mean longitude as follows:

$$L_t = L_m + \nu - l'$$ [6.14l]

where L_m = the geometric mean longitude of the Sun,
referred to the mean equinox of date

= $280°$ 27' 57".850 + (100r + $0°$ 46'11".270) T + 1".089 T^2 [6.14m]

l' = the mean anomaly of the Sun (see [6.8c])

To compute the true anomaly ν, Meeus (1982, p.122) suggests using the following (the expression for e has been updated to J2000.0):

$$\nu = 2 \arctan \left[\sqrt{\frac{1 + e}{1 - e}} \tan \frac{E}{2} \right]$$ [6.14n]

where e = the eccentricity of the Earth's orbit

= 0.016708320 - 0.000042229 T - 0.000000126 T^2 [6.14o]

E = the eccentric anomaly of the Earth in its orbit

= l' + e sin E   [6.14p]

Equation [6.14p] is computed by iteration until convergence. Equations [6.14l] - [6.14p] can be computed at the beginning of the evening.

This procedure neglects terms, known as the E-terms, which depend on the eccentricity and longitude of perihelion of the Earth's orbit. These E-terms are about 0".34, and are constant throughout the year for a particular star, changing very slowly over a period of centuries. The E-terms are evaluated by:

(in longitude) $\Delta\lambda$ = k e sec β cos(ω - λ)   [6.15a]
(in latitude) $\Phi\beta$ = k e sin β sin(ω - λ)   [6.15b]

where k = the constant of aberration = 20".49552   [6.15c]

e = the eccentricity of the Earth's orbit (see [6.14o])
ω = the longitude of perihelion of the Earth's orbit

= $102°$ 56' 18".046 + $1°$ 43' 10".046 T   [6.15d]

Longitude (λ) and latitude (β) are measured from the ecliptic (not the equator).

The effect of the E-terms on (α, δ) can be computed as follows:

$$\Delta\alpha = c \, \Delta C + d \, \Delta D$$ [6.15e]

$$\Delta\delta = c' \, \Delta C + d' \, \Delta D$$ [6.15f]

where c and d are given by [6.14e and f]

$$\Delta C = + k \; e \; \cos \; \omega \quad \cos \; \epsilon \qquad\qquad\qquad [6.15g]$$

$$\Delta D = + k \; e \; \sin \; \omega \qquad\qquad\qquad [6.15h]$$

These quantities are added to α and δ, respectively, to effect the correction for E-terms.

By convention, catalog mean places of stars already contain the E-terms, until 1984 (see Section J below). Theoretically, one should remove the E-terms from the catalog position, compute precession, nutation, and proper motion, then recompute the E-terms using the new equinox and epoch, and apply them to the new coordinates. Since the E-terms are relatively constant over short periods of time (decades), the maximum error per century created by treating the E-terms as constant instead of following the above procedure is given by Gurnette and Woolley (1961, p.145) as 0s.0001 in $\Delta\alpha \cos \delta$ and 0".002 in $\Delta\delta$. These errors are so small that the E-terms can be treated as being constant in ephemerides published before 1984.

4. Methods of Computing Diurnal Aberration. Meeus (1982, p.71) correctly defines the apparent place of a star as "its position on the celestial sphere as it is actually seen from the center of the moving Earth, and referred to the instantaneous equator, ecliptic, and equinox". From this perspective, diurnal aberration need not be computed to find the apparent place of a star. However, telescopes do not function well at the center of the Earth. "Apparent position" is used here in the sense of pointing a telescope located on the Earth's surface (the topocentric position), and, therefore, to obtain high pointing accuracy, corrections for diurnal aberration need to be performed.

Gurnette and Woolley (1961, p.49) express the velocity of an observer on the surface of the Earth as v r cos L_c, where v (= 0.46 km/sec) is the equatorial rotational velocity of the surface of the Earth, r is the geocentric radius of the Earth at the observer's site expressed in units of the Earth's equatorial radius, and L_c is the geocentric latitude of the observer. This corresponds to a constant of diurnal aberration of

$$\frac{v}{c} \; r \; \cos \; L_c = 0".320 \; r \; \cos \; L_c = 0s.0213 \; r \; \cos \; L_c$$

This can be resolved into the following corrections (apparent minus mean) to (α, δ):

$$\Delta\alpha = 0s.0213 \; r \; \cos \; L_c \; \cos \; h \; \sec \; \delta \qquad\qquad [6.16a]$$

$$\Delta\delta = 0".320 \; r \; \cos \; L_c \; \sin \; h \; \sin \; \delta \qquad\qquad [6.16b]$$

where h = the hour angle = local sidereal time - α [6.16c]

Since the equatorial radius of the Earth (6378140 m) differs from the polar radius (6356755 m) by only about two parts in 600, and any Earth-based observer's geocentric radius is likely to fall in between the two, r can be taken to be 1 independent of the observer's position, and still keep the correction for diurnal aberration accurate to 0".1. This also means the distinctions among geocentric, geodetic (geographic), and astronomical latitude can be ignored.

E. PARALLAX

1. The Physical Basis of Parallax. Since the positions of the Sun, Moon, planets, and stars are published in geocentric or heliocentric coordinates, and since the observer is not at the position of reference for the published coordinates, these bodies usually appear to be displaced in the sky. When viewed by an observer on the Earth's surface, the displacement depends directly on the distance of the observer from the reference point and inversely on the distance of the body from the reference point. The parallax of a close star caused by the orbital motion of the Earth about the Sun is depicted in Figure 6-9.

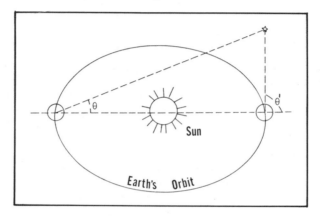

6-9 Heliocentric Parallax

2. Magnitude of Parallax Corrections. Stellar positions are typically referred to the Sun, but the heliocentric parallax caused by the Earth's distance from the Sun is less than 0".1 for all but the closest stars, which have a parallax of about one arc second. The geocentric parallax, caused by the observer's being separated from the center of the Earth by its radius, is negligible for stars.

For the Sun and planets, both the heliocentric and geocentric parallax corrections are on the order of a few arc seconds, and may be approximated by first order theory with sufficient accuracy. The close proximity of the Moon to the Earth requires the use of exact formulae to compute the Moon's geocentric parallax, and the much closer proximity of artificial satellites requires that their positions be computed exactly from a known geocentric ephemeris. Because most astronomers have no interest in pointing a telescope accurately at the moon or artificial satellites, these cases are ignored in the discussion below. Furthermore, most astronomers have little interest in those few close stars that would require a stellar parallax correction, so in most cases, it, too, can be ignored. The discussion of stellar parallax is presented for completeness.

3. Methods of Computing Stellar Parallax. Gurnette and Woolley (1961, p.64) give the corrections to the mean (catalog) position (α, δ) of a star due to heliocentric parallax as follows (the constant of parallax has been updated to J2000.0):

$$\Delta\alpha = \pi \frac{Y \cos \alpha - X \sin \alpha}{\cos \delta} \qquad\qquad [6.17a]$$

$$\Delta\delta = \pi (Z \cos \delta - X \cos \alpha \sin \delta - Y \sin \alpha \sin \delta) \qquad [6.17b]$$

where π = the annual parallax of a star
  = 8".794148/D, where D is the heliocentric distance of the star in A.U.

(X,Y,Z) = the geocentric coordinates of the Sun

Often, neither the heliocentric distance of a star nor its annual parallax are known, so the correction for parallax cannot be made. This usually occurs, however, only when the parallax is too small to matter. Many star catalogs contain values of heliocentric parallax when it is of the order of 0".1 or larger.

Meeus (1982, p.85) uses the following method to compute (X,Y,Z):

$$X = R \cos L_t \qquad\qquad\qquad\qquad\qquad\qquad [6.18a]$$

$$Y = R \sin L_t \cos \varepsilon \qquad\qquad\qquad\qquad\qquad [6.18b]$$

$$Z = R \sin L_t \sin \varepsilon \qquad\qquad\qquad\qquad\qquad [6.18c]$$

where R = the distance from the Sun to the Earth in A.U.

L_t= the Sun's true longitude referred to the mean equinox of date (see equation [6.14j])

ε = the mean obliquity of the ecliptic for that date (see equation [6.14k])

Meeus (1982, p.83) computes R as follows:

$$R = \frac{1.0000002 (1 - e^2)}{1 + e \cos \nu} \qquad\qquad\qquad [6.18d]$$

where e = the eccentricity of the Earth's orbit (see [6.14o])

ν = the true anomaly of the Earth in its orbit (see [6.14n])

As an alternative method, Gurnette and Woolley (1961, p.64) suggest using the star constants c, d, c' and d' as follows:

$$\Delta\alpha = \pi (Yc - Xd) \qquad\qquad\qquad\qquad [6.19a]$$

$$\Delta\delta = \pi\,(Yc' - Xd') \qquad\qquad [6.19b]$$

where c, d, c', and d' are given by equations [6.14e] - [6.14h]. This permits corrections for annual parallax to be included with the aberration terms of the reduction from mean to apparent place, as follows:

$$\Delta\alpha = (C + \pi\,Y)\,c + (D - \pi\,X)d \qquad\qquad [6.20a]$$

$$\Delta\delta = (C + \pi\,Y)\,c' + (D - \pi\,X)d' \qquad\qquad [6.20b]$$

These equations can be simplified if the annual parallax is small enough, as follows:

$$\Delta\alpha = C\,(c + d\,\pi\,k1) + D\,(d - c\,\pi\,k2) \qquad\qquad [6.21a]$$

$$\Delta\delta = C\,(c' + d'\,\pi\,k1) + D\,(d' - c'\,\pi\,k2) \qquad\qquad [6.21b]$$

$$\text{where } k1 = \frac{R \sec \varepsilon}{20''.49552} \qquad\qquad [6.21c]$$

$$k2 = \frac{R \cos \varepsilon}{20''.49552} \qquad\qquad [6.21d]$$

in which 20".49552 is the constant of aberration (J2000.0). This method does not require that X, Y, or Z be computed.

Although the method using [6.21], by itself, is slightly more efficient than the method using [6.17], when the other arithmetic operations in [6.21a,b] are taken into account, the real savings accrue when both annual aberration and annual parallax are computed, as is usually the case. Thus the preferred method of computing both annual aberration and annual parallax is to combine them using the latter method.

4. Methods of Computing Solar or Planetary Parallax. For bodies such as the Sun, the planets, or comets, published catalog positions are referred to the center of the Earth, so there is no need to compute heliocentric parallax.

Gurnette and Woolley (1961, p.63) give the following equations for computing geocentric parallax (updated to J2000.0):

$$\Delta\alpha = \pi\,(\rho \cos \varphi' \sin h \sec \delta) \qquad\qquad [6.22a]$$

$$\Delta\delta = \pi\,(\rho \sin \varphi' \cos \delta - \rho \cos \varphi' \cos h \sin \delta) \qquad\qquad [6.22b]$$

where π = the horizontal parallax of the object = 8".794148/D where D is the geocentric distance of the object in A.U.

ρ = the geocentric distance of the observer in units of the Earth's equatorial radius

φ' = the geocentric latitude

h = the topocentric hour angle of the object

δ = the topocentric declination of the object

The topocentric quantities h and δ may be replaced by their geocentric counterparts h' and δ', which are the quantities listed (δ') or derived (h') from the catalog or ephemeris used to find an object. This is because the distances to the Sun and the planets makes the quantities (h - h') and (δ - δ') insignificant. Gurnette and Woolley (1961, p.57) give equations to compute the geocentric latitude if the geodetic latitude is known. This kind of conversion is necessary only for objects less than about 1/4 A.U. from the Earth, such as the Moon.

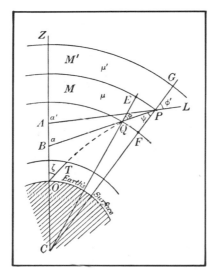

6-10 Refraction of Light in Passing Throught the Earth's Atmosphere
Courtesy of Cambridge University Press

F. REFRACTION

1. The Physical Basis of Refraction. As light leaves interplanetary space, considered to be a vacuum with index of refraction μ = 1, and penetrates the Earth's atmosphere on its way to a telescope, it passes through regions of the atmosphere with different μ. This is depicted in Figure 6-10. To compute the effective refraction for a plane-parallel atmosphere, only the index of refraction at the Earth's surface is required. Since the Earth's atmosphere is not strictly plane-parallel, the resulting errors at low altitudes can be on the order of seconds of arc. This, combined with other factors, such as temperature inversions, make it impossible to compute refraction corrections from fundamental considerations of the physical theory of refraction. This is in contrast to precession, nutation, aberration, and other phenomena, which can be computed on the basis of theoretical considerations.

Instead, models of atmospheric refraction are usually based on extensive observations. These observations are often published in the form of a set of tables, the most widely-used being those of Pulkovo Observatory (Orlov, 1956). These tables are quite lengthy, so when building a computerized control system, it is not realistic to try to place the tables on a mass storage device and access them in the required amount of time. Two different mathematical models of refraction are described below. These models differ widely in structural, mathematical, and computational complexity, with the more computationally intensive model being more accurate.

The theory of refraction is complicated by the fact that different colors exhibit different amounts of refraction, so star images are actually vertical spectra. Woolard and Clemence (1966, p.84) list the following differences in refraction angle between red and blue light for varying zenith angles z:

z (degrees)	Delta R
30	0".35
45	0".60
60	1".04
75	2".24

Refraction is also complicated by the fact that turbulence in the atmosphere changes the refraction of a given volume of air with time. This effect makes the apparent position of a star change so rapidly that it smears the star's point-like image into a "seeing disk" for the naked eye and all instruments except those with very high time resolution. This "seeing" effect is ignored in the discussion below, since there is nothing one can do about it without a very expensive star tracker and highly optimized servos. One final effect is differential refraction, which tends to distort extended objects. This effect can be overcome using a pair of rotating prisms, but is usually ignored in all but large telescopes.

2. Magnitude of Refraction Corrections. Gurnette and Woolley (1960, p.55) give values of $J = (\mu \sin(Z - R))/\sin Z$ for an ideal spherical atmosphere, where μ is the index of refraction of the atmosphere at sea level, Z is the true zenith distance of a star, and R is the angle through which the star's light is refracted, for various angles of Z. The values of J were derived using mean observed values of R and μ. Using $\mu = 1.00029241$ (at a wavelength of 5780 Angstrom, from Garfinkel, 1967, p.248), the author has computed the corresponding refraction angle R. These values are reproduced below:

Z (degrees)	J	R
0	1.000000	-
30	1.000000	-
60	1.000001	1'.7
70	1.000003	2'.7
75	1.000005	3'.7
80	1.000011	5'.5
82	1.000016	6'.7
84	1.000026	8'.6
86	1.000046	11'.8
88	1.000095	18'.1
89	1.000148	23'.7
90	1.000242	34'.5

Note that the values for the angle of refraction R in the table above are in arc-minutes. It is clear that even for relatively small zenith distances, refraction can be quite large, compared to other sources of error. Therefore, except for systems of modest accuracy, refraction corrections are essential.

3. Methods of Computing Refraction. Smart (1979, p.73) gives the effect of refraction on the (α, δ) coordinates of a star as:

$$\Delta\alpha = R \; \frac{\sec^2 \delta \; \sin h}{\tan z \, (\tan \delta \; \tan \varphi + \cos h)} \qquad\qquad [6.23a]$$

$$\Delta\delta = R \; \frac{\tan \varphi - \tan \delta \; \cos h}{\tan z \, (\tan \delta \; \tan \varphi + \cos h)} \qquad\qquad [6.23b]$$

where R = the angle of refraction = Z - z

   h = the hour angle of the observed body

   z = the apparent zenith distance

   Z = the true zenith distance

   φ = the observer's latitude

The angular units of $\Delta\alpha$ and $\Delta\delta$ are the same as those of R.

Many astronomers use a single simple equation to compute the refraction angle R over the entire range of zenith distances $0 \leq z \leq 90°$. Such methods typically do not offer high accuracy at large zenith distances, but are quite adequate for most applications. For example, the Almanac for Computers (p.B13) lists the following:

$$R = \frac{P}{273 + T} \left\{ 3.430289 \, [z - \arcsin (0.9986047 \sin 0.9967614 \, z)] - 0.01115929 \, z \right\}$$

where R is the refraction correction in minutes of arc

   z is the apparent zenith distance in degrees

   T is the temperature in degrees Celsius

   P is the atmospheric pressure in millibars

This equation is accurate to 0'.1 for altitudes greater than 15°, 1'.0 above 3°, and 3' near the horizon. To obtain more accuracy, Eisele and Shannon (1975) fit the equation

$$R = A \tan Z + B \tan^3 Z$$

at the points Z = 0°, 45°, and 85°, where Z is the true zenith distance.

For $85° \leq Z \leq 90.°6$, the equation

$$R = A\left[e^{-B\ (90\ -\ Z)} + e^{\frac{-B}{5}\ (90\ -\ Z)}\right]$$

was used. The resulting equations are as follows:

$$R = \frac{17\ P}{460\ +\ T}\ (57.626039\ \tan z - 0.05813517\ \tan^3 z) \qquad\qquad [6.24a]$$

for $0 \leq Z \leq 85°$, and

$$R = \frac{17\ P}{460\ +\ T}\ 871.94412\ \left[e^{-0.53520501\ (90\ -\ Z)} + e^{-0.107041\ (90\ -\ Z)}\right] \qquad [6.24b]$$

for $85° \leq Z \leq 90°.6$,

where R = the refraction angle in arc seconds

  Z = the true zenith distance in degrees

  P = the atmospheric pressure in inches of Hg measured at the Earth's surface

  T = the atmospheric temperature in °F measured at the Earth's surface.

Overall accuracy of this set of equations over the range $0 \leq z \leq 90°.6$ is quoted as 3.7 arc seconds, with larger excursions detected at $Z = 85°$.
  Although this level of accuracy does not compare with the 0.1 arc second accuracy that is possible in the other astronomical corrections used to compute apparent place from mean place, it is adequate for all but the most demanding of real-time telescope control applications. Garfinkel (1967) describes a polytropic model of the atmosphere and an algorithm for computing refraction to high accuracy. He claims overall error of less than one arc-second, with most computations being in agreement with the standard tables within 0.1 arc seconds, even for zenith distances greater than 90°. Garfinkel's computer program contains roughly 300 lines of FORTRAN statements, and performs several iterations within many different loops. Its author estimates that it would take on the order of ten seconds to run this program on a typical minicomputer (Rodin, 1982). This would translate to about 30 seconds on a typical microcomputer programmed in assembly language using floating point hardware, and roughly 10 to 20 times that if done in interpretive BASIC. Since refraction must be re-computed with changing altitude as the object of interest rises and sets, to obtain 0".1 accuracy, refraction should be computed at least once per second when observing near the horizon. Therefore, this requirement for computation time makes Garfinkel's algorithm impractical for all but the most powerful microcomputers.

The problem in using Garfinkle's algorithm in a real-time control system is not simply the execution speed of the control computer. Equation [6.24b] can be used to estimate the changes in P or T which would change R by 0".1 . Using z = 90° as the worst case, [6.24b] reduces to

$$R = \frac{17\ P}{460\ +\ T} \times 871.94412 \times 2$$

With T = 50° F and P = 30 inches of Hg, dR/dP = 57".6 per inch of Hg, so R changes by 0".1 when P changes by 0.0017 inches of Hg. This kind of pressure change can happen within a minute or so. Similarly, under the same conditions, dR/dT = -3".9 per 1° F at T = 50° F,   so R changes by 0".1 when T changes by 0.03° F. Again, changes of this magnitude can happen very quickly. Since either P or T or both can change in a period of time comparable to that required to compute refraction directly from Garfinkel's algorithm, this method is not satisfactory for field use.

This also brings to light another problem. Atmospheric pressure cannot be measured with an error less than about 0.01 inches of Hg, and temperature cannot be measured with an error less than about 0.25° F without very expensive instrumentation. This means that even if Garfinkel's algorithm is capable of computing refraction to 0".1 accuracy, field instrumentation will probably limit the accuracy of refraction calculations to 1-5 arc seconds. Thus the equations of Eisele and Shannon are well-suited to the real-time telescope control environment because of their minimal computation requirements and the accuracy of typical temperature and pressure field instruments.

If a means could be found to compute Garfinkel's algorithm quickly and to measure P and T very accurately, telescope pointing accuracy under normal field conditions could be improved. Wellnitz (1983) indicates that changes in atmospheric pressure at high altitudes are always accompanied by the appropriate changes on the ground, since the pressure measured at the Earth's surface is simply the hydrostatic pressure of the air mass above the sensor. However, a given temperature measured at the Earth's surface does not imply a particular atmospheric temperature distribution along the line of sight, since temperature inversions and inhomogenieties in azimuth temperature change rapidly. Despite this, Wellnitz claims that the error introduced by these effects is only about 0".05 near the zenith, growing to about 1" RMS between 10° and 40° elevation. This means that refraction algorithm accuracies of about 0".5 would be useful under many observing conditions. Wellnitz also confirms the limits imposed by temperature and pressure field instrumentation.

G. ORBITAL MOTION

The motion of two stars in a binary star system causes their observed positions in the sky to change with time, as shown in Figure 6-11. If the stars are relatively far apart, the changes in the stars' positions will be large enough to be observable. The rate of change in apparent position depends upon the separation between the stars and the distance of the binary system from the observer.

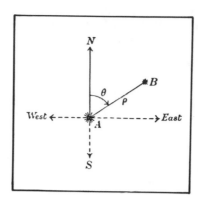

6-11 Changing Star Orientations Due to Orbital Motion
Courtesy of Cambridge University Press

Most binary systems are at a distance which is sufficient to keep the changes in positions of the two stars less than 0".1 in any appreciable time. This is because in order to move rapidly in an orbit, the stars must be so close together that the angle the orbit subtends from the Earth is small. Conversely, for the orbit to subtend an appreciable angle on the sky, the stars must be so widely separated that they move very slowly in their orbits.

In either case, once the desired star is located (by using a recent catalog), the effects of its orbital motion can usually be ignored. In those few cases where it cannot be ignored, orbital motion can be considered to produce a position change which varies linearly over time in the course of a year, and is usually noted in the star catalogs in a form that is readily entered into the computer for a trivial correction calculation.

H. PROPER MOTION

The proper motion of a heavenly body is its motion relative to the "fixed stars" (objects known not to move significantly with respect to each other over long periods of time). Many star catalogs give values for $\Delta\alpha$ and $\Delta\delta$ due to proper motion over some period of time, such as a year or a century, for those stars whose proper motions are known. Such motion is depicted in Figure 6-12. Most stellar proper motions of significant size are known, and for these stars the correction for proper motion is trivial. For those stars listed in fundamental catalogs, or in catalogs based on the fundamental catalogs, the mean place includes a correction for proper motion.

The proper motion of other (closer) bodies, such as planets and comets, is the basic phenomenon of interest to those preparing ephemerides for these objects, and thus proper motion is included in ephemerides of these bodies. Therefore, the designer of a computerized telescope control system need not be concerned with corrections for proper motion of solar system objects.

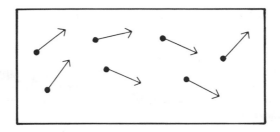

6-12 Proper Motions of Stars

I. REDUCTION FROM MEAN TO APPARENT PLACE

The preceeding sections have described methods of computing corrections to the mean (catalog) position of an object to obtain its apparent position as one would see it from a point on the Earth's surface. Proper motion, orbital motion, stellar aberration, refraction, and parallax cause changes in the direction in which a star is actually observed, but do not affect the frame of reference. Precession, nutation, and polar motion cause changes to the fundamental coordinate system, but do not affect the direction in which a star is observed. The corrections for precession, nutation, and annual aberration are large enough to cause significant errors if the corrections are not applied in the proper order and the cross terms are neglected.

Gurnette and Woolley (1961, p.150) quote an analysis by Porter and Sadler which indicates that the most accurate method of reducing mean to apparent place is to apply the correction for aberration to the mean place, then to use the resulting coordinates for the precession correction, then to apply the nutation correction to the precessed coordinates. The corrections for the heliocentric parallax of stars, proper motion, and orbital motion are small enough to be applied at any convenient point. The correction for refraction is applied last. The following table summarizes the process of reducing a mean place, represented by $(\alpha, \delta)m$, to an apparent place, represented by $(\alpha, \delta)a$:

STEP	CORRECTION	INPUT	OUTPUT
1a	Annual aberration	$(\alpha, \delta)m$	$(\alpha, \delta)1$
1b	Stellar parallax		
	(included as part of Step 1a)		
2	Precession	$(\alpha, \delta)1$	$(\alpha, \delta)2$
3	Nutation	$(\alpha, \delta)2$	$(\alpha, \delta)3$
4*	Orbital motion	$(\alpha, \delta)3$	$(\alpha, \delta)4$
5*	Proper motion	$(\alpha, \delta)4$	$(\alpha, \delta)5$
6	Diurnal aberration	$(\alpha, \delta)5$	$(\alpha, \delta)6$
7	Planetary parallax	$(\alpha, \delta)6$	$(\alpha, \delta)7$
8	Refraction	$(\alpha, \delta)7$	$(\alpha, \delta)a$

* Usually not needed

In this table, each correction is treated as if it were a computer software subroutine, in which a set of input coordinates is transformed into a set of output coordinates by computing a correction and applying the correction to the input coordinates. The input coordinates for Step 1 are the mean place right ascension and declination, and after applying all the corrections, the output coordinates of Step 8 are for the apparent place of the object. It is these latter coordinates that are used to point the telescope. Polar motion has been ignored in this table because of its negligible effect and our inability to model it. Because it was found above to be computationally more efficient to combine the computation of heliocentric stellar parallax and annual aberration, these corrections are listed together.

In a computerized control system, those corrections which need be computed only once per evening should be at the top of the list, then the other corrections should be ordered by their frequency of computation, with the lowest frequency corrections being closer to the top of the list. This is because the output of each step is the input to the succeeding step. This constraint places diurnal aberration near the bottom of the list, instead of at the top, as recommended by Porter and Sadler. However, the error resulting from this is typically negligible in telescope pointing applications.

Note that the process depicted in this table is "serial", with each correction being a sequential step in which the result of the previous step undergoes a transformation to obtain the input for the succeeding step. A different approach is the one used before electronic digital computers became available, and which is still used today because of this tradition. This latter approach is to compute Besselian day numbers, which, when used with star constants computed using the star's mean place coordinates, form a series of eight terms in right ascension and seven terms in declination. This "parallel" approach is popular because the Besselian day numbers depend only on the time an observation is made, not on either the object's or the observer's coordinates. Thus the Besselian day numbers can be computed in advance for the entire year, and published in an ephemeris. The only calculations the observer must do are to compute the star constants, then apply the Besselian day numbers for the observation date, and compute the two series for right ascension and declination.

This approach is still useful for one doing the calculations by hand or using a pocket calculator, but it has no advantage over the serial approach when used in a small computer to control a telescope in real-time. In the latter case, some Besselian numbers must be recomputed throughout the evening, then the whole series must be re-evaluated. Since some corrections need be computed only once per evening, while others must be re-computed several times during the night, there is little to be gained by this approach.

Another approach along the same lines is to use independent day numbers, which are used in a different series for position.

J. CHANGES IN THE 1984 EPHEMERIDES

The method of preparing the catalogs that astronomers use as the sources for the coordinates they enter into telescope control systems has changed recently. These changes include new constants for computing some of the corrections given in previous sections, but the equations given in this chapter already reflect these changes, so there is no need to bring them up to date.

Kaplan (1981) and Seidelmann and Kaplan (1982) list a set of changes in the method of preparing the Astronomical Ephemeris starting with the ephemeris for 1984. These were brought about by resolutions adopted by the International Astronomical Union in 1976, 1979, and 1982, and are summarized below:

1. The new fundamental epoch is J2000.0, corresponding to January 1.5, 2000, or Julian Day number 2,451,545.0. The previous fundamental epoch (1950.0) was 1950 January 0.5 . A Julian century of 36525 ephemeris days will continue to be used as the time unit for most calculations.

2. The new constant of general precession in longitude, per Julian century, is 5029".0966 . The previous precession constant for epoch 1900.0 was 5025".64 .

3. The equinox of the Fifth Fundamental Catalog (FK5) has replaced the FK4 equinox as the origin of right ascension. This produced a shift of 0s.06 in 1984 plus a correction that is time dependent.

4. The new constant of aberration is 20".49552. The previous constant was 20".496 . The E-terms of aberration are no longer included in catalog mean places of stars.

5. The new solar parallax constant is 8".794148, compared to the previous value of 8".794. The new equatorial radius of the Earth is 6378140 m (the previous value was 6378160 m), with a new flattening factor of 0.00335281 = 1/298.257, compared to the previous value of 0.00335285.

6. The new constant of nutation is 9".2025. The constant at epoch 1900.0 was 9".2235. A new theory of nutation, based on a non-rigid model of the Earth, has been adopted. The new model uses two 106-term series to evaluate nutation.

7. Other changes, including the definition of new time scales and new astronomical constants, do not affect telescope pointing algorithms, but do affect the way in which ephemerides will be generated from 1984 onward.

The corrections described in this chapter are used to convert mean catalog positions to apparent position in the sky. In the following chapter, corrections for telescope characteristics which affect pointing accuracy are discussed. Both sets of corrections must be made for high accuracy telescope control.

Chapter 7. **SOURCES OF SYSTEMATIC ERROR II
—MECHANICAL CORRECTIONS**

The previous chapter dealt with the problem of where in the sky an object appears to be placed, and how to compute this apparent position from the coordinates given in a catalog or ephemeris. If a perfect telescope were set to the apparent position coordinates, the object would appear in the center of the field. Any pointing errors seen in the eyepiece would be due only to errors in computing the apparent position. The corrections that are computed to obtain the apparent position are based on factors external to the particular telescope being used, and thus are computed in the same way for all observers.

The following paragraphs discuss the problem of how to point an imperfect telescope accurately. These corrections are computed in a manner that can be used with any telescope of a particular mounting type, but the values of the error parameters are unique to each individual telescope, since they are characteristics of that telescope.

In a typical telescope control system, the apparent position is what is presented to the control computer as the desired position. In closed-loop systems, the position feedback shaft angle encoder produces a number corresponding to the angular position of the axis. This number is converted to a meaningful coordinate, such as declination, by applying a calibration algorithm to the raw encoder readout. The corrections discussed in the following paragraphs are incorporated into the calibration algorithm to produce numbers representing the true position of the telescope. The control computer then compares the apparent position of the desired object with the position of the telescope (determined from the encoder readings), and generates commands to the motors to minimize the difference between the object and telescope positions. In open-loop systems, the control software keeps track of where it "thinks" the telescope is pointing, so there is no position comparison to make. Since there are no encoder readings in open-loop systems, the desired position of the telescope is adjusted to compensate for the telescope's mechanical errors.

The mechanical corrections of interest are as follows:

1. Zero offset (the $0°$ point in hour angle or declination)
2. Polar axis misalignment (equatorial mounts)
   Azimuth axis misalignment (alt-az mounts)
3. Non-perpendicularity of the axes
4. Collimation errors
5. Tube flexure
6. Mount flexure
7. Servo lag
8. Gearing errors
9. Bearing errors
10. Drive train torsion errors

Since the size of each correction depends so much on each individual telescope, each type of correction can be any size, from arc seconds to degrees.

A. TELESCOPE POINTING CORRECTIONS--EQUATORIAL MOUNT

1. Zero Offset. Closed-loop systems using shaft encoders must determine the encoder reading that corresponds to the zero point of that axis. The simplest form of the encoder calibration algorithm is a linear model, in which the hour angle (or declination) reading is obtained by multiplying the number produced by the encoder by a fixed factor that is the number of arc seconds per least significant bit of the encoder count, then adding a constant. Included in this constant is the offset of the zero position of the encoder relative to the direction of zero hour angle. This correction, shown in Figure 7-1, is unnecessary in open-loop systems, which are oriented by pointing at a star and entering its coordinates, or by using limit switches and a clock to determine the telescope's initial position.

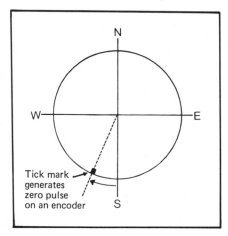

7-1 Zero Offset Error in Azimuth

Associated with zero offset is the scale factor for converting from encoder counts to an angle. This factor can be measured accurately with an autocollimator, or by sighting a distant (stationary) object, rotating the telescope through 360°, and sighting the object again. The scale factor can be found by noting how many encoder counts occur during the 360° rotation.

2. Polar Axis Alignment. Careful measurements of star positions with any equatorially-mounted telescope will indicate that the polar axis of the telescope is not exactly parallel to the Earth's rotation axis. The angle between these two axes cannot be expressed as a constant correction to hour angle or declination, since the corrections to the position (h, δ) depend on h and δ. However, this angle, once measured, can be resolved into components of the elevation error (M_{el}) and azimuth error (M_{az}) of the polar axis, as shown in Figure 7-2. Wallace (1975, p.284) uses the following equations to compute corrections Δh and $\Delta \delta$ to the telescope position (h, δ):

$$\Delta h = \tan \delta \ (M_{el} \sin h \ - \ M_{az} \cos h) \qquad\qquad [7.1a]$$

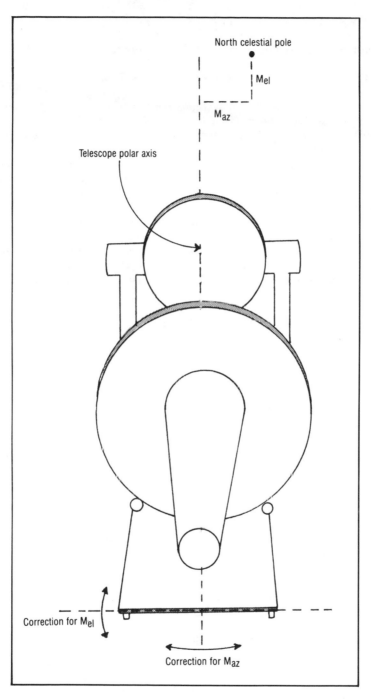

7-2 Corrections for Polar Axis Misalignments

$$\Delta\delta = M_{el} \cos h + M_{az} \sin h \qquad\qquad [7.1b]$$

These equations were derived using a small angle approximation for M_{el} and M_{az}.

B. TELESCOPE POINTING CORRECTIONS--ALT-AZ MOUNT

1. **Zero Offset.** In closed-loop systems, the zero offsets of the encoders on an alt-az mount are included in constant terms in a linear encoder calibration model. This is exactly analogous to the equatorial mount zero offset, and therefore is not necessary in open-loop control systems.

2. **Azimuth Axis Alignment.** The alt-az mount analogy to polar axis misalignment is zenith misalignment. This is shown in Figure 7-3. Imagine the rectangular coordinate system (x,y,z) has its origin at the intersection of the telescope's elevation and azimuth axes, and its x-y plane is parallel to the horizon such that the x-axis points north, the y-axis points west, and the z-axis points to the zenith. A telescope azimuth plane which is tilted can be represented by the coordinate system (x'',y'',z'') which has been rotated through angle **a** in elevation about the x-axis to form coordinate system (x',y',z'), and then rotated through angle **b** in azimuth about the z'-axis. Typically, angle **a** is small, but angle **b** can have any value.

The matrix representing the first rotation is:

$$\begin{bmatrix} 1 & 0 & 0 \\ 0 & \cos a & \sin a \\ 0 & -\sin a & \cos a \end{bmatrix}$$

while the matrix representing the second rotation is:

$$\begin{bmatrix} \cos b & \sin b & 0 \\ -\sin b & \cos b & 0 \\ 0 & 0 & 1 \end{bmatrix}$$

with the resulting matrix product representing the complete set of rotations on (x,y,z) to obtain (x'',y'',z'') as follows:

$$\begin{bmatrix} \cos b & \cos a \sin b & \sin a \sin b \\ -\sin b & \cos a \cos b & \sin a \cos b \\ 0 & -\sin a & \cos a \end{bmatrix} \qquad [7.2]$$

When this resulting matrix is applied to the coordinate system (x,y,z), the expressions for the coordinates (x'',y'',z'') are as follows:

$$x'' = x \cos b + y \cos a \sin b + z \sin a \sin b \qquad\qquad [7.3a]$$

$$y'' = -x \sin b + y \cos a \cos b + z \sin a \cos b \qquad\qquad [7.3b]$$

$$z'' = -y \sin a + z \cos a \qquad\qquad [7.3c]$$

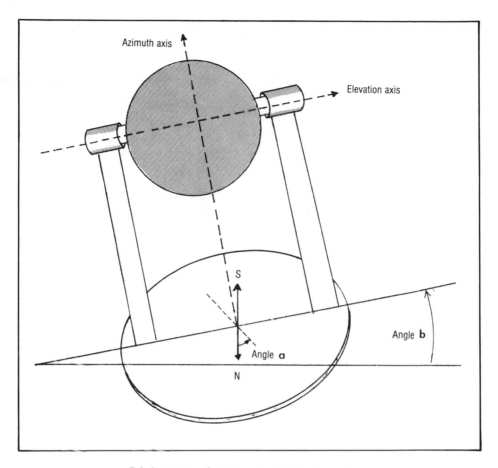

7-3 Corrections for Azimuth Axis Misalignments

Using the small angle approximations sin a = a and cos a – 1, equations [7.3] become

$$x'' = x \cos b + (y + za) \sin b \qquad\qquad [7.4a]$$

$$y'' = -x \sin b + (y + za) \cos b \qquad\qquad [7.4b]$$

$$z'' = -ya + z \qquad\qquad [7.4c]$$

If the true azimuth A and zenith distance Z are computed for a star from the apparent place, but the star is measured at azimuth A" and zenith distance Z" with all other error sources taken into account, the angles a and b can be found by rearranging [7.4] into the form

$$a = \frac{z - z''}{y} \qquad\qquad [7.5a]$$

$$\sin b = \frac{x''(y + za) - xy''}{x^2 + (y + za)^2} \qquad\qquad [7.5b]$$

and by using the spherical-to-rectangular coordinate system transformations

$x = \sin Z \cos A \qquad x'' = \sin Z'' \cos A'' \qquad\qquad [7.6a,a'']$

$y = \sin Z \sin A \qquad y'' = \sin Z'' \sin A'' \qquad\qquad [7.6b,b'']$

$z = \cos Z \qquad\qquad z'' = \cos Z'' \qquad\qquad\qquad [7.6c,c'']$

By sighting only a single star, computing its apparent position (A,Z) from its catalog mean coordinates, and measuring its position at (A'',Z''), the angles **a** and **b** can be found. Additional measurements can be used to improve the accuracy of the values measured for **a** and **b**.

Once **a** and **b** are found, the true (A,Z) of the telescope is found from the position (A'',Z'') maintained in the program (open-loop) or read from the shaft angle encoders (closed-loop). The matrix [7.2] may be transposed to give

$x = x'' \cos b - y'' \sin b$

$y = x'' \cos a \sin b + y'' \cos a \cos b - z'' \sin a$

$z = z'' \sin a \sin b + y'' \sin a \cos b + z'' \cos a$

When the same small angle approximation is applied, the result is:

$x = x'' \cos b - y'' \sin b \qquad\qquad [7.7a]$

$y = x'' \sin b + y'' \cos b - z''a \qquad\qquad [7.7b]$

$z = x''a \sin b + y''a \cos b + z'' \qquad\qquad [7.7c]$

Knowing **a** and **b** from previous measurements, and (A'',Z'') from the telescope axis angle encoders, one can compute the (A,Z) where the telescope is actually pointing using equations [7.6] and [7.7]. Once equations [7.7] are evaluated, Z is obtained from z = cos Z, then A is found from x = sin Z cos A. To expedite computing of these equations, values for sin b and cos b can be computed once and stored when **a** and **b** are found from measurements.

3. Equatorial to Alt-Az Conversion. Catalogs and ephemerides generally give positions in right ascension and declination coordinates. The encoders attached to the axes of equatorial mounts give readings in terms of these coordinates, so they are the most convenient coordinates for a computerized system controlling an equatorially-mounted telescope.

A control system for an alt-az mount, however, computes motor commands in the altitude (elevation) - azimuth coordinate system. Depending on the mechanical design of the mount and tube or truss, and on how various error sources are modelled, one might choose to work in either alt-az or equatorial coordinates. Thus a conversion from one set of coordinates to the other is usually required at some point.

a. Conversion Equations. Smart (1979, p.35) gives the following equations for transforming (α, δ) into alt-az coordinates:

$$\cos Z = \sin \varphi \sin \delta + \cos \varphi \cos \delta \cos h \qquad [7.8a]$$

$$\cos A = \frac{\sin \varphi \cos Z - \sin \delta}{\cos \varphi \sin Z} \qquad [7.8b]$$

where A = azimuth, measured westward from the south

h = local hour angle

φ = observer's latitude

δ = declination

Z = zenith distance = $90°$ - altitude

The numerator of [7.8b] has the numerator terms reversed in sign from Smart, who measures azimuth westward from north.

b. Driving Rates. In contrast to the equatorial mount, which must be driven in one axis only at a constant rate (excluding minor corrections), both axes of an alt-az mount must be driven at rates that vary over a wide dynamic range. Smart (1979, p.51) gives these rates as:

$$dZ''/dh = 15 \sin A'' \cos \varphi \qquad [7.9a]$$

$$dA''/dh = -15 (\sin \varphi + \cot Z'' \cos A'' \cos \varphi) \qquad [7.9b]$$

where Z'', h, A'', and φ are as before, and the rates dZ''/dh and dA''/dh are in units of "/sidereal second. Equation [7.9b] differs from that given in Smart, because he measures azimuth westward from north. Note that as Z'' approaches zero, cot Z''grows without bound, forcing dA''/dh to do the same. Therefore, there is a cone around the zenith of angular size determined by the maximum azimuth drive rate within which celestial bodies cannot be tracked. Figure 7-4 is a plot of dZ''/dh versus h and Figure 7-5 is a plot of dA''/dh versus h for bodies of different declinations. From these plots, it is seen that the drive rates needed to cover a useful fraction of the total sky vary considerably. This places severe requirements on the motors used in the drive, since they must be capable of delivering the required torque over a wide range of speeds. This is in contrast to equatorial mounts, in which motor speed requirements are far less severe, since motor speeds during tracking are relatively constant. Note that drive rates are computed for the tilted coordinate system.

c. Field Rotation Corrections. One final aspect of alt-az mounts is that the field of view rotates in the image plane as the telescope tracks a celestial body. This is because the image plane of an alt-az telescope stays aligned with the local vertical, whereas the image plane of an equatorial telescope stays aligned with the hour circle through the object being tracked. Since the stars in the field are constant in right ascension, their relative hour circle alignments will not change. Thus the field as a whole in an alt-az telescope appears to rotate at the rate of change of the parallactic angle.

As shown in Figure 7-6, Smart (1979, p.34) describes a spherical triangle whose apexes are the celestial pole (P), the zenith (Z), and the star (X). Angle XPZ is the hour angle h, angle XZP is the azimuth A (measured using Smart's method), and the remaining angle is the parallactic angle, the angle of interest (PXZ). Side PZ is the

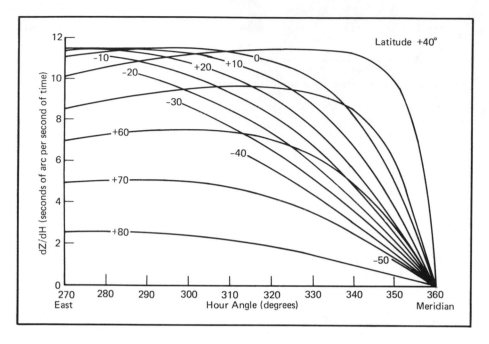

7-4 dZ/dH versus Hour Angle for Various Declinations

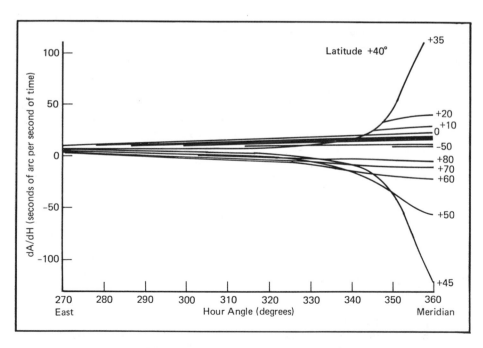

7-5 dA/dH versus Hour Angle for Various Declinations

co-latitude ($90°$ - φ, where φ is the observer's latitude), side PX is the co-declination ($90°$ - δ), and side ZX is the zenith distance of the star. From the law of cosines,

$$\cos PXZ = -\cos XPZ \cos XZP + \sin XPZ \sin XZP \cos PZ$$

Let C denote angle PXZ. Then

$$\cos C = \cos h \cos A + \sin h \sin A \sin \varphi \qquad\qquad [7.10]$$

The minus sign disappears from the first term when our convention of measuring A west from south is used.

Differentiating both sides with respect to h,

$$dC/dh = \frac{15}{\sin C} [\sin h \cos A'' (1 - \sin \varphi \, dA''/dh) + \cos h \sin A'' (dA''/dh - \sin \varphi)] \quad [7.11]$$

where dC/dh is in units of "/sidereal second. Again, tilted (A", Z") coordinates are used.

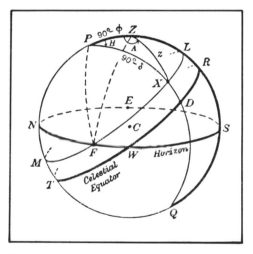

7-6 The Geometry of the Parallactic Angle (PXZ)
Courtesy of Cambridge University Press

If one is doing simple photometry on a star centered in the field, field rotation can be ignored. However, if one is photographing the entire star field, a separate field rotation motor must be used to rotate the camera in synch with the rotation of the image. The rate for driving this motor, dC/dh, must be recomputed frequently enough to prevent field stars from smearing significantly on the plate. If the image field is four inches in diameter, and stars must be kept stationary to within 0.0001 inches on the edge of the field, the field cannot rotate more than about 10". The rates at which the image plane must be driven to counteract the field rotation are given in Figure 7-7.

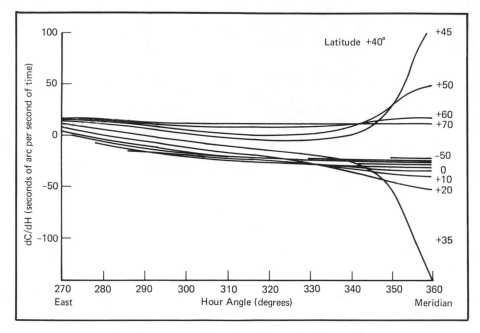

7-7 dC/dH versus Hour Angle for Various Declinations

C. INTRINSIC TELESCOPE CORRECTIONS

The previous two sections dealt with the orientation of the telescope mount on the face of the Earth. The corrections in this section are concerned with intrinsic features of the mount itself. Each correction is analyzed in the coordinate system which is most natural. Transformations from the coordinate system chosen for the analysis of the correction to the equatorial or alt-az coordinate systems are not detailed for each correction, since these transformations have been discussed above.

1. Non-Perpendicular Axis Alignment. The alignment errors discussed above concerned the polar axis in an equatorial mount, and the azimuth axis in an alt-az mount. If the complimentary axis in each case is not exactly perpendicular, errors will result which require correction.

In an equatorial mount, assume $90^\circ + p$ is the angle between the polar and declination axes, as shown in Figure 7-8. If the telescope's optical axis crosses the equator at point C, the star is transiting the meridian at B, and point A has the same true declination as B and lies on the arc described by the declination axis, then angle ACB is p, angle ABC is a right angle, the side BC is the declination δ, and side AB is the hour angle correction Δh (see Figure 7-16). Using Napier's rule,

$$\sin BC = \tan AB \tan (90^\circ - ACB)$$

$$\sin \delta = \tan \Delta h \cot p$$

$$\tan \Delta h = \tan p \sin \delta$$

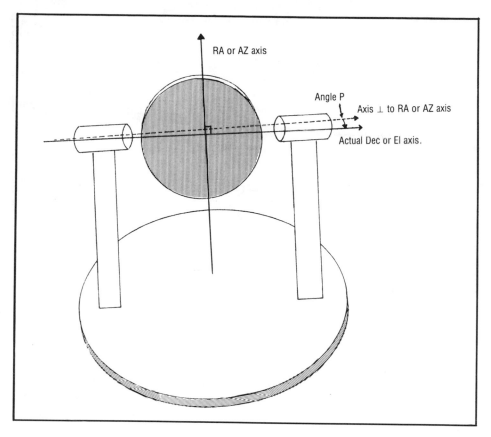

7-8 Non-Perpendicular Axes

Since Δh and p are small angles, hour angle positions should be corrected by

$$\Delta h = p \sin \delta \qquad\qquad\qquad [7.12a]$$

Since side AC > side BC, there exists a point D on arc AC where DC is the declination δ. Again, using Napier's rule,

$$\sin \Delta h = \cot p \tan \Delta \delta$$

$$\tan \Delta \delta = \tan p \sin \Delta h$$

Since $\Delta \delta$, p, and Δh are all small angles, $\Delta \delta = p \Delta h$ [7.12b]

By analogy, for the alt-az mount,

$$\Delta A = p \sin (\text{altitude}) = p \cos Z \qquad\qquad [7.13a]$$

$$\Delta Z = p \Delta A \qquad\qquad\qquad [7.13b]$$

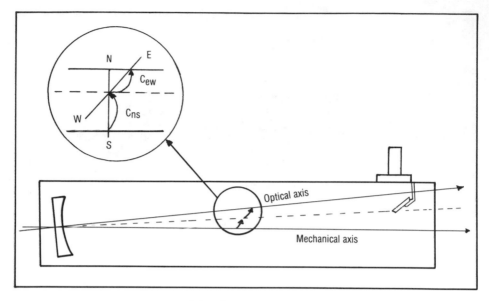

7-9 Collimation Errors

2. Non-Alignment of Mechanical and Optical Axes. Collimation errors between the mechanical and optical axes have two sources: (1) a static component, due to simple constant offsets built in when the telescope was constructed and the optics aligned, and (2) dynamic components, which arise from flexure (droop) of the tube or truss, and of the mount. The dynamic collimation errors are treated in the following two sections.

Static collimation errors are small constant angles which affect the hour angle and declination in an equatorial mount. As shown in Figure 7-9, to correct an east-west collimation error angle of C_{ew} measured on the celestial equator, Wallace (1975, p.295) uses

$$\Delta h = C_{ew} \sec \delta \qquad [7.14]$$

A north-south collimation error angle of C_{ns} is the correction to declination. In an alt-az mount, if C_{ew} is measured at the horizon,

$$\Delta A = C_{ew} \csc Z \qquad [7.15]$$

A "north-south" (actually, zenith distance) collimation error angle of C_{ns} is the correction to zenith distance. The formulation of these corrections is based on the assumption that the optical components of the telescope are in good alignment among themselves, and that the alignment of the optical axis as defined by the optical system as a whole is what accounts for the collimation error with the mechanical axis, which is defined by the bearings. If the optical components are not aligned among themselves, the result is usually visible as unduly large coma or other optical anomalies, which can be detected, diagnosed, and corrected relatively easily.

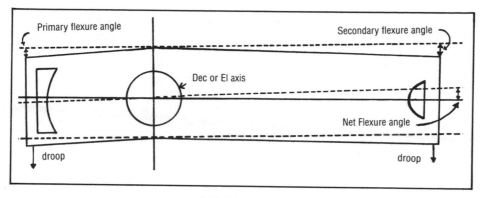

7-10 Tube Flexure

3. Tube Flexure. The cylindrical tube (on small telescopes) or the Serrurier truss (on larger telescopes) that supports the optical components bends or droops due to gravity by amounts in zenith distance which vary with the zenith distance. The actual amount of the flexure depends on the masses of the truss members and optical components, the stiffness of the truss tubing material, the tubing diameter and wall thickness, and the geometrical design of the truss. Such a structure cannot be modelled in real-time on a microcomputer. For example, the model of the KPNO 4-meter Serrurier truss requires three seconds to execute on a CDC 6400 computer (Abdel-Gawad, 1969, p.A-1). That is roughly equivalent to a few minutes on an Apple II microcomputer, during which time the flexure can change significantly.

Tube flexure is depicted in Figure 7-10. The table below (based on Abdel-Gawad, 1969, p.33) shows the flexure of the upper truss (F_u), an empirical approximation of the upper truss flexure using

$$\Delta Z = - F_u(90^\circ) \sin Z, \qquad\qquad [7.16]$$

the net flexure of both the upper truss (holding the secondary mirror) and the lower truss (holding the primary mirror) (F_n), and an approximation of the net flexure using [7.16] for F_n. All angular deflections are in arc seconds.

Z (deg)	F_u	$F_u(90^\circ) \sin Z$	F_n	$F_n(90^\circ) \sin Z$
0	0.000	0.000	0.000	0.000
15	1.190	1.188	0.159	0.096
30	2.390	2.295	0.210	0.185
45	3.240	3.246	0.260	0.262
60	3.980	3.975	0.320	0.320
75	4.410	4.434	0.330	0.357
90	4.590	4.590	0.370	0.370

This table shows that equation [7.16] is useful in modelling net tube or truss flexure. Abdel-Gawad (1969, p. A-2) gives a listing of a FORTRAN program useful in predicting the value of $F_n(90^\circ)$ for a Serrurier truss. Wallace (1975, p. 295) uses the same sin Z model, but quotes the secondary support flexure at f/8 in the Anglo-Australian telescope as 23" at the horizon. The large difference between the KPNO 4-meter deflections quoted in the table above and the AAT deflection indicates that the value of $F_n(90°)$ should be found for each telescope. This can be done using the program listed by Abdel-Gawad, although a direct measurement of tube flexure probably would be more accurate. To reduce the effects of residual errors from the calibration of the control system, $F_n(90°)$ should be minimized first by using good design and construction techniques, then the remaining flexure can be modelled. According to Abdel-Gawad (1975, p.71), this is more accurate than relying on the computer to compensate for a bad design dynamically, because the residual error varies with the size of the total error.

The effect of flexure is easy to compute for an alt-az mount. To compute the change in (h, δ) in an equatorial mount due to flexure,

$$\Delta h = \Delta Z \sin C \qquad\qquad\qquad [7.17a]$$

$$\Delta \delta = \Delta Z \cos C \qquad\qquad\qquad [7.17b]$$

where C is the parallactic angle found in [7.10]. Therefore, [7.8] - [7.10] and [7.16] must be evaluated for both equatorial and alt-az mounts, unless tube flexure is small enough to ignore.

4. Mount Flexure. When equipment of large mass is attached to the telescope mount and is not counterbalanced, or if the mount is not stiff enough to support the telescope mass properly, the mount will flex by a noticeable amount. An example of mount flexure is shown in Figure 7-11. The amount and direction of flexure varies with the design and construction of the mount, and must be measured empirically for each telescope. This can be done by recording pointing residuals during observing operations, then analyzing them for variation with hour angle or declination.

Wallace (1975, p.295) has determined by such an empirical method that the horseshoe flexure in hour angle h of the AAT is given by

$$\Delta h = -18''.9 \sin h \sec \delta \qquad\qquad\qquad [7.18]$$

Since the AAT is capable of at least 2".5 RMS pointing accuracy, the mount flexure must be modelled to obtain this level of accuracy.

Mount flexure is more of a problem in equatorial mounts, in which the force vectors due to the weight of the telescope tube assembly shift with changing hour angle. In alt-az mounts, the weight borne by each part of the mount does not change with telescope position, so mount flexure does not change. Any mount flexure is, therefore, embedded automatically in the shaft encoder calibration. The mount flexure should be modelled in equatorial mounts when high pointing accuracy is required, but can be ignored in alt-az mounts.

5. Servo Lag Errors. In a linear one-term servo, the motor speed command on either axis is directly proportional to the error signal calculated for that axis. If the servo is over-damped, as most telescope control systems are, or if it is critically damped, the telescope approaches, but never equals, the desired position. This lag can be determined by measuring the servo performance (error signal) in response to a step function. The time constant τ of the servo can be measured, then used in the

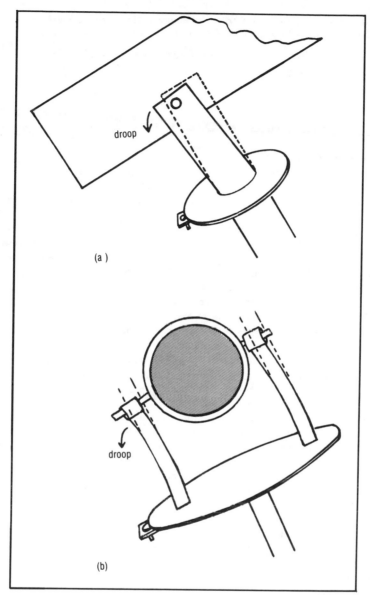

(a)

(b)

7-11 Mount Flexure

equation for an underdamped servo,

$$\text{lag error} = I\ e^{-t/\tau}$$

where I is the amount the motor speed changes at the instant the step function is applied, time t is measured from the time the new error signal is computed, and τ is the measured time constant. The constant exponential term to be used in the servo lag model is determined by using for t the time between error signal calculations (which are, presumably, performed at fixed time intervals), so $\tau' = e^{-t/\tau}$ is a constant. The servo lag is then

$$lag = I \tau' \qquad\qquad\qquad [7.19]$$

where I is the motor speed correction before the servo lag error correction is applied.

6. Position Encoder Errors. The encoders used most frequently for high accuracy shaft angle position feedback in telescopes are optical. Errors in the master pattern, or errors in depositing the pattern onto the encoder substrate, can cause random "tooth-to-tooth" errors of up to about 10% of the angular size of one encoder count.

These errors are usually small enough to be ignored. This is because optical encoders always give the correct integral number of counts per revolution. If the encoder has the intrinsic resolution to be connected directly to the telescope shaft without gear reductions, errors of up to 10% of the desired resolution are not important. If the encoder is connected to the telescope shaft through a reducing gear train, the size of the pattern errors is reduced by the same gear reduction factor as the angular size of an encoder count, so again, the error can be ignored.

Small sensor misalignments inside optical encoders typically do not produce errors. Absolute encoders use gray codes to prevent false readings. Incremental encoders often use two tracks for both count and direction sensing. The logic circuitry prevents false counts. If a sensor is in the wrong position, the encoder will not function. Therefore, if an optical encoder works, sensor misalignments can be neglected.

Other types of encoders may have a nonlinear response curve (e.g., a differential capacitor or a resolver), or some other kind of error. The error modelling that is required depends on the device. If you have an angle position encoder that is not optical, find someone with an encoder more accurate than yours, and calibrate yours against his. A power spectrum analysis (a one-dimensional Fourier transform) of the resulting curve can be used to find the periodic dependence (if any) to use in modelling the encoder errors. A much simpler approach is to use encoders specified at or better than the required accuracy, to avoid the need to model the encoder errors.

7. Gearing Errors. Figure 7-12 shows two main types of gear shape errors: (1) random tooth-to-tooth error, and (2) periodic error. In addition, when a drive train consists of multiple gears, there is backlash when the drive gear reverses direction. Backlash can be reduced by using precision-cut split gears which are spring loaded. The other errors can be modelled if very high accuracy is needed.

In a closed-loop system, if a shaft encoder is placed in the gear train between the motor and the telescope axis, any errors due to gears between the motor and the encoder can be ignored, but errors caused by gears between the encoder and the telescope axis appear in the graph of telescope axis position (as measured by calibration stars corrected to apparent position) versus encoder reading. This is the encoder calibration curve. Since any errors that appear in this curve must be modelled in some manner, it is best to place the smallest possible number of gears

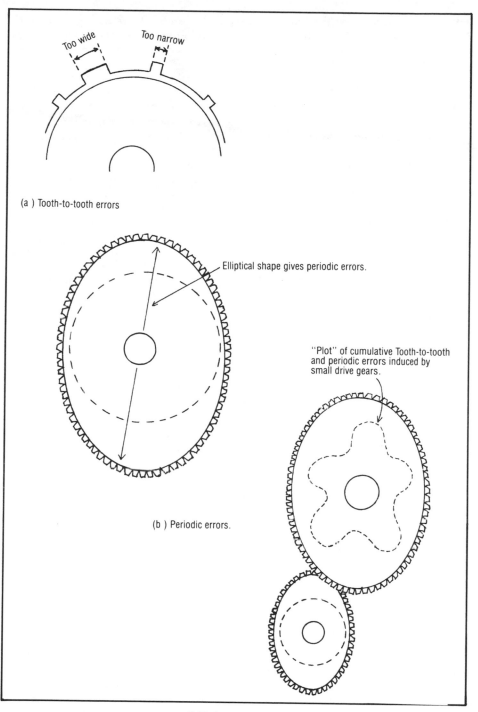

(a) Tooth-to-tooth errors

Elliptical shape gives periodic errors.

"Plot" of cumulative Tooth-to-tooth and periodic errors induced by small drive gears.

(b) Periodic errors.

7-12 Gear Errors

between the encoder and the telescope axis, and to ensure there are no clutches or other means to detach the encoder from the telescope axis. The encoder gears should be marked before assembly and calibration so that upon subsequent disassembly, the calibration is not lost.

Periodic gear errors appear in the encoder calibration curve as sine terms, whose amplitudes are found empirically. If there are several gears in the drive train, a power spectrum analysis of the curve will show an amplitude for each gear's periodic error. To avoid complicating the gear error model, the number of gears in the drive train should be kept as small as possible. This approach also reduces the total backlash.

To model small high frequency, regularly occuring gear errors, one can keep track of gear positions and use the residuals in the calibration curve to detect the tooth-to-tooth gear errors. As in the case of tube flexure, the modelling accuracy depends on the size of the systematic error being modelled. Rather than try to implement software corrections for sloppy gears, a better approach is to use precision gears, especially in the stages of the drive gearbox closest to the driven telescope axis, since the errors in gears closer to the motor are divided down by the gear ratio between the given gear and the telescope axis. If the power spectrum of error residuals still shows peaks at the gear rotation frequencies, a simple sine wave gear error model can be implemented.

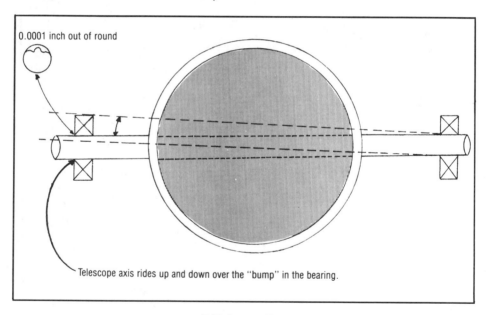

7-13 Bearing Errors

8. Bearing Errors. Bearing runout has the effect of periodically changing the gear reduction slightly between the motor and the driven shaft. The tolerance on ball or roller bearing race runout on good bearings is typically held to about 0.0001 inches or less. As shown in Figure 7-13, if one altitude bearing in an alt-az mount has a runout of 0.0001 inch and the other bearing has the same runout in the opposite direction, and the two bearings are separated by 40 inches (e.g., a 30-inch telescope), the runout produces an apparent non-perpendicularity between the altitude and azimuth axes of about 1". From equation [7.13], this results in a pointing error that grows large at small zenith distances. If the runout is in a direction parallel to the azimuth axis, it is indistinguishable from non-perpendicularity between the axes. If the runout is in a direction perpendicular to the azimuth axis, it is indistinguishable from a zero offset in azimuth. Runout in any other direction can be resolved into components perpendicular to and parallel to the azimuth axis. Either way, the altitude bearing runout errors are a periodic form of non-perpendicularity and zero offset errors.

Runout in the declination bearing of an equatorial mount is treated much the same as in the altitude bearing, except that equation [7.12] is used in modelling the runout that looks like axis non-perpendicularity. Runout in the right ascension bearing is a periodic form of polar axis misalignment.

Only the runout in the bearings supporting the principal telescope axes have any consequence. Bearing errors in the drive train produce no noticeable telescope pointing effects.

Bearing errors are dynamic, in that the axis running through the bearing exhibits the error only when a ball or roller is travelling over the "hill" in the race that has the runout. These errors are small and difficult to isolate, so they probably aren't worth worrying about until larger errors have been modelled sucessfully. As in the case of gearing errors, a power spectrum analysis will indicate what periodic terms and coefficients to include in the model. An accurate dial indicator will help in determining what fraction of the residuals is due to bearing errors.

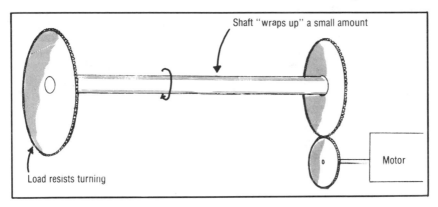

7-14 Drive Train Torsion Errors

9. Drive Train Torsion Errors. If one end of a rod of length L is held fixed and a torque T is applied to the other end with the direction of the vector T along the longitudinal axis of the rod, the rod will twist through angle Θ according to the formula

$$T = K L \Theta \qquad\qquad [7.20]$$

where the coefficient K accounts for the stiffness, cross sectional area, and diameter of the rod. This is shown in Figure 7-14. Equation [7.20] is the expression for a Hook's law relationship, provided the torsion angles remain within the rod's elastic limit. If a shaft angle encoder is placed in the drive train of a telescope, since the gears amplify the motor torque to very large values, there is a (variable) difference between the shaft angle encoder position and the telescope axis position produced by the mass of the telescope and the torque of the motor, as amplified by the gearbox. The amount of twist in the drive train gear shafts will vary, especially in alt-az mounts, in which the motor speeds vary considerably. This is because a motor's torque decreases rapidly with increased speed, and the motor acceleration varies with time.

This source of error can be reduced to tolerable levels simply by using short thick drive shafts with very high K coefficients and by placing the encoder as near to the telescope axis as is possible. This error can also be controlled by using a separate gear system to drive the encoder from the telescope axis. This latter approach also has the advantage of allowing one to choose convenient gear ratios for both driving the telescope axis with the motor and reading the telescope axis position using the encoder.

D. REDUCING THE EFFECTS OF SYSTEMATIC ERRORS

The easiest way to eliminate most of the errors discussed in the preceding sections is to provide adjustments on the telescope to counteract them. This works for static errors, such as position alignment errors (e.g., polar axis alignment on the refracted pole), axis perpendicularity errors, and collimation errors, but it is difficult to compensate for dynamic errors, such as flexure, in this manner. For smaller telescopes and in applications not requiring high accuracy, compensating for errors with mechanical adjustments is the best approach. There is no point in burdening a small computer with software corrections that can be made unnecessary with a few simple adjustments. However, it is difficult to provide adjustments in larger telescopes that do not detract from the overall stiffness of the mount, and it is easier to measure an error to a certain accuracy than it is to make a compensating mechanical adjustment to the same accuracy. Therefore, in larger telescopes, and in applications requiring high accuracy (say 30" or better), the preferred approach is to model the errors in software.

1. **Mechanical Adjustments.** The first mechanical adjustment is to align the polar axis. Until this is done, measurements of the other errors do not make sense. One way of aligning the polar axis is to use the "drift" test. In this procedure, a star near the meridian and celestial equator is centered and tracked, with only RA tracking adjustments being allowed. The eventual movement in Dec suggests which way the azimuth of the mount should be changed. A similar procedure can be applied to a star also on the celestial equator, but near the eastern horizon instead of the meridian for an indication of mount elevation adjustment. For highly accurate setting, this procedure is a bit slow, as it requires long drifts to achieve high accuracy. Figure 7-15 depicts polar axis alignment errors.

For microcomputer controlled telescopes, a modification of the drift test can be used that is much faster and quite accurate. This method was used to align the DFM Engineering mount on the Automatic Photoelectric Telescope at Fairborn Observatory East. The mount is first rough aligned using whatever convenient methods are at hand. The scale factors relating steps or counts to movement in RA

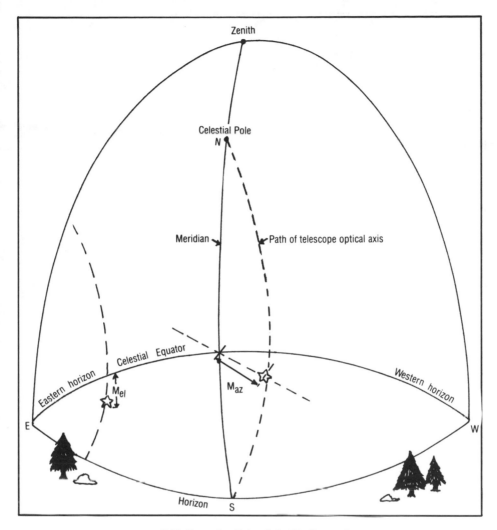

7-15 Measuring Polar Axis Misalignments

and Dec are then determined. A star of known position located near the celestial equator and meridian is centered on the crosshairs and its position (current epoch) is entered. The telescope is then commanded via the microcomputer to move to a star near the pole and on the meridian (within 20 degrees of the pole and a few degrees of the meridian is adequate). When the telescope stops, presumably somewhat away from the star, it is centered using the Dec control on the paddle and the azimuth adjustment of the mount. This is then repeated using a star on the equator near the eastern horizon, and the star near the pole is chosen with a similar RA. After moving to the star near the pole, the star is centered using the Dec paddle control and the elevation adjustment of the mount. This can be repeated a time or two if desired.

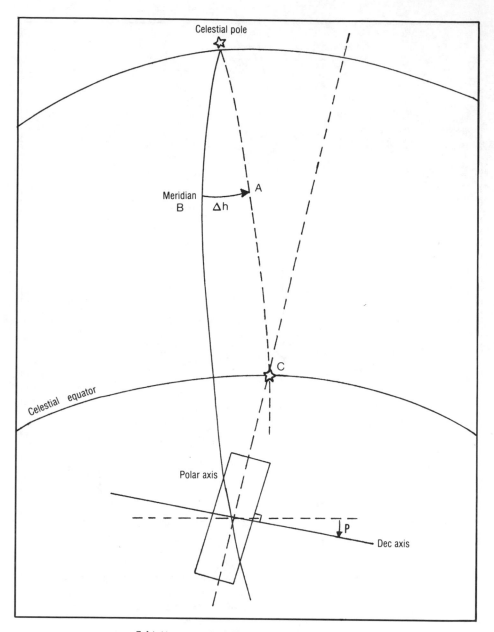

7-16 Measuring Non-Perpendicularity of the Telescope Axes

The errors that remain after polar alignment tend to be mixed together. If pointing residuals are carefully analyzed, errors proportional to RA or Dec, or a trigonometric function of RA or Dec, can be found. For example, Figure 7-16 shows how to measure pointing errors related to non-perpendicularity of the RA and Dec axes. As given in equation [7.12a], this causes errors in RA proportional to the sine of the declination. In Chapter 14, Trueblood discusses his proposed method for error correction in the Winer Mobile Observatory trailer telescope.

2. Compensating for Mechanical Behavior in Software. Although mechanical adjustments can be made in small telescopes to correct for static errors, in those applications requiring dynamic error compensation, or in large telescopes which do not permit mechanical adjustments, error correction can be performed in software. Kibrick (1984) describes a large effort underway at Lick Observatory to use software to compensate for a periodic error in the RA worm gear, and to damp low-frequency telescope vibrations of the 120-inch telescope at Lick. Figure 7-17 shows the error in the RA worm gear appearing every two minutes (of time). An optical encoder was attached to the large worm gear using a friction drive. When the encoder count read by the control computer indicates that the spot on the worm that has the error is about to enter the tooth mesh with the worm gear, the software sends commands to the RA motor to change speed to compensate for the error.

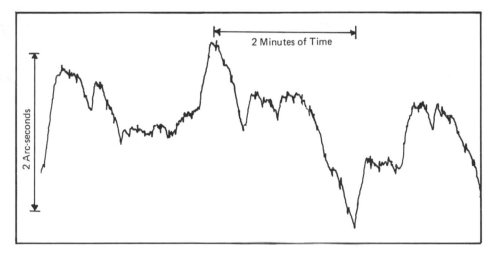

7-17 Tracking Error vs. Time in the Lick 120-inch Telescope
Courtesy of The Lick Observatory

Figure 7-18(a) shows the other problem--excessive ringing of the truss structure due to sudden starts or stops, or wind gusts. Although better mechanical design could have reduced this problem to a tolerable level, a new truss is too expensive on a telescope this large. A less expensive solution is to compensate for the low-frequency ringing in software. The pattern of ringing was carefully recorded and analyzed, to ensure that an accurate model of it could be embedded in the software. When a command is given that would produce mechanical ringing, the software issues commands to the motors to produce motions to counteract the ringing, according to the software model. The model is stored as a table of constants which are used in an equation (developed empirically) to compute the correction. Results of this effort are shown in Figure 7-18(b). Both the initial excursion and the time it takes for vibrations to damp are reduced considerably, making the telescope more compliant to commands issued by the computer, and better able to handle a series of short quick motions. Further work is expected to achieve even better results.

This concludes our discussion of the theoretical aspects of telescope control. The remainder of this book is devoted to detailed discussion of several examples of telescopes controlled by computers.

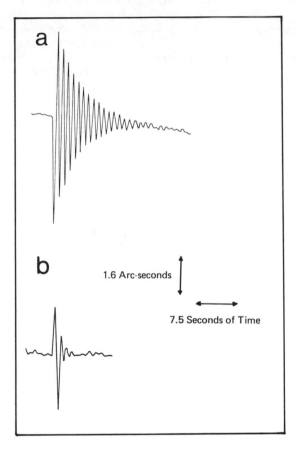

7-18 Telescope Vibration Compensation in Software
Courtesy of The Lick Observatory

EXAMPLES OF TELESCOPES CONTROLLED BY MICROCOMPUTERS

Chapter 8. CONTROL SYSTEMS BASED ON THE APPLE HOME COMPUTER

A. THE TOMER PORTABLE TELESCOPE

A few years ago, Tomer (Cox and Sinnott, 1977) built a trailer mounted 12-inch Cassegrain telescope that used an analog servo motor drive without position feedback. A few difficulties were encountered when Tomer connected his Apple computer to this telescope, but the project eventually succeeded (Tomer and Bernstein, 1983). The result, however, was cumbersome, with complex interface hardware, and even more complex interrupt handling software written in assembler language. This meant that the software was not easy to modify to serve a particular observing program (Tomer, 1984).

A new project, which was recently completed, was undertaken to develop a drive that was easy to interface to the Apple computer, and which could be controlled using software written entirely in interpretive BASIC. The new system relies on stepper motors run open loop without angle feedback encoders. It uses a hand paddle and hard-wired logic to send pulses to the steppers in manual mode. A "local/remote" switch determines whether the hand paddle electronics or the computer sends pulses to the steppers. A high precision worm gear is used on each axis, with mechanical differential gear boxes driving the worms. This approach permits a low-cost system to be built of reasonable accuracy, capable of bringing most objects into the field of a low-power eyepiece.

1. **Drive Train.** The strategy pursued by Tomer was to use one motor to "stop the sky" in RA, then make all other motions with respect to, and within, this moving reference frame. A single synchronous AC motor driven by a conventional drive corrector performs this function. Two steppers are used on each axis: one for fast (slew) motion, and the other for fine (set and guide) motion. These motors are connected to a differential gear using two different gear ratios. Thus three motors are used in RA (with two differentials), and two in Dec (with a single differential). In RA, the AC motor and the fine motion stepper drive the first differential, the output of which serves as an input to the second differential. The RA slew stepper serves as the other input to the second differential, the output of which drives the RA worm. All differentials give an automatic 2:1 reduction. A DC motor is used to move the secondary mirror for focussing.

Figure 8-1 is an overview of the trailer and telescope, while Figure 8-2 is a closeup of the RA bearing, polar axis, and counterweight assembly. The telescope has RA, Dec, and azimuth axes for easy polar alignment in remote locations. Jack stands permit levelling the azimuth bearing. The telescope is stowed in a box in the trailer for towing, and is raised up using a system of pulleys to tilt the polar axis for the observer's latitude. Figure 8-3 shows the fast (slew) and slow (set and guide) rate Dec steppers connected by toothed belts to the differential. The belts are used for gear reduction, and to provide smoothing of the stepper pulses (full steps are used, instead of half steps) and damping of mechanical resonances that could stall

8-1 Tomer "Phoenix IV" Trailer Mounted Telescope

8-2 Closeup of the RA Axis and Counterweight Assembly

8-3 Declination Drive Train

the motors. A similar arrangement is used in RA, with the addition of the AC synchronous motor to provide the sidereal rate.

The RA differential drives a worm and 400-tooth worm gear, while a 360-tooth worm gear is used in Dec. The gear reduction to the worm from the 30 oz.-in. steppers was initially 2:1 for the slew motor. With this small ratio, the motors tended to stall. The reduction ratio was finally increased to 6:1, which proved adequate. This added gear reduction raised the no-ramp (control paddle) step rate from a dead stop from about 300 full steps per second up to over 1000 full steps per second. When the computer ramps the motors, the top speed of 1500 steps per second yields a slew rate of roughly one degree per second. The fine stepper on each axis has an additional 5:1 reduction over that of the slew stepper, yielding a step size of 0".3 in Dec, and 1".5 in RA.

2. Electronics. The large box in Figure 8-4 holds the power supply and all the stepper controller and logic cards. Resting on this main control box is the inverter/drive corrector for the AC synchronous motor, which recently was incorporated into the main box. A closeup of the electronics box is shown in Figure 8-5. The cards are arranged to allow the free flow of cooling air to circulate inside the box.

The approach to the electronics is to provide two autonomous systems -- one manual and one computerized. A series of stepper pulses is generated in the hand paddle by three separate adjustable oscillators. Switches on the hand paddle select which oscillator output is connected to the hand paddle pulse output line. The Apple bus interface cards generate pulses at a rate determined by numbers loaded by software into registers on these cards. Bus switch IC's on a separate pc card select either the hand paddle pulse train or the computer pulse train, depending on the position of the hand-operated "local/remote" toggle switch. Since the hand paddle is not connected to the computer, no computer interface or software are needed to read the hand paddle inputs.

8-4 "Phoenix IV" Electronics

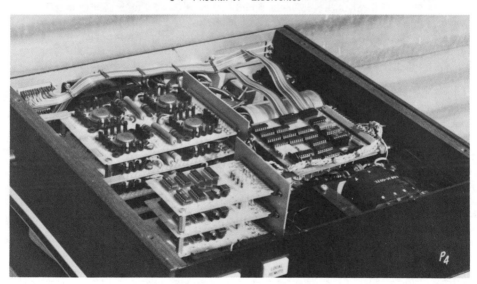

8-5 Closeup of the Stepper Control Box
Showing the Power Supply and Stepper Driver Boards

Figure 8-6 shows the hand paddle rate generator schematic diagram. Three separate 74LS123 one-shots are used to generate the slew, set, and guide rates. Since each has its own 20k ohm potentiometer to adjust the oscillator frequency over a 10:1 range, all three rates can be adjusted independently. The nominal frequency of each oscillator is 100 kHz. 74LS93's are used to divide down the 100

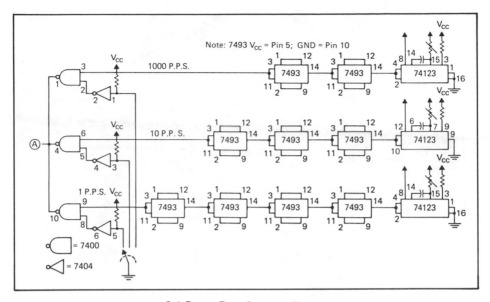

8-6 Tomer Rate Generator/Selector

kHz to 1000 Hz, 100 Hz, and 1 Hz. A switch on the hand paddle selects one of these rates to be sent to the stepper controller cards. Since the selected rates can be sent to either the slew or the fine stepper, each axis has six possible speeds. This has proved to be quite useful, since it provides an appropriate set or guide speed for most eyepiece magnifications. The other hand paddle switch directs the pulses to the 74LS157 bus switches on the computer/paddle interface card, which is shown in Figure 8-7. Tomer acknowledges the help of Joe Bell in the design of this board.

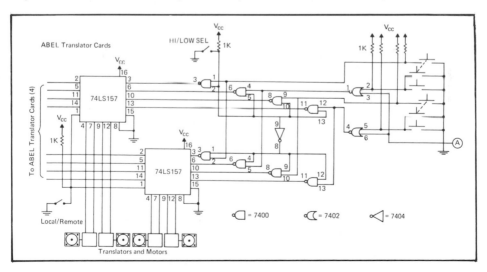

8-7 Tomer Computer/Paddle Interface

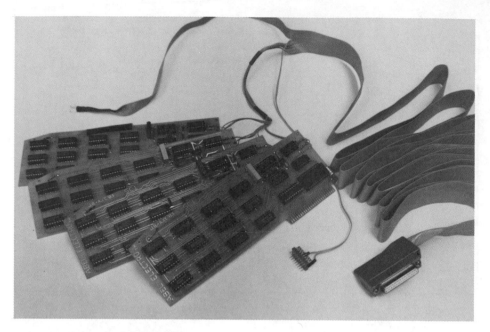

8-8 Abel Stepper Interface Cards for the Apple Computer

The computer interface cards shown in Figure 8-8 were designed and built by engineers at Robert Abel and Associates in Hollywood to drive motion picture animation cameras. They offered to Apple owners many of the features available on smart stepper drivers designed for the STD bus (see Chapter 9). These features included automatic ramping, a "busy" check, and a "panic stop". The computer loaded the direction, the step rate, and the total number of steps, then signaled the card to "go". The card handled the rest, by sending out a stream of TTL pulses at the appropriate rate. Unfortunately, these cards are no longer for sale.

The 74LS157 bus switches shown in Figure 8-7 receive step rate pulses and direction bits from either the hand paddle or the Abel stepper driver cards inside the Apple and route them to the four driver/translator cards, one for each motor. The driver cards are patterned after those used by DFM Engineering (Melsheimer, 1983). The parts count of the DFM circuit was reduced by using 2N6284 power Darlingtons. Also, a direction bit was added to each board.

3. Software. Using smart computer interface cards has enabled the software to be simplified, and to be written entirely in interpretive BASIC. The various cases handled by the RA and Dec routines are shown in Figures 8-9 and 8-10, while the routines themselves are listed in Figures 8-11 and 8-12. Not included in these listings are corrections for refraction and precession which are currently being implemented.

The control paddle inputs are handled entirely in hardware, which has three advantages. The first is that no software is needed to handle control paddle inputs, so as soon as the hardware was built, the telescope could be used. Software was then added incrementally without forcing the telescope to be out of service for extended periods of time. The second advantage is that the control paddle is a true

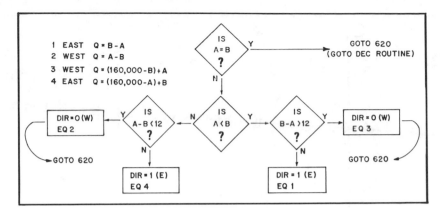

8-9 RA Control Algorithm

manual override. Software that has run amuck, or failed computer hardware, will not leave the telescope unusable. Finally, any residual pointing error can be tweaked out manually without the need for special software to tell the computer to leave the current RA and Dec alone, despite the fact the telescope is not now pointing where the computer left it after a slew.

The result is that the control software is extremely simple. The telescope is pointed at an object and its coordinates are entered. From that point on, the computer "knows" where the telescope is pointed when it is in control. As shown in Figure 8-9, if a new position is entered on the computer keyboard, or called up from a star catalog on disk, the new RA is compared with the current RA. If they are the same, the telescope has the correct RA, and the software branches to the Dec control routine. If not, the new and current RA are compared to find which direction to move and by how much. The number of steps to move is computed and executed, then the software branches to the Dec routine while the intelligent Apple stepper controllers are executing the RA command.

The various cases that arise in controlling the telescope in Dec are shown in Figure 8-10. In this figure, X is the current Dec and Y is the new Dec. Type 1 moves are those confined to the quadrant between the north pole and the celestial equator, Type 2 moves are those which cross the celestial equator, while Type 3 moves are confined to the quadrant south of the celestial equator. As shown in Figure 8-12, these cases are separated and dealt with individually in Tomer's software. Note the simplicity of the software which sends pulses to the motors (lines 430-500 in RA and lines 850-930 in Dec). This is a result of using the Abel intelligent stepper interface cards to generate the pulses to the driver cards.

A display generated by the software is shown in Figure 8-13. This display shows the current and new telescope coordinates and other status items, and permits commands to be entered to slew to new coordinates and other maneuvers.

4. Development History. Since the original DC servo motor drive worked with the Apple computer, the development strategy was to build a small test bed mount to determine if the new system would perform better than the existing one. Four stepper motors were purchased from a local surplus electronics store, the electronics were designed and built, and the system was assembled. The test bed system worked well, and the simplified control software was developed rapidly.

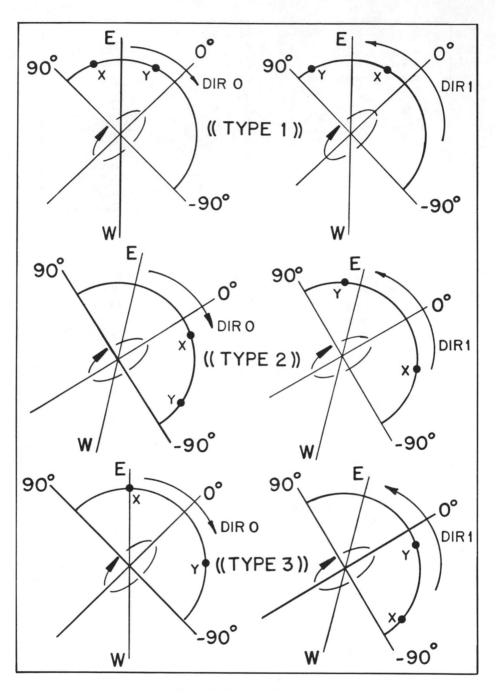

8-10 Declination Control Cases

```
 10   REM   ********************************************************
 20   REM   ********************************************************
 30   REM
 40   REM                   INITIAL CALCULATIONS OF POSITION
 50   REM
 60   REM   ********************************************************
 70   REM   ********************************************************
 80   REM
 90   X =   INT (DD * 200 + (DM * 200 / 60) + (DS * 200 / 3600))
100   REM         INITIAL DEC POSITION IN MOTOR PULSES
110   Y =   INT (ED * 200 + (EM * 200 / 60) + (ES * 200 / 3600))
120   REM         NEW DEC POSITION IN MOTOR PULSES
130   A =   INT (RH * 430) + (RM * 50) + (RS * .833333333333)
140   REM         INITIAL RA POSITION IN MOTOR PULSES
150   B =   INT (SH * 430) + (SM * 50) + (SS * .833333333333)
160   REM         NEW RA POSITION IN MOTOR PULSES
170   REM
180   REM   ********************************************************
190   REM   ********************************************************
200   REM   *                                                      *
210   REM   *                   THE R.A. ROUTINE                   *
220   REM   *                                                      *
230   REM   ********************************************************
240   REM   ********************************************************
250   REM
260   M2 = 49344: REM        -- SLOT 5 --
270   REM
280   REM
290   IF A = B THEN  GOTO 620: REM      NO MOVE AT ALL     <------
300   IF A < B THEN 360
310   IF A - B < 36000 THEN 340
320   DIR = 1:Q = (72000 - A) + B: REM                 EQUATION # 4
330   GOTO 430
340   DIR = 0:Q = A - B: REM                           EQUATION # 2
350   GOTO 430
360   IF B - A > 36000 THEN 390
370   DIR = 1:Q = B - A: REM                           EQUATION # 1
380   GOTO 430
390   DIR = 0:Q = (72000 - B) + A: REM                 EQUATION # 3
400   REM
410   REM
420   REM
430   B2 =   INT (Q / 256):B1 = Q - 256 * B2
440   POKE M2,200: POKE M2 + 1,0: REM         STANDARD PULSE INTERVAL
450   POKE M2 + 2,B1: POKE M2 + 3,B2: REM     LOAD PULSE COUNTS
460   POKE M2 + 5 + DIR,1: REM                LOAD THE DIRECTION
470   POKE M2 + 4,1: REM                      >>>> GO <<<<
480   FOR QQ = 1 TO 100 STEP 2
490   FOR ZZ = 1 TO 10: NEXT ZZ: REM          STD RAMP RATE
500   POKE M2,200 - QQ: NEXT QQ
510   REM
520   REM
```

8-11 RA Control Routine

After tests demonstrated the soundness of the approach, Tomer decided to implement the system on his 12-inch telescope.

The first problem that was encountered was that the 30 oz.-in. steppers had adequate torque for the test bed, but when the slew steppers were connected directly to the inputs of the differentials driving the worms on the 12-inch telescope, the steppers tended to stall. Although each differential provides a 2:1 reduction between the slew stepper and the worm, this was not adequate. An additional 2:1 reduction was installed, providing an overall reduction of 4:1. This improved performance, but the motors still tended to stall on occasion. The

```
530   REM   *********************************************************
540   REM   *********************************************************
550   REM   *                                                       *
560   REM   *                                                       *
570   REM   *              THE DEC SUBROUTINES                      *
580   REM   *                                                       *
590   REM   *********************************************************
600   REM   *********************************************************
610   REM
620   IF NE$ = "Y" THEN NE = 0: REM     RESPONSE, DEC INITIAL POSITION
630   IF NE$ = "N" THEN NE = 1: REM     RESPONSE, DEC INITIAL POSITION
640   IF NA$ = "Y" THEN NA = 0: REM     RESPONSE, DEC GO TO    POSITION
650   IF NA$ = "N" THEN NA = 1: REM     RESPONSE, DEC GO TO    POSITION
660   REM
670   IF NE + NA = 2 THEN   GOTO 720: REM         TYPE 1 MOVE HERE
680   IF NE + NA = 1 THEN   GOTO 770: REM         TYPE 2 MOVE HERE
690   IF NE + NA = 0 THEN   GOTO 800: REM         TYPE 3 MOVE HERE
700   REM
710   IF NE + NA =  > 3 THEN NE = 0:NA = 0: GOTO 9600
720   IF X > Y THEN DR = 0:Z = X - Y: GOTO 840
730   REM              ..........SOUTH MOVE
740   DR = 1:Z = Y - X: GOSUB 840: GOTO 9600
750   REM              ..........NORTH MOVE
760   REM
770   IF NE = 1 THEN DR = 0:Z = X + Y: GOTO 840
780   DR = 1:Z = X + Y: GOTO 840
790   REM
800   IF X < Y THEN DR = 0:Z = Y - X: GOTO 840
810   REM              ........NORTH MOVE
820   DR = 1:Z = X - Y: GOTO 840
830   REM              ........SOUTH MOVE
840   REM
850   Y2 =   INT (Z / 256):Y1 = Z - 256 * Y2
860   M1 = 49360: REM     -- SLOT 4 --
870   POKE M1,200: POKE M1 + 1,0: REM          STD INTERVAL
880   POKE M1 + 2,Y1: POKE M1 + 3,Y2: REM      LOADING THE PULSE COUNT
890   POKE M1 + 5 + DR,1: REM                  LOAD THE DIRECTION
900   POKE M1 + 4,1: REM                       >>>>> GO  <<<<<
910   FOR QQ = 1 TO 100 STEP 2
920   FOR ZZ = 1 TO 10: NEXT ZZ: REM           STD RAMP RATE
930   POKE M1,200 - QQ: NEXT QQ
```

8-12 Dec Control Routine

additional 2:1 reduction was changed to 3:1, for an overall reduction of 6:1. This proved to be entirely adequate, but reduced the slew rate from over 3° per second to just over 1° per second.

Another problem linked to the first was the mesh of the worm with the worm gear. Although the gears are cut to high accuracy (10" or better), a tight mesh increases the amount of friction in the drive. Tomer found that by permitting as much as 1' of "slop" in the mesh, the friction could be reduced to tolerable levels. This means the total pointing error is at least 1', but this is well within Tomer's system performance expectations.

A third problem that was encountered was heat dissipation from the stepper driver/translator cards. With all four steppers powered up at all times, the inside of the stepper electronics box got quite warm. The 12-volt, 10-amp power supply runs at 8 amps, which generates additional heat. Tomer tried powering only the slew motors, but found they were capable of backdriving the fine steppers through the differentials when the latter were not producing holding torque. The solution to the heat problem was to add more ventilation slots to one end of the case.

8-13 Tomer "Phoenix IV" Control System Display

The final set of problems were minor electrical faults. Arcing was occuring between the hand paddle case and the telescope trailer. The solution was to relocate some wiring and add proper grounding to all metal parts. Additional small filter capacitors were added to the electronics boards. Large 500 μF capacitors were added to the 12-volt supply line on each board to prevent the voltage from sagging during motor steps. Transient clippers (metal oxide varistors) were placed across the coil of the AC synchronous sidereal rate motor and several switch contacts to prevent the motors from stepping erroneously when switches were closed. Finally, the hand paddle switches were debounced to prevent rapid and erroneous direction reversals when using the guide speeds.

All projects encounter problems, and this one was no exception. All the problems were solved, and Tomer considers this project to be a successful demonstration of how proper hardware design can make the software development manageable.

Tomer's system demonstrates how stepper motors can be used in conjunction with smart interface cards and a small personal computer to produce a useful telescope control system. Such open loop systems are simple and inexpensive, and can be used in applications which do not require extremely high accuracy, or when final centering is performed by the observer. This system consistently gives absolute pointing accuracy on the order of a few arc minutes or better.

B. THE DFM ENGINEERING 0.4-METER TELESCOPE

1. **Overview of the Telescope.** Many smaller research observatories are still struggling with outdated and inadequate telescopes and telescope control systems. A modern research telescope is computer controlled, allowing the astronomer merely to enter the desired coordinates and epoch of the object to be observed. The telescope then automatically slews to the object quickly and efficiently. Tracking is highly accurate, so that stars do not drift out of small diaphragms on long integrations in photometry, and long exposure photographs need less manual guiding. Centering of stars is quick and crisp, and the best modern telescopes have no perceptible vibration, even after abrupt stops.

8-14 The DFM Engineering Small Telescope Mount

The 0.6-meter telescope at Grinnell College is a good example of a modern microcomputer controlled telescope. This telescope is one of a line of moderately sized telescopes made by DFM Engineering. These have included the MIRA 0.9-meter, the University of Colorado 0.5-meter, and (with L & F Industries) the 2.4-meter telescope recently installed on Kitt Peak. The DFM Engineering telescopes use friction disk drives in both right ascension and declination to eliminate backlash and provide fast response to computer commands. The forks and mounts are unusually rigid so that the rapid accelerations and stops possible with computerized telescopes do not induce annoying vibrations. The highly sophisticated control systems on the larger DFM Engineering telescopes use two 6502 microprocessors in custom computers developed specifically for accurate closed loop telescope control. DC servo motors are used on these larger telescopes for slewing, and stepper motors are used for fine motion. Position feedback information is supplied by optical encoders on each axis.

Such telescopes are ideal instruments for modern astronomical research, which often employs computers of various types to control both the telescope and its data gathering instruments. Telescopes in this class have the disadvantage of high price. For example, the Grinnell 0.6-meter telescope and control system cost about $175,000, including installation and checkout. However attractive such a telescope may be, many smaller colleges and advanced amateurs simply cannot afford such a telescope, and have had to settle for telescopes without computer control, accurate tracking, and exceptionally rigid mounts. Realizing that a gap existed between the small manually controlled telescopes currently available for under $10,000 and somewhat larger microcomputer controlled telescopes costing $130,000 and more, DFM Engineering designed and now produces a moderately sized microcomputer controlled research telescope in the $25,000 range. The complete telescope consists of one of several options for optics, the Model CCT-16 small telescope mount, and the Model TCS-16 telescope control system.

2. Optics and Tube Assembly. A 0.4-meter classical Cassegrain system was chosen after careful consideration. The cost of optics and mounts are roughly the cube of the telescope aperture, and it was felt that 0.4 meters was about the largest size telescope possible if the cost was to be kept within $25,000. A 0.4-meter telescope can accomodate a wide range of research programs at smaller observatories.

The primary mirror cell has a bolt pattern for attaching instruments at the Cassegrain focus. The light shielding and the perforation in the primary mirror are computer optimized to provide the desired unvignetted and fully illuminated field. The focal ratio of the primary mirror is 1/3 to reduce the length of the tube to minimize wind loading and to reduce the size of the dome or shelter. The secondary mirror is large enough to provide the unvignetted field with sufficient extra diameter to ease the optical figuring. It has a magnification of 4x, for an effective focal ratio of f/12. An electric motor moves the secondary mirror to change the position of the focal plane.

The primary and secondary mirrors are pyrex. More expensive materials such as quartz provide only marginally better performance in this size telescope at considerably more expense. The primary effect of the non-zero temperature coefficient of the pyrex is to change the position of the focus. Unless the telescope tube itself is temperature compensated, the focus shift with temperature is more a function of the length change of the tube than the focal length change of the optics. The larger DFM telescopes use invar rods to maintain the spacing of the primary and secondary mirrors, which eliminates most of the focus shift with temperature. The invar spacer technique would add considerable expense and weight to an optical tube assembly of this size. The optics are hand figured and tested using an optical flat and an interferometer, and are diffraction limited. The standard coating is

8-15 Closeup Showing the Declination Drive Assembly

aluminum with silicon monoxide overcoat. The primary mirror has a nine-point back support and four-point radial support.

The tube assembly is made from machined aluminum castings and a rolled and seamed aluminum tube. This construction method produces a stiff but lightweight tube with sufficient thickness at the attachments for the bolts. The focus housing and spiders are cast as one piece with the integral top ring to provide maximum stiffness for the secondary mirror. A sliding counterweight is provided to balance different instrument weights. The counterweight is located on the tube opposite the Declination drive and partially balances the weight of the drive.

3. Mount and Drive Assembly. To achieve high performance, it was critical that the generously sized friction disk drives and highly rigid mounts of the larger DFM Engineering telescopes be retained in the design, yet at greatly reduced cost. The result is the CCT-16 small telescope mount, shown in Figure 8-14 with the optional Meade 10-inch Schmidt-Cassegrain optics. This system was recently installed at the Fairborn Observatory East.

The mount is an equatorial fork made primarily from machined aluminum castings. The fork is a hollow box section to increase the stiffness over that of standard production forks, such as those used in commercial Schmidt-Cassegrain telescopes. An instrument clearance of 14 inches is provided, along with a carrying capacity of 20 pounds.

The polar axle is conical with the larger end having a diameter of eight inches. It is supported on a sealed ball bearing on the south end, while the north bearing is formed by the 15-inch diameter drive disk, which rests on two rollers supported by ball bearings. One of the rollers is powered to form the RA drive. This entire assembly is enclosed in an outer conical aluminum shell. This "cone within a cone" design adds stiffness to the mount.

Friction between the hardened steel disk and roller forms the final stage of the gearing with a "gear" ratio of 24:1. The drive roller has two electric clutches which alternately engage two different gear ratio worm gear sets to provide a high gear ratio for the slow motion drive and a lower gear ratio for the slew drive. A single stepper motor drives both worms through separate toothed belts to provide the remainder of the gear ratio. The worm gear set used in the slow motion gear path is very precisely ground for maximum tracking accuracy, and is spring loaded into contact to produce zero backlash. Any error in this gear set is divided by the final friction drive gear ratio to give a very smooth and accurate drive. The Dec drive, shown in Figure 8-15, is similar to the RA drive. The dual ratio gear boxes are identical. Only the mounting and loading of the friction drive stages are different.

The stepper motors have 200 full steps per revolution with a slewing speed of 3.5 degrees per second, and a set speed of 10'.5 per second. This configuration yields a tracking accuracy of ± 5" over five minutes, and ±20" per hour, excluding errors due to refraction and alignment.

The pedestal has both azimuth and elevation adjustments. The azimuth adjustment uses a push-push pair of screws mounted in a block attached to the sole plate. A block is mounted on the north face of the pedestal which is pushed by the screws to rotate the pedestal in azimuth. The elevation adjustment is located on the south end of the polar axis and has a range of 25 to 48 degrees latitude. The mount may be used outside of these latitudes by building the pier with a tilt in the north-south direction.

4. Control System. Perhaps the greatest challenge was making significant simplifications in the control system hardware and software so that the cost could be reduced from about $55,000 per system, typical of the larger DFM Engineering telescopes, to approximately one tenth of this for the 0.4-meter system. To achieve this, DFM Engineering went from a closed loop system using optical encoders, two motors per axis, and somewhat complex electronics and software, to an open loop system without encoders, only one motor per axis, and greatly simplified electronics and software. This open loop approach can be surprisingly accurate on a friction drive system. On long slews across the sky, accuracies of a few arc minutes have been achieved, while for short local moves of just a few degrees, accuracies are consistently better than 30". For many applications, these accuracies are more than adequate.

In photometry, for instance, after a slew clear across the sky, an object will be within the field of view of most photometers, and hence not difficult to locate. After a short move, such as between variable and comparison stars, the object will be within the diaphragm and will require only centering. While an accuracy of a few arc seconds may be typical of larger and more expensive telescopes, many observatories would be perfectly content with somewhat lower accuracy at a much lower cost. By employing an open loop system, not only were the expensive shaft angle encoders eliminated and the software greatly simplified, it was possible to use only one motor per axis. To permit the system to be open loop, this motor had to be a stepper motor, so that the computer could keep track of position by "counting" the steps commanded. However, using a single stepper motor for both slewing and tracking presents a problem. Because of the finite step size of steppers and their limited top speed, it is difficult to get both a fast slew and a small step size out of a single motor. This problem was overcome by using 10".0 steps during slewing and 0".2 steps during tracking, automatically switching between the two with electric clutches.

Also, by using an open loop system, the computational load on the computer was reduced by at least an order of magnitude. In the Grinnell College 0.6-meter system, two 6502 microprocessors are kept busy full time reading encoders, making calculations, and updating a complex display. There is little additional computer capacity left for instrument control, so the two processors are totally dedicated to telescope control. They operate from a fixed program in EPROM and can be treated as a peripheral device from another computer. In the 0.4-meter open loop control system, the computational burden can be handled easily by a single 6502 microprocessor, with spare capacity for instrument control and data processing. In fact, it was possible to use a low cost, off-the-shelf microcomputer and place all the software on a single disk, instead of placing it in EPROM's. As there is considerable reserve capacity, the same computer can also be used for instrument control. Further cost savings were realized by using existing stepper driver cards and many existing software utilities.

The control system is based on the Apple IIe or the Apple II+ with a language card. A single 5.25-inch floppy disk drive and controller are required. A DFM I/O interface circuit board plugged into the Apple provides the hardware interface. A multiconductor cable runs from the interface board to the motor driver/power supply. Separate cables run from the motor driver to the hand paddle and to the telescope. The pushbutton hand paddle has N, S, E, and W direction buttons, and SET, SLEW, and STOP buttons. The TCS-16 control system is shown in Figure 8-16.

The motor driver/power supply contains two high speed bipolar stepper motor drivers and their center-tapped 60-volt DC power supply. A 5-volt regulated power supply is provided for the stepper motor driver logic, and a 24-volt regulated supply is used to operate the clutches. Relays are provided to control the clutches.

Telescope control requires nine inputs and seven outputs. The interface board contains three 6522 PIA's, buffers, and associated circuitry, including a real time clock-calendar. One 8-bit parallel input port, a similar output port, and their four control lines are available for user control of instruments.

The software consists of a multi-tasking mixture of a high level language called XPL and assembly code. The high level language is very similar to Pascal, but was written for real time control and allows direct access to memory and calls to assembly language subroutines. The assembly language code handles interrupts and performs the high speed real time control activities. A user function is included which handles entry and exit, and allows a user to write an instrument control program. The user function is executed by entry from the menu. Procedures are supplied which perform slewing to coordinates, setting tracking rates, reading the real time clock, reading the input port, writing to the output port, reading the telescope coordinates, and reading to and writing from the floppy disk. These functions give the user software sufficient flexibility to command the telescope to perform complex maneuvers required by the instruments, and to store instrument data on the floppy disk for later processing. The menu is displayed on the lower portion of the monitor screen, and the telescope position data (RA, Dec, and hour angle), along with the sidereal time, are displayed on the upper portion of the screen.

8-16 The TCS-16 Small Telescope Control System

The system is designed to control a telescope mounted on the CCT-16 small telescope mount. Electric clutches are used to couple the two sets of gearing to the telescope. Normally this is done using a form "C" relay with the track gearing connected when the relay is de-energized. The telescope must be in reasonable balance, as the drive system is disconnected for approximately 40 milliseconds during gear ratio changes, and any movement of the telescope during this period will result in a position error.

Position information is obtained by counting steps to the motors. In track mode, the counts are given a weight of 0".2, while the counts are weighted at 10" in the slew mode. Software is provided to allow initializing, entering manual position corrections, and slewing to coordinates. Correction for atmospheric refraction is calculated and applied to the coordinates to improve the pointing.

The software is menu-driven and allows five choices for proper operation. The first step in each session is to initialize the software by entering where the telescope is pointing. A method which works quite well is to use a three step procedure. With power on and the control program loaded, the hand paddle is first used to slew the telescope to the zenith. A spirit level attached to the telescope tube is used to determine the zenith direction. The second step is to initialize the system with the current Universal date and time. The latitude and longitude of the observatory are built into the software and only require reentry if the telescope is moved to a substantially different location. The initialization program then sets the current telescope position to an hour angle of zero hours, and the Dec is set to the latitude of the site. The sidereal time is calculated using the current Universal time and date. The final step is to slew the telescope and center it on a bright star either using the hand paddle or by executing a "Slew to Coordinates" menu command. The actual coordinates of the star are entered using the "ZPOINT" command. This completes the initialization. "ZPOINT" may be used whenever necessary to correct any accumulated pointing error.

The telescope's RA, hour angle, and Dec, and the sidereal time, are displayed on the monitor. The menu commands are:

1. INITIALIZE	Enter UT and date
2. ZPOINT	Enter actual coordinates of star
3. SLEW TO COORDINATES	Enter coordinates of desired position
4. RATES	Enter desired tracking rates
5. USER	Enter user program

One gravity-operated limit switch is provided with the telescope, since the software does not check for over-travel motions. This makes it possible, however, to command the telescope to move to undesirable positions. A STOP button is provided on the hand paddle to cancel slew commands.

The two telescopes described in this chapter vary markedly in their approach to both mechanical and software designs, yet they both use the Apple II as the control computer. This demonstrates the wide range of control designs that can be implemented on a common home computer. In the next two chapters, other telescope control systems are discussed which use computers not as well known as the Apple II.

EXAMPLES OF TELESCOPES CONTROLLED BY MICROCOMPUTERS

Chapter 9. AN STD BUS CONTROL SYSTEM

A. THE STD BUS

The STD bus was originated in 1978 by Prolog and Mostek. These two companies reached an agreement on the STD bus so that from the beginning, there would be a second source for STD cards. They wisely did not copyright or patent the STD bus, purposely putting it in the public domain. The STD bus has had one of the most rapid growths of any bus, and in 1985, it was supported by almost 100 manufacturers and over 700 different cards--a truly phenomenal growth for only seven years!

The STD bus is not widely used by hobbyists, however, and this has its good and bad points. On the bad side, it means that the STD bus is not well known. On the good side, STD manufacturers do not vie with each other trying to produce the cheapest possible cards for an overly competitive hobby market. The STD manufacturers tend to be highly concerned with maintaining a reputation for high quality and reliability.

Compared to most other buses, the STD bus is simple and straightforward. It contains lines for addresses, data, bus control, and power. It would be difficult to have a much simpler, yet general-purpose bus. While the bus was designed for the 8080/Z-80 class of microprocessors, others are available on STD bus CPU cards, including the 6502, 6800, 6809, and a few CMOS CPUs (ones with TINY BASIC as part of the CPU card itself).

Surprisingly, there is only one book we are aware of on the STD bus, and this is Titus et al. (1982). This book does a very good job of describing the bus in detail, and providing guidance on address decoding, interrupts, and other detailed matters for those who want to design their own cards or understand the details of commercially available cards. This book is must reading for anyone interested in the STD bus.

The STD bus is just one of a number of buses that work well for controlling telescopes with microcomputers. This chapter gives some details on a particular bus only as an example. Trueblood prefers the Q-bus, and there is much to be said for the S-100, IBM PC, and Apple buses.

B. DEVELOPMENT HISTORY

The Fairborn Observatory was founded in January, 1979 by Genet for the specific purpose of making photoelectric observations of variable stars. Photoelectric observations commenced in August of the same year after construction of an observatory, 0.2-meter Cassegrain telescope, and a photoelectric photometer. The initial telescope and photometer at the observatory have been described (Genet, 1979). Observations were made and recorded manually, but a Radio Shack TRS-80 was used to make the reductions and analyze the data. Observations were initially concentrated on RS CVn type binary stars in a cooperative program with other small observatories. D. S. Hall at Dyer Observatory (Vanderbilt University) provided overall direction on this observing program.

By late 1979, it was decided that the role of the TRS-80 microcomputer should be extended to that of logging the data directly. The necessary interfaces and software were developed to accomplish this, as described in Genet (1980). The TRS-80 did not control either the telescope or the photometer, but it did record the data, provided prompts to and received commands from the observer, and provided analysis of the data just seconds after it was collected. The system was quite reliable and gave almost trouble-free performance for several years. The final version of this data-logging system is described in book-length detail (Genet, 1982b).

After observing with the 0.2-meter telescope for some time, construction was started on a larger 0.4-meter telescope in June of 1980. First light was obtained in August of 1981. The primary quartz mirror was ground, polished, and figured by Genet, and the structure itself was mainly built from steel obtained from the local scrap yard.

9-1 0.4-m Telescope Overview

The primary objective of building the 0.4-meter telescope was to provide the capability for dual-channel photometry. By using two photometers to observe both a variable and comparison star simultaneously, it is potentially possible to eliminate most of the undesirable effects of the occasional poor seeing in the midwestern United States, as well as to increase the data rate. The telescope was

9-2 0.4-m Tailpiece Showing the Starlight-1 Photometer

designed so that it could be operated manually without a computer control system, and its initial use was for single-channel photometry using the Starlight-1 photometer. An overview of the entire 0.4-meter telescope is given in Figure 9-1, while a closeup of the EMI photometer on the tailstock is shown in Figure 9-2.

In the Fall of 1981, the decision was made to develop a computer-controlled dual-channel photometer along with a computerized control system for the 0.4-meter telescope. Two early decisions were later regretted. These were the use of the TRS-80, and a rather "informal" (non-bus) layout of the peripheral electronics. Although the main capabilities of the TRS-80-based system began to become operational by July of 1982, there were constant problems with random system crashes and with connector and cable reliability. While perhaps "staying with it" might have resulted in eventual success and elimination of some of these persistent problems, in a program review in September of 1982, it was decided that other alternatives should be given serious consideration.

After considerable research and careful consideration, the STD bus was selected as the most appropriate for our particular circumstances. A number of manufacturers were helpful with suggestions and donations of cards and equipment to the observatory. What emerged was a reliable and capable telescope/instrument control system. It is a bit sad, perhaps, that not long after this system became operational (in some of its capabilties) that it was replaced with the fully automatic photoelectric telescope system. It was more important to sleep at night than to make dual-channel observations with the STD system. Even though the STD control system is no longer used, it is a good example of a bus-oriented telescope control system.

Before considering the details of the control system, the overall approach that was taken will be considered. This will be done in light of the general advice given earlier on real-time control systems.

C. OVERALL APPROACH

Operator convenience is one of the most important characteristics of any real-time control system--telescopes being no exception. We felt that the greatest possible operator convenience would be to get the observer out of the cold and to automate some of the more mundane aspects of a typical variable star photoelectric run. The system was designed to allow this capability. However, initial operation was from inside the observatory in the evolutionary spirit of system development. Control of the telescope when standing right next to it was by a control paddle for telescope movements, and a small hexadecimal keypad and video monitor for other control functions.

The hardware/software approach chosen was the use of a single master CPU running in a high-level language to control a number of transparent slave microprocessors in areas where high speed or external "smarts" were required. The system was implemented as an STD system primarily employing off-the-shelf commercial cards.

Except for the PROM machine-language programs supplied by the manufacturers on the smart STD cards, all software was written in interpretive MBASIC running on the CP/M operating system. While the capability of compiling the interpretive MBASIC exists (to speed it up), the interpretive MBASIC proved to be faster than required. (The reserve fallback was comforting, however--especially considering the CPU utilization figures for an LSI-11 given in Chapter 14.) The program was modular, and was developed with cards in place, logic probe in hand, and stepper motors powered up nearby.

Adaptability for future requirements was kept in mind. A 24-slot Scanbe motherboard/card cage provided space for future expansion, and the 30-amp Lambda 5-volt switching power supply had ample reserve. The rear connector panel was modularized so that connector strips could be added or changed easily.

System reliability was enhanced by using a proven bus and commercial high-quality pc cards as much as possible. The system was placed in an insulated and heated control room. Maintainability was enhanced by organized and detailed documentation (most of it supplied by the STD card manufacturers), extensive labeling, and ease of access. System maintainability, system construction, software development, and a number of other things have all been helped by the small size of the system. All the electronics, including the power supplies, but excluding the terminal and disk drive, were in a single box that weighed less than 40 lbs., and could be placed under one slightly large arm. The TRS-80-based control system resided in two relay racks and was anything but portable.

D. SYSTEM BLOCK DIAGRAM

An overall block diagram of the system is shown in Figure 9-3. The STD Control Unit contained all the STD cards in a card cage, all the power supplies for the system, a rear panel, where most of the system cables were connected, and a front panel that contained the system reset switch, low and high voltage pilot lights, and

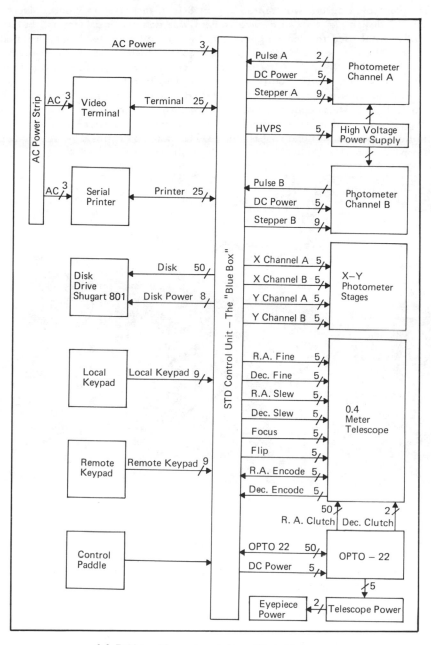

9-3 Fairborn Observatory System Block Diagram

the clock/calendar continuous display and setting switches. As the STD Control Unit was contained within a blue colored aluminum box, it was sometimes simply called the "blue box".

The control program was stored on an 8-inch diskette and loaded from a Shugart 801 disk drive. Software development was done with a terminal, but the terminal was not used for normal operation. Instead, all operator inputs are made using either the control paddle near the telescope or using one of the two small hexadecimal Enlode keypads. One of the keypads was on the observatory floor, while the other was in the control room near the terminal.

The telescope was controlled with cables coming directly from the back of the STD Control Unit. These cables ran out to the nearby telescope (in the next room) and plugged into various stepper motors and encoders. High power control, such as for the telescope clutches and roof motors, was handled by Opto-22 power modules located near the STD Control Unit, which is shown in Figure 9-4.

9-4 STD Control Unit

Each of the two photometer heads of the type shown in Figure 9-5 was connected to the STD Control Unit by cables, and to a common high-voltage power supply. Finally, the four linear steppers on the two X-Y stages were connected by cables to the STD Control Unit. These X-Y stages were used to position and/or scan the two photometer heads.

E. STD CONTROL UNIT

A functional diagram of the STD Control Unit is shown in Figure 9-6. The order of the STD cards shown in this figure is not exactly that of the actual system, since some have been rearranged to make cabling easier. Perhaps this system can most easily be understood by discussing the function of each of the cards, and this is done below.

1. Pulsar Computer: The Pulsar is a single-board STD bus computer made in

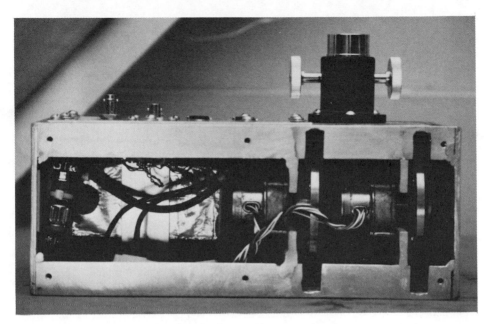

9-5 STD Bus System Photometer Head

Australia by Pulsar Electronics Ltd., and distributed in the United States by Infinity, Inc. (see Appendix I). The CPU is a 4 MHz Zilog Z-80. On board are 64KB of RAM and a 2KB Monitor ROM. A disk controller (up to four disks), two RS-232 serial ports (terminal and printer), and a clock/calendar with battery backup round out this amazing little board. It is really impressive what those "Aussies" can pack on a 4-inch by 6-inch board! The Pulsar CPU is Slot No. 1 of the 24-slot Scanbe motherboard and card cage, and the Pulsar communicates with all the other boards using the lines on this STD bus motherboard. As might be expected, the Pulsar easily handles CP/M, and it comes with many helpful utilities.

    2. MCPI Relay Control Board: The Mullins Computer Products, Inc. (MCPI) board contains eight reed relays, along with appropriate address decoders and a data latch. It was used to control the high voltage power supply for the PMTs, leaving plenty of reserve for future expansion.

    3. and 4. Whedco Intelligent Stepper Controllers: Each of the two identical Whedco stepper controllers is a complete computer in its own right with its own preprogrammed "canned" software stored in ROM. It can operate in a positioning mode, a constant velocity mode, or a single step mode. The positioning mode is the most interesting. The Whedco controller automatically ramps a stepper up to slewing speed, and then ramps it down to a stop exactly at the commanded position. The initial start speed, acceleration rate, and maximum slew speed are all software settable, allowing one intelligent controller to handle a number of different steppers with different inertias, friction loads, and allowable top speeds, etc. With the two Whedco boards, any two steppers can be controlled at the same time.

    5. Circuits and Systems Counter-I/O Board: This board contains three 16-bit programmable counters, and four 8-bit parallel I/O ports. Two of the counters were used to count the pulses coming from the two photometer heads. The third counter was used as a programmable divide-by-N to set the stellar rate for the RA fine stepper. This avoided tying up one of the Whedco controllers in such a mundane task. Three of the 8-bit ports were in an "output" mode and were used to (1) select which two steppers would be driven by the Whedco controllers, (2) enable the stepper driver chips, and (3) set the stepper direction. The fourth port was used as an "input" port, and it sensed overflow flags from the counters, and busy and ready flags from the Whedco stepper controllers.

6. STD Patch Card: Since the Whedco controllers were used to control a number of different steppers, provisions had to be made for logically switching the pulse streams and for making interconnections between the various cards. This was taken care of by a single wire-wrap board -- the only one in the entire system. All other boards were pc boards.

7. Quad Stepper No. 1: There were four identical quad stepper cards in the control system. Each card contained four Hurst hybrid stepper controller/driver chips. The Hurst chips provided the direct switching logic for controlling four-phase steppers, and also contain drivers with sufficient power to drive low-powered stepper motors. Pulse signals were provided by a jumper from the STD Patch Card. Only power was drawn from the STD bus itself. This fairly simple card was not available commercially, so Doug Sauer layed out a pc card. Quad Stepper No. 1 drove the geared Hurst RA fine and Dec fine steppers directly, and drove the Slo-Syn (Superior Electric) RA slew and Dec slew steppers through the dual stepper power driver on Board 8.

8. Dual Stepper Driver: Eight Darlington transistors provided the muscle for the RA and dec Slo-Syn slew motors. The four-phase signals were provided from two of the Hurst controller chips on Quad Stepper No. 1.

9. Quad Stepper No. 2: The four Hurst controller chips on this board provided the four-phased signals for the four linear Hurst stepper motors that drove the two X-Y stages. These stages positioned the two photometer heads.

10. Quad Stepper No. 3: The four steppers controlled by this card were the diaphragm and filter wheels in each of the photometer heads. These steppers were small, geared Hurst steppers.

11. Quad Stepper No. 4: Only two of the four controllers on this board were utilized for secondary mirror focus (a linear Hurst stepper), and flip mirror control. This left two stepper controllers for future possibilities.

12. MCPI DC Input: This board provided eight opto-isolated DC inputs. All eight of these inputs were used to sense limit switches. If any of these switches was activated, it resulted in the stopping (at maximum deceleration) of all steppers, and the removal of the high voltage, etc.

13. Enlode Clock/Calendar Card: This card was another smart card with its own on-board microprocessor. Besides the normal time-keeping functions, it provided alarm and timed-interrupt functions. One unique feature was a remote indicator that automatically displayed the time and date, or if desired, any other commanded data, such as RA and Dec.

14. Electrologic Z-501 and Opto-22: These two cards formed a team. The Z-501 provided up to 24 control signals to the remotely-placed Opto-22 board. On the Opto-22 board were provisions for up to 24 plug-in control modules. These can be for AC power, DC switching, DC input sensing, or even AC input sensing. These control modules are industry standards, and are available from Opto-22 (their originator) and other companies. The Z-501 controlled the modules, and the modules in turn controlled the slew clutches, roof motor, and other heavy loads.

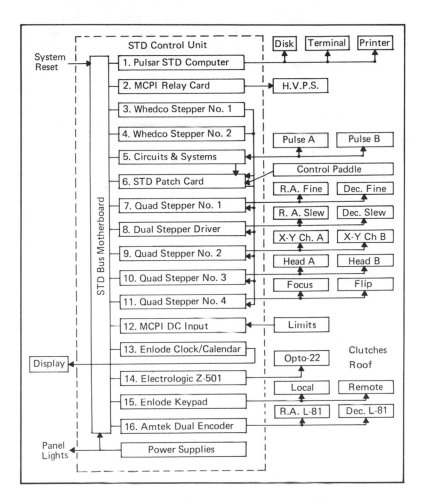

9-6 Fairborn Observatory STD Control Unit Functional Diagram

15. Enlode Keypad Encoder: The keypad encoder can handle up to four different keypads. However, only two Enlode hexadecimal keypads were used in this system. One was located next to the telescope for manual telescope control (the remote keypad), while the other was located in the control room, and was used for semiautomatic control (the local keypad). The remote keypad had first priority, on the theory that the observer at the telescope should not be overridden by someone in the control room.

16. Amtek Dual Encoder: The Amtek dual encoder is another smart card with its own microprocessor (Z-80) and a program stored in PROM. It can handle two incremental encoders at the same time, keeping continuous track of their positions. Two Litton Model-81 optical shaft angle encoders were used to sense position in RA and Dec, and the quadrature signals from these two sensors were fed directly to the Amtek dual encoder card.

By using a well-recognized bus with commercial off-the-shelf cards, and by paying some attention to division of equipment along functional lines, a reliable control system resulted. Genet gratefully acknowledges the considerable help in the development of this system by George C. Roberts and Douglass J. Sauer.

EXAMPLES OF TELESCOPES CONTROLLED BY MICROCOMPUTERS

Chapter 10. THE FAIRBORN OBSERVATORY EAST
YOKE-MOUNTED TELESCOPE

A. INTRODUCTION

The telescope control system described in this chapter is an open loop type with a single stepper per axis. Since the system is open loop, there are no shaft angle encoders or other telescope position sensors. A fixed gear ratio is used in each axis, and the necessary speed range results from the use of stepper ramping and special overvoltage (bi-level) stepper drivers. The electronics were designed by Louis Boyd, and are simpler than his earlier design of the automatic photoelectric telescope (APT) currently in operation at Fairborn Observatory West in Phoenix. Details of the second generation APT have been discussed previously elsewhere (Boyd, Genet, and Hall, 1984a). The schematics, diagrams, and software listings which appear in this book have been furnished courtesy of the Fairborn Observatory.

The yoke-mounted telescope is operated by a single board computer which costs only $300. One need not spend a lot of money to develop a computer controlled telescope. The computer used in this example is based on the Motorola 6809 microprocessor. While the 6809 is a good choice, and the Microware OS9 operating system and Basic09 language are well-suited for complex real-time control tasks such as telescope control, other microcomputers, operating systems, and languages would also work well with the electronics described in this chapter.

While the second generation system is presented in this chapter as a simple manual control system for operation by a human observer, the hardware can be used to build a system which is capable of fully automatic operation. Chapter 13 describes how additional software procedures and a photoelectric "eye" can be added to provide closed loop feedback which permits the computer to "see" how well it is aligned on the stars. In fact, this telescope is usually operated in the fully automatic mode. However, we are not concerned here about automatic control, but instead will focus on the many details needed to achieve basic manual operation.

B. TELESCOPE MOUNT AND DRIVE

When the mechanical portion of this system was designed, the decision was made to use existing piers and optics. The north and south piers were large concrete structures that were originally used to support the massive 0.4-meter telescope at Fairborn Observatory East in Ohio. The optics consisted of a 10-inch Schmidt-Cassegrain set donated by K. Kissell. A yoke configuration was chosen to utilize the two piers and maintain a low moment of inertia with a symmetrical mount. An overall view of the mount, drive, and optics is shown in Figure 10-1.

The yoke was constructed of welded 3-inch square steel tubing with 1/8-inch thick walls, while light-weight 3-inch channel was welded to form the "inner gimbal" that holds the optical assembly. It was imperative that the centerlines on the stub axles be exactly aligned--both in the RA axis and the Dec axis. As the

10-1 Overall View of the Yoke-Mounted Telescope

10-2 Declination Stepper Motor and Berg Reduction Gear

system was assembled in a simple shop without any sort of alignment aids, we used an approach that others in a similar predicament might find useful. After the yoke was welded together, round holes were torched out at each end for the axle stubs. A long piece of straight 1.5-inch diameter rod was then placed through the holes and welded in place--assuring that the stubs sticking out were in line. The rod was then cut on the inside of the yoke, and this smaller piece was in turn used in the same manner for the stub axles on the inner gimbal. Hollow thick wall tubing might have been more appropriate, as it would have allowed passing various cables out through the axis of rotation.

Drives in both axes consist of large diameter thin aluminum disks driven by chains and sprockets. These in turn are driven by steppers through a gear reduction unit consisting of a single antibacklash worm gear in a housing which seals the gears against dust and dirt. The 32-inch diameter aluminum disks (1/8-inch thick) are driven by 1/4-inch pitch industrial chain from 1-inch diameter sprockets. This gives the final drive stage a 32:1 reduction. The worm gear unit is a 120:1 reduction unit from Winfred Berg. The worm wheel is split in two and spring loaded against itself to eliminate backlash. The Berg unit is driven by Slo-Syn M061-FD02 steppers in the half-step mode (400 steps per revolution). The Berg reduction gear and Slo-Syn stepper motor for Dec are shown in Figure 10-2. The total number of half-steps per revolution for the telescope is 32 x 120 x 400, or 1,536,000 half steps per revolution. As there are 360 x 60 x 60 or 1,296,000 arc seconds per revolution, each half step corresponds to 0".84375.

Chain drives have some distinct advantages and disadvantages. The advantages are that they are very easily built without the need for special machinery or careful alingments--ideal for the casual "homebrewer". Also, they have no perceptable backlash and they are reasonably low in cost. For small telescopes, such as that discussed here, the chain is stiff enough that the total system stiffness is not seriously degraded, but this would not be the case for much larger telescopes. The sprocket and chain combination does cause some periodic error. While this may not be objectionable in Dec, the RA periodic error might not be appreciated by those engaging in astrophotography, although it is insufficient to concern photometrists.

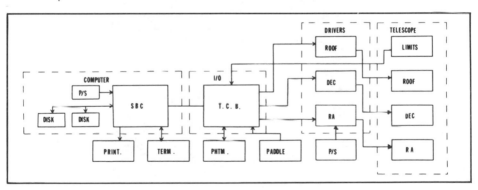

10-3 Control System Block Diagram

C. CONTROL SYSTEM HARDWARE

1. **System Block Diagram.** The overall layout of the control system electronics can be seen in the system block diagram shown in Figure 10-3. The equipment, besides the steppers and limit switches on the telescope itself, has been placed on three aluminum chassis plates, one each for the computer, telescope control board, and the stepper driver board. There are also a number of auxilliary items such as a control paddle, terminal, etc.

10-4 The PT69 Single Board Computer

10-5 Compete Computer

There is nothing particularly sacred about the way the system was modularized. In this case, it was made so that the stepper drivers and their power supply could be placed close to the telescope, while the computer could be placed some distance away, with the telescope control board directly beside it.

2. PT69 Computer. The PT69 single board computer was purchased from Peripheral Technology as a completely assembled and tested board. The cost is slightly less than than $300. The board is only 5.5 x 6.5 inches in size, as shown in

Figure 10-4. Included on the board are the 6809E processor, a floppy disk controller, two RS-232 ports with adjustable baud rate, a clock/calendar using the MC146818 chip, 4K bytes of EPROM, and 56K bytes of RAM. Needed to complete the computer is a power supply, two disk drives (5.25-inch floppies), and a chassis or cabinet. The home-brew computer used at the Fairborn Observatory is shown in Figure 10-5.

One minor "plug in" modification was made to the PT69 computer (it can easily be returned to the original configuration at any time). The change was to replace the peripheral interface adaptor (PIA) chip with a "chip carrier" containing two smaller chips on it. The reason this was done was to provide four input and four output ports, instead of the two standard PIA ports.

The system described in this chapter would work well with many other computers. All that is really required are provisions for access to the data lines and enough address lines to provide the port select signals. While an on-computer clock/calendar with battery backup is certainly handy for telescope control, this could be added as part of the external circuitry.

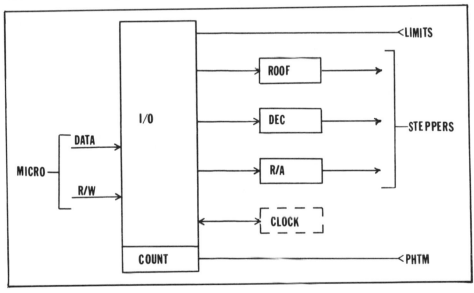

10-6 Telesdope Control Board Functions

3. Telescope Control Board. The telescope control board provides three vital functions, and three optional ones, as can be seen in Figure 10-6. The vital functions are the buffering or latching of the control computer signals, control of the RA stepper, and control of the Dec stepper. The optional functions are control of a third stepper which changes a photometer filter, a counter for use with external instrumentation (such as a photometer), and a clock/calendar. The version of the telescope control board discussed here implements all the vital and optional functions, except for the clock/calendar, as there is already one on the PT69 computer board.

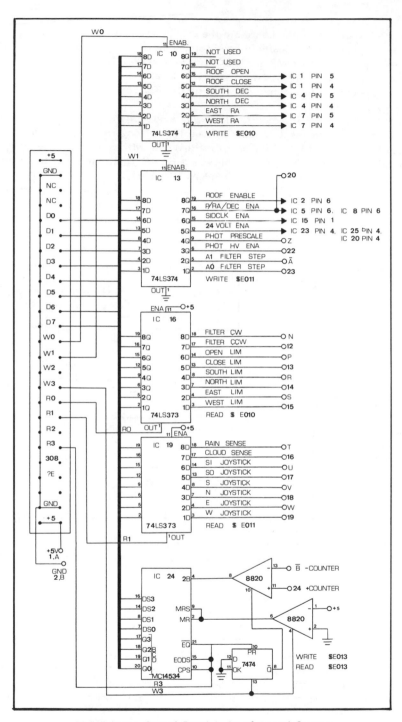

10-7 Telescpe Control Board A--Interface and Counter

The microcomputer interface portion of the telescope control board is shown in Figure 10-7, along with the optional counter. On the left of the schematic is the 50-pin connector that runs from the PT69 (the connections would be different for other computers). Coming from the computer are the eight data lines, four read port select lines and four write port select lines, +5 volts DC, ground, and an already somewhat divided down clock signal at 308 kHz. This clock signal is just passed on to the RA section, to be discussed later, for the sidereal rate drive.

The top chip in Figure 10-7, IC 10, provides inputs to the three stepper motors used in this system (there are provisions for a fourth stepper if one were desired). The outputs are normally high, and the appropriate outputs are brought momentarily low for the stepper or steppers one wishes to move.

IC 13 is another output chip, and it provides latched outputs to control a variety of functions. Two lines are used to move the photometer filter stepper motor, enable high voltage for the photometer, and switch in a photometer count prescaler. These photometer functions are optional, and these lines could easily be used for other functions. Another function is "enable" line for the 24-volt overvoltage, which turns on the high drive voltage to the steppers during high-speed operation. This is a necessary, but not sufficient condition for applying the overvoltage. Once the sidereal clock is enabled (high), the sidereal rate pulses are provided to the RA stepper motor. When the photometer/RA/Dec stepper enable line is high, these steppers are enabled. Finally, there is the roof stepper enable line (separate so it can be disabled unless the telescope is in its home "stowed" position).

IC 16 is an input port that handles all the limit and position sensing switches for the entire system. While a microcomputer controlled telescope can be operated without limit switches, one software problem could cause the telescope to encounter an obstruction or wind up the cables. The "home position" limit switches also define an initial position for the telescope.

IC 19 is another input port that handles two control paddle inputs and two auxilliary inputs. The auxilliary inputs in this system are weather sensors which, when activated, cause automatic system shutdown. These are not needed for a manually operated system, as most astronomers can tell when it is raining on their telescope and know what action to take!

Finally, IC 24, the two halves of the 8820 chip, and one half of the 7474 chip provide a simple counter with differential (balanced line) input. Writing to the 7474 starts the counting from the input, while the reading of the first digit stops the count. After the other digits are read, another count can be taken.

Shown in Figure 10-8 is the roof control portion of the telescope control board. IC 1 is a binary up/down counter. Roof open pulses cause the counter to count up, while roof close pulses cause the counter to count down. The three least significant bits of the counter output are fed to a 3-to-8 line decoder/multiplexor (IC 2), which, in conjunction with the three input 7410 NAND gates (IC's 3, 6, 9, and 12), provides the four-phase pulses required by the roof control stepper motor.

IC 23 is a retriggerable monostable multivibrator used as a one-shot. As long as pulses are continuously received from "roof open" or from "roof closed" and the 24-volt enable is high, then the Q output (pin 8) will be high and the roof high voltage (24-volt slew overvoltage) will be enabled, but if there are no very recent steps or if the 24-volt enable is low, then the overvoltage will not be enabled. The reason for all this is that providing a stepper with a five times overvoltage for high speed operation is just fine, as long as the stepper continues to step along at a

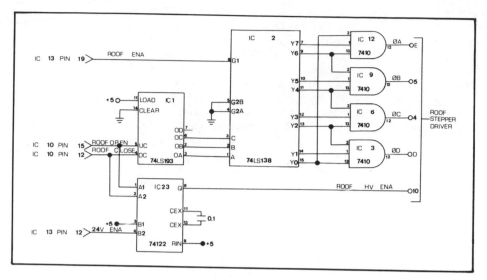

10-8 Telescope Control Board B--Roof Control

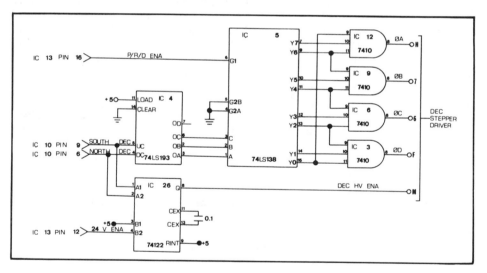

10-9 Telescope Control Board C--Dec Control

reasonable speed. However, if this same overvoltage were applied when the stepper was standing still, it would quickly overheat. While the software should turn off the overvoltage at the instant the stepper stops stepping, IC 23 makes sure that even if this were forgotten, that it would be turned off within a fraction of a second anyway -- protecting the hardware from software errors.

The Dec control portion of the telescope control board is shown in Figure 10-9. The schematic is identical to that of the roof control, except that the IC numbers are different, as are the inputs and outputs. As the functions are identical, no additional explanation is needed.

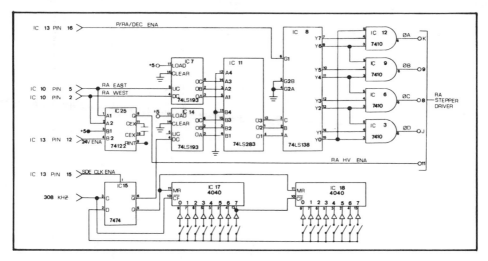

10-10 Telescope Control Board D--RA Control

The RA control portion of the telescope control board is shown in Figure 10-10. IC's 7, 8, and 25, and the 7410's perform exactly the same functions as IC's 1, 4, and 7 in the roof and Dec controls. The added chips provide a constant sidereal rate in RA. IC's 15, 17, and 18 form a DIP-switch settable binary divider. This allows the 308 kHz constant clock from the PT69 computer to be divided down to the exact rate needed for sidereal tracking. IC 14 is an up/down counter that performs for the sidereal rate signal the same function provided by IC's 1, 4, and 7. IC 11 is a four bit binary full adder, and it "adds" or combines the sidereal rate signal with any RA motion commanded by the PT69. When RA motion is west, the two are added; when RA motion is east, it is more appropriate to think of them as being subtracted.

In theory, one could provide the sidereal rate pulses and any commanded motions entirely from the computer without this external hardware on the telescope control board, but in practice, this is very cumbersome in software, and it also would add a greater processing burden to the computer.

Shown in Figure 10-11 is the layout of the chips on the telescope control board (component side). The board itself is shown in Figure 10-12 on its chassis. The cable from the computer enters from the left. The cable to the stepper control chassis leaves from the front, and jacks for the control paddle and photometer are provided. Table 10-1 lists the pinout of the telescope control board, while Table 10-2 summarizes the port commands.

4. Stepper Drivers. The schematic for the stepper driver boards is shown in Figure 10-13. The only unusual feature of the circuit is the use of the TIP 120 and TIP 125 to turn on and off the 24-volt overvoltage, and the use of diodes to switch between the 5- and 24-volt supplies.

All the parts except for the 10-watt zener diode are available at Radio Shack. The parts were wired point-to-point on a perforated circuit board. The driver board is shown in Figure 10-14.

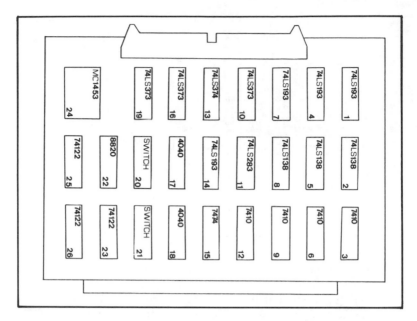

10-11 Telescope Control Board Chip Layout (Component Side)

10-12 Telescope Control Board on its Aluminum Panel Chassis

TABLE 10-1
TELESCOPE CONTROL BOARD PINOUT

1 +5 Buss	A +Buss
2 Ground	B Ground
3 NC (Common to C)	C NC (Common to 3)
4 Roof Phase C	D Roof Phase D
5 Roof Phase B	E Roof Phase A
6 Dec Phase C	F Dec Phase D
7 Dec Phase B	H Dec Phase A
8 R A Phase C	J R A Phase D
9 R A Phase B	K R A Phase A
10 Roof HV Pulse	L NC
11 R A HV Pulse	M Dec HV Pulse
12 Filter CCV Limit SW	N Filter CW Limit Swicth
13 Roof Close Limit SW	P Roof Open Limit Switch
14 North Limit Switch	R South Limit Switch
15 West Limit Switch	S East Limit Switch
16 Cloud Sensor	T Rain Sensor
17 Joystick Speed O	U Joystick Speed 1
18 North Joystick	V South Joystick
19 West Joystick	W East Joystick
20 Photo. Stepper Enable	X NC
21 NC	Y NC
22 Phot. HV Enable	Z Phot. Divide by 10
23 Phot. Stepper Addr O	-A Phot. Stepper Addr 1
24 Counter Input +	-B Counter Input -
25 NC	-C NC
26 NC	-D NC
27 NC	-E NC
28 NC	-F NC

TABLE 10-2
PORT ADDRESSES AND FUNCTIONS

Address	Bit	Pin	Read Limit Switches	Bit	Pin	Write Stepper Pulses	Bit
$E010	0	15	West	1		R.A. West	1
	1	S	East	2		R.A. East	2
	2	14	North	4		Dec. North	4
	3	R	South	8		Dec. South	8
	4	13	Roof Close	16		Roof Close	16
	5	P	Roof Open	32		Roof Open	32
	6	12	Filter CCW	64		Unassigned	
$E011			Misc. Inputs			Misc. Outputs	
	0	19	West Joy Stick	1	23	AO Filter Stepper	1
	1	W	East Joy Stick	2	A	A1 Filter Stepper	2
	2	18	North Joy Stick	4	22	Photometer HV Enable	4
	3	V	South Joy Stick	8	Z	Photometer Prescale	8
	4	17	Joystick Speed	16		24 Volt Enable	16
	5	4	Joystick Speed	32		Sidereal Clock Enable	32
	6	16	Cloud Sensor	64	20	Phot., R.A., Dec. Enable	64
	7	T	Rain Sensor	128		Roof Enable	128

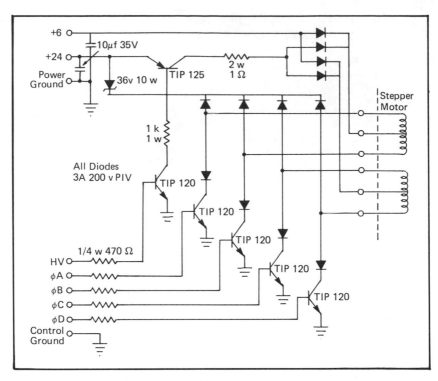

10-13 Stepper Driver Schematic Diagram

10-14 Assembled Stepper Driver Board

10-15 Control Paddle

5. Hand Paddle. Manual operation of the system is made simpler with the control paddle shown in Figure 10-15. The top four buttons determine direction (two can be pressed simultaneously, such as north and east, to move the telescope northeast). With none of the buttons pressed, the telescope moves at its slowest speed. The buttons on the bottom row are used with the direction buttons to obtain medium (set) and high (slew) speed operation. One of the bottom buttons can also be used as a panic "stop everything" button.

D. ELEMENTARY SOFTWARE

1. Introduction. The software needed to run a microcomputer controlled telescope can be quite simple or very complex. Obviously, the more functions accomplished by the software, the more complicated it gets. The most important design decisions in any computer control project involve the allocation of functions to hardware and software, and the tradeoff between the two.

Perhaps the most elemental software is that needed to move the telescope about the sky in slow motion. This is essentially performing the same function with the computer that one has on fairly simple non-computerized telescopes. In this section, the most elemental control software will be presented. The software is written in a way that individual control functions can be identified and separated from other functions by making each function a module called a "procedure". The more advanced modules needed to do such functions as slew to coordinates and precess coordinates are presented in Chapter 13, along with the modules needed to make the operation of the telescope completely automatic.

QJOY2 is the main control procedure for manual control of the telescope without the display of coordinates. This procedure is not used very often, as the telescope is normally run in an automatic mode, and even in a manual mode, one would usually prefer to slew to coordinates. However, it has the merits of being very simple.

The QJOY2 procedure "runs" or "calls" two other procedures-- DIGITAL and MOVE, which in turn calls RAMP. This is a high speed assembly language procedure that performs the telescope slewing and all other telescope movements. These procedures are summarized in Table 10-3.

TABLE 10-3
SOFTWARE PROCEDURE FUNCTIONS

Name	Functions
QJOY2	Main control, decodes control paddle, and sets incremental move steps
DIGITAL	Reads Control Paddle
MOVE	Breaks overall move into two seperate moves and develops command for RAMP procedure
RAMP	Assembler procedure to ramp steppers up and down

Before considering each procedure, it is appropriate to see what they do overall. After the procedures are loaded, one simply types in the command "RUN QJOY2" and picks up the control paddle and operates the telescope. The first thing QJOY2 does is run DIGITAL, and DIGITAL returns to QJOY2 with the parameter PADDLE, which is the value representing the keys pressed on the control paddle. From the combination of keys pressed, QJOY2 calculates two parameters--MOVEX and MOVEY--which contain the number of steps and direction to move in an X-Y (RA-Dec) coordinate system. West is +MOVEX, while north is +MOVEY.

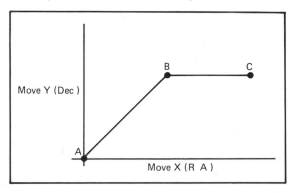

10-16 General MOVE Case

MOVE is a very general procedure that can take any MOVEX and MOVEY input and cause the telescope to make the desired move. The general move case is shown in Figure 10-16. To move from A to C, MOVE breaks the move into two segments, the AB and BC segments. In the AB segment, both the RA and Dec steppers operate together, simultaneously taking an equal number of steps. After momentarily stopping at point B, the remaining steps on just one of the steppers (the case shown is RA) is executed. Each one of these two moves is made by MOVE running the RAMP procedure. Thus generally MOVE will call the RAMP procedure twice. After the second move has been made, MOVE realizes that the move is completed, and returns to QJOY2, which runs DIGITAL again to see if there is a new command.

If one of the direction buttons on the control paddle is held down--say the north button--then the telescope makes a series of moves to the north. For guide speed, each motion is a very short one separated by a brief pause, and the telescope does not move very fast. For slew speed, the moves are much longer, and the telescope has a chance to get up to full speed and move along some distance before it stops again. This series of discrete movements allows (when other procedures are added) coordinates to be updated and other actions to be taken between each move.

Each procedure will now be considered in more detail.

2. QJOY2. The flow diagram for procedure QJOY2 is shown in Figure 10-17. As can be seen, the first thing QJOY2 does is to run DIGITAL. It then sets or resets MOVEX and MOVEY to zero, and MULT to its smallest value. MULT is the number of steps that will be taken. A check is then made to see if the highest order "STOP" key has been pressed. If it has, all execution is stopped.

A check is then made to see if the slew or set buttons have been pressed. If they have, MULT is increased from its minimal value. If one of these buttons has not been pressed, then it is assumed that the system is in the guide mode and the MULT parameter is left at its smallest value.

A check is then made to see if any of the direction buttons has been pressed. If none has been pressed, then MOVEX and MOVEY are left equal to zero. If one or two have been pressed, then MOVEX and MOVEY take on the appropriate sign. The final value of MOVEX and MOVEY is the product of the signed direction information and the number of steps, MULT.

After executing a small delay (to give a very brief pause between each discrete move), procedure MOVE is run.

3. DIGITAL is a very simple procedure. The control paddle is read by doing a PEEK at location E011 Hex. A button press brings lines low which are normally high. This gives an inverted value for the key, which must be subtracted from 255 to obtain the value corresponding to the key pressed. Since bit 7 is not defined, the resulting number (stored in PADDLE) needs to be between 0 and 127 regardless of whether bit 7 is high or low. A flow diagram of a procedure that will implement this is shown in Figure 10-18.

4. MOVE. The flow diagram for MOVE is shown in Figure 10-19. The inputs to MOVE are the parameters MOVEX and MOVEY, which are the number of steps and the direction the telescope is to be moved. In the general case, MOVE divides the total move into two separate moves and determines the parameters for calling RAMP. These parameters are known as Elements 0 through 5. Elements 0 - 2 are the number of steps to be taken, specified as a three-byte number. Element 3 is a constant that defines the acceleration to be used in ramping the motor in the

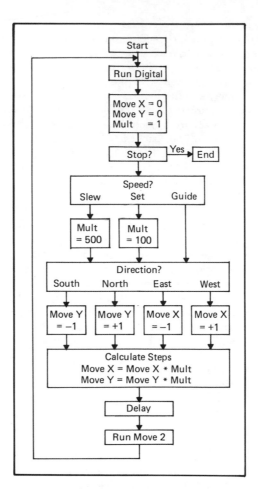

10-17 QJOY2 Flow Diagram

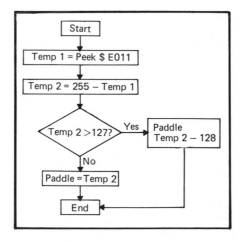

10-18 DIGITAL Flow Diagram

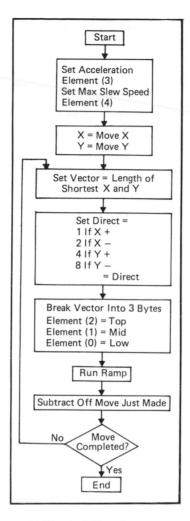

10-19 MOVE Flow Diagram

assembly language RAMP procedure. Element 4 sets the maximum slew speed in RAMP. Element 5 is a motor direction command.

Once these elements are calculated, MOVE runs RAMP. When RAMP returns to MOVE, the move which was just completed is subtracted from what was originally requested. If the move has not been completed, the process is repeated.

5. RAMP is the assembly language procedure which moves the chosen steppers in the commanded direction the proper number of steps. It accelerates to maximum slewing speed, and after the proper number of steps have been executed, decelerates to a stop.

6. JOY4. After using the simple QJOY2 discussed above for some time, various enhancements were made over time. These included a table of stars and other objects, the ability to move the telescope to any coordinates entered, the ability to "tweak" the telescope position without changing coordinates, and a list of objects with preset coordinates. A call to a precession routine was also added, as was a move to the home (storage) position.

The details of the algorithms used to control the yoke-mounted telescope are presented in Appendix K, while the listings of the procedures are given in Appendix L. They can be taken together as a group and used to build an open loop telescope control system capable of pointing a well-made telescope with an error less than a few arc minutes. As discussed in Chapter 13, these procedures can also be combined with others to form the basis of a completely automatic telescope control system which uses a photometer to make rapid and accurate photometric measurements of variable stars without human intervention.

Chapter 11. **FAIRBORN OBSERVATORY TELESCOPE DESIGN CONSIDERATIONS**

A. INTRODUCTION TO SECTION V

This section of the book describes the approach to telescope control taken at the Fairborn Observatory. It is, quite literally, "A Tale of Two Cities", in this case Fairborn, Ohio and Phoenix, Arizona. The Fairborn Observatory's east and west facilities are located in these two cities respectively. Genet, P. Scott Hawthorn, Douglass J. Sauer, and Lloyd W. Slonaker, Jr. did the work in the east, while Louis J. Boyd did the work in the west.

In this first chapter of Section V, some of the practical considerations involved in the approach taken at the Fairborn Observatory to telescope control are described. These are matters that ended up being of concern to us, and we pass them along to you as things worth considering. We must admit that in many cases, we only became concerned about these after we had learned the hard way about the serious consequences of unconcern! Many of the ideas developed in this chapter also appeared in a **Byte** magazine article on real-time control (Genet, Boyd, and Sauer, 1984). We thank the editors of **Byte** for permission to exerpt from the original material.

In Chapters 12 and 13, a complete description is given of the automatic photoelectric telescopes (APT's) at the Fairborn Observatory. There are three of these telescopes. The original one was developed by Boyd over a four-year period, and is a 10-inch system located in Phoenix. The remaining two APT's are located in Fairborn. One of these is a 10-inch Schmidt-Cassegrain system on a home built yoke mount, while the other is a 10-inch Schmidt-Cassegrain system on the first of the new DFM Engineering fork mounts that were specifically designed for small computer-controlled telescopes. To understand these systems, an introduction is given to APT's in Chapter 12. This is followed by a description of the control algorithms and software in Chapter 13. Chapter 10 in the previous section described the hardware (both mechanical and electronic) for the APT's.

We turn, for the remainder of this chapter, to practical considerations gleaned from experience at the Fairborn Observatory.

B. OPERATOR CONVENIENCE

Although some microcomputer-controlled telescopes are only intended for fully automatic operation, most control systems will be operated by an astronomer through the computer. A good control system should be easily operated by an average astronomer without specialized training. It is tempting, of course, to have lots of dials, knobs, and pretty red lights, but simplicity is preferable. It is also tempting to use specialized computer terms and jargon, but plain language oriented to the operation of the telescope is more appropriate. It is not easy to develop a good astronomer/telescope control system interface, so if the real intent of a project is total automation, the equipment and project can be simplified somewhat if no attempt is even made to provide for a human observer.

The first data-logging and photometer-control systems at the Fairborn Observatory (Genet, 1980, 1982) used a hand-held hexadecimal keypad and remote video monitor for all communications between the astronomer and computer. All instructions and requests were displayed on the monitor in plain English, and the operator did not even need to know that a computer was involved. While a telescope can be controlled from a hexadecimal keypad, it is so unlike the traditional telescope control paddle that significant relearning must take place. Also, it is questionable if stars can be moved into position and centered as quickly with a hexadecimal keypad as they can be with a more traditional control paddle. For these reasons, a normal control paddle should be used in computerized telescope control systems. As a control paddle is not useful for entering numeric data, a separate numeric keypad may also be appropriate.

In some cases, it would be possible to place the computer and its terminal/keyboard inside the observatory dome for the astronomer to enter responses directly. However, not only is the observatory not a hospitable environment for electronic equipment, trying to look through a telescope while cradling a computer in your arms (or running across the room to enter each response) is not conducive to productive observing! The way to make it most convenient for the operator is to have a fully automatic control system that operates the telescope while the astronomer sleeps. For the "part time" astronomers that have a regular job during the day, a telescope that totally operates itself is convenient indeed!

C. HARDWARE/SOFTWARE TRADEOFFS

There is a natural, but often expensive and occasionally disastrous tendency to make the hardware as simple and inexpensive as possible, and to place the burden on the software. The hardware costs are both immediate and obvious, but the software development and life cycle costs are off in the future and less obvious-- especially to the optimists among us. Many real-time control projects have been abandoned in failure when it finally became clear that the software programming would take an order of magnitude more time and money than originally anticipated.

Removing the burden from software development and placing it on the interface hardware is what standardized buses with plug-in slots are all about. Such bus-based microcomputers can initially cost more than those without standardized slots for expansion. However, this investment can open up the world of ready-to-go, plug-in interface cards. These cards come fully assembled and debugged. The better ones have excellent documentation and have been "burned in" (tested with full power applied for several hours to weed out bad components).

Many of these interface cards are highly intelligent, with their own built-in microprocessors and machine language programs in EPROM. The software work has been done by the card manufacturer, and its cost spread out over hundreds or thousands of users. Smart interface cards are often used by control system designers to do the fast, repetitive, and mundane tasks, such as controlling stepper motors and reading shaft angle encoders. By delegating these standard tasks to the off-the-shelf smart cards, it is often possible to program most or all of the "custom" software in a convenient high level language such as BASIC, and in some systems, the complexities of interrupts may be reduced, or the need for interrupts may even be eliminated entirely. By picking the right bus and making a careful selection from the off-the-shelf interface cards available for this bus, the cost and time spent in custom software development can often be reduced by an order of magnitude. This can be the difference between the success and failure of an entire project.

As chips have become more capable over time, it has been possible to pack an increasing number of functions in a given amount of space. A number of manufacturers realized that by placing an entire microcomputer on a board along with the chips needed to implement most real-time control functions, that some very low-cost control systems would be possible. If one of these single board control systems will do the job for you, and future flexibility is not critical, then there will not be a cheaper solution. The second generation Fairborn Observatory APT hardware previously discussed in Chapter 10 is an example of such a single board computer and non-bus interface. It is almost an unbelievably simple system from the hardware viewpoint, suggesting that under some circumstances, a non-bus approach truly has merit.

For an example of hardware/software tradeoffs, consider the microcomputer control of stepper motors. At one extreme, the microcomputer can calculate the bit pattern needed to drive each individual phase of each of the stepper motors. Typically, this function would be implemented in machine language and embedded in a larger BASIC program. To slew the telescope, it is necessary to "ramp" the steppers up to a high speed, and then ramp them down again. A chip or two added to the hardware can simplify the software considerably. For instance, Hurst Mfg. makes a one-chip stepper controller/driver, with TTL pulses for stepper motion and direction on the input, and a direct connection to small steppers on the output. Sigma Instruments makes a two-chip set (which is used with a few discrete components) that takes a pulse stream of some length as the input, and provides a ramped pulse stream as the output.

At the other smart extreme, there are intelligent stepper controllers, of which the Whedco STD card is a prime example. From a high level language such as BASIC (or from a low level language), the Whedco stepper controller can be told the start speed for a stepper, the ramping acceleration, top velocity, and initial position. Anytime thereafter, a simple command, such as to move 1,235,476 steps CCW, will be executed exactly, including the ramp up, slew, and ramp down-- bringing the stepper to a smooth stop exactly where desired. Both hardware and software flags are available to signal "movement complete". An onboard register keeps accurate track of the total motion, and hence the current position. Unramped moves and single steps can be commanded. In-motion stops, either ramped or immediate, can be commanded at any time. While the Whedco controller has an onboard Z-80 and its own program in EPROM, all this is totally transparent to the user, and the controller just appears as an output (and input) port to the user to which the high-level commands are issued.

As mentioned earlier, the Fairborn Observatory APT's use two steppers, one each on the right ascension and declination axes. By only allowing the telescope to go in one of eight directions at any one time, and breaking up the move between two stars into two separate segments (with the telescope coming to a momentary stop between the segments), it is possible to get by with somewhat simpler software. Here is a case where the overall concept (telescope movement) was made slightly more complex, and the total system efficiency was reduced (imperceptably) by adding a "waypoint" stop in every move -- all to simplify the hardware and software.

D. SINGLE BOARD COMPUTERS AND BUSES

Selecting the right bus-based or single board microcomputer for any given real-time control task is relatively straightforward if prejudices not related to real-time control can be set aside. If the task is reasonably well defined and will not be changing in the future, if a number of essentially identical systems are to be made at low cost, and if a single board computer will do the job, then use it.

The single board computer used for the APT is based on the Motorola 6809 and is made by Peripheral Technology in Marietta, Georgia. Fred Brown, president of this small company, says that his philosophy is to put everything needed on one board and to make it at the lowest cost. What this board has is a 6809E processor, 56KB of RAM and 4KB of EPROM, a floppy disk controller, two serial and two parallel ports, and a real-time clock. If this meets your needs, then for less than $300 you can get a fully assembled and tested system, but if you need more, then you might consider a bus-based system with plug-in slots. For convenience, buses will be discussed in three wide categories: (1) hobby, (2) personal computers, and (3) industrial control. Hobby buses would certainly include the venerable and ever popular S-100, which has done much to make the 8080 and Z-80 processors well known, and the SS-50 bus which remains the favorite of the 68XX fans. Categorization as a "hobby bus" simply means that besides the fine complement of ready to go commercial systems and cards, these buses are patronized by those hardy individualists, the home brewers. Personal computers with a wide selection of plug-in cards for their slots are really limited to the Apple II and the IBM-PC and their very close clones. If the real-time control task at hand is not overly difficult, or if one of the systems mentioned above is available and adequate to the task, then it makes a lot of sense to use it. Honeycutt (1977) developed one of the first microcomputer control systems based on the S-100 bus, while Tomer controls his mobile telescope with an Apple II (Tomer and Bernstein, 1983). Skillman (1981) developed an automatic telescope that is controlled by a combination of Apple II and KIM microcomputers.

If the real-time control job is a tough one, or if one desires to tap the full potential of what is available off-the-shelf for real-time control, then it is necessary to turn to the buses used by industry in their control systems. Good examples of industrial buses are the STD Bus, Q-Bus, and Multibus.

The STD Bus has a large number of different manufacturers (about 100) and a wide selection of different cards (nearing 1000). The card selection is particularly rich in the area of such real-time control tasks as smart stepper controllers, optical encoder interfaces, and DC servo motor interfaces. While there are 16-bit STD systems, this bus is primarily an 8-bit bus that favors the Z-80 microprocessor and the CP/M operating system. For greater crunching power, the Q-Bus and Multibus are good choices. For those who have been brought up on DEC operating systems, languages, and features, the Q-Bus will be the natural choice. While the first LSI-11 microcomputers were somewhat slow compared to their minicomputer brethren, speeds have increased considerably with the LSI-11/23, the LSI-11/73, and now the Micro/VAX. While cards for the Q-Bus were somewhat expensive a few years back, competition has brought prices tumbling down.

E. HARDWARE APPROACHES

There are essentially three approaches that can be taken to the hardware. Each of these is discussed below.

First, commercial off-the-shelf electronics can be used in all or most of the system. This usually implies that a bus approach be used--preferably a bus with a good selection of real-time control cards (such as the S-100, STD, or Q-Bus). This approach has a number of advantages. First, fully developed, tested, and functioning cards can be used in the system. Second, commercial cards that have been in use for some time tend to have their "bugs" worked out, and they are more reliable than almost any new "first time" cards. This is especially so if the cards are from a well-known supplier of cards to industry (cards built for the hobby market are not necessarily as reliable). Third, such cards usually come with rather complete documentation--obviating the need to write your own documentation. Fourth, for "smart" off-the-shelf cards, the software is already written and stored in ROM--thus additional effort need not be invested to program the intelligent card. Finally, using off-the-shelf commercial cards can often save time, money, and grief. It is difficult to design, build, test, debug, and document a board cheaper than one can be bought. This is due, of course, to the economies of scale inherent in commercial production.

The second approach that can be taken to hardware is to fabricate to an established design. This saves design time, and can reduce, but not eliminate, the debugging and documentation time. This approach is reasonable when no commercially available card exists for a critical function. Sometimes, however, one is forced to do the best with what is available, and fabricating to an established design makes some sense. The third and final hardware approach is designing, building, testing, debugging, and documenting cards entirely from scratch. When, for some critical function, no commercial card is available, and no design is available, then one has no choice but to go through the entire process. Plenty of time and money should be set aside to do a good development job, as otherwise the performance and reliability of the total system will be jeopardized because of one poorly done card. For non-bus systems, making custom cards is likely to be the rule. If the design is done very carefully (such as that done by Boyd, Genet, and Slonaker), then this approach can work well, as exemplified by the current Fairborn Observatory APT control system. If not thought out well, it quickly becomes a nightmare.

F. INTERFACE SOFTWARE

The complexity of software in telescope control systems makes the approach to software development of critical importance. Methods of reducing the software complexity with good system design and the use of appropriate hardware have already been discussed. By selecting the right development and run-time environments, the effort of building and debugging the software that remains in the system can be minimized.

Assembly language has its uses. An excellent use in real-time control is by the manufacturers that put it in the EPROMs on the smart interface cards. On rare occasions, a high level language has a task to do that is simply beyond its speed capabilities. A small and limited assembly language subroutine is appropriate in this case.

Really good high level languages are a pleasure to use. The Fairborn Observatory has gravitated towards the use of two BASIC languages. On our Z-80 based CP/M systems, we ran Microsoft MBASIC in both its interpretive and compiled versions. Programming in the interpretive version was convenient, and if greater speed was needed, we could always use the compiled version. MBASIC and all its look-alikes are probably the most widely used computer languages in the world, and thus need no description here.

Our current 6809-based systems use Microware's "Basic09" BASIC interpreter running under the OS9 operating system. For the true high level language control system afficionado, this language is such a delight to use, that a short digression to mention its origins and salient features is in order. When Joel Boney and Terry Riter at Motorola started the design of the 6809 microprocessor, they contracted with Ken Kaplin at Microware to develop a language and operating system that would fully utilize the many unique and valuable features of this most advanced of the 8-bit microprocessors. What Microware came up with was a language that supported structured programming (sort of a cross between BASIC and Pascal), that was completely modular (allowing independent programs to work together), and that included an interactive compiler with very high level debugging functions. The ability to develop the software for a complex control system as a number of independent programs has proven to be very valuable.

Having selected system and hardware approaches and an operating system and language, software development can begin once sufficient hardware is on hand. Just as an evolutionary approach to the hardware makes sense--starting as simple as possible and adding the frills later--so it makes sense to start with the simple and move to the more complex with the software. Breaking the program into little modules, where each module accomplishes a specific function, makes the entire task much easier. Each function can be checked out while the program is fresh in one's mind and the bugs worked out right away. It is sometimes helpful to write a small, highly simplified program on the side to check out the major features of some function, e.g., to see if the equipment does what you expect it to. Unlike normal, non-real-time programming, programming a control system usually requires that the control system itself be in operation, that a logic probe and circuit schematic diagrams be close at hand, and the expected effects of the software on the hardware be constantly checked.

Sometimes it is possible to check out some specific function with just a small part of the system in operation, or even with a special piece of equipment used just for checking software. For instance, in the 0.4-meter telescope control system, two small steppers and short connection cables were used to check out the software as it was written. For example, a command in declination, instead of moving the telescope, could turn a little stepper right beside the terminal.

Besides having special cables and peripheral equipment just for software development, it is quite helpful if a second and identical computer can be set aside for development of the software and hardware while the primary computer stays hooked up to the telescope and is used to control it. Later on, the second computer can be used for data analysis, word processing, and as a backup to the primary computer.

G. ADAPTABILITY

No matter how carefully you plan out your system on paper (and it should be most carefully planned!), there will be changes--usually quite a few of them--before you get the system operational. No matter what the configuration of the system is when it first becomes operational, it will be modified later (sometimes extensively so). This suggests that the prudent thing to do is to plan on change from the very beginning so that changes can be made as gracefully as possible.

One approach to having an adaptable system is to use a standardized bus. If some of the hardware doesn't work, only a card's worth needs to be tossed. As suggested earlier, if you are making your own card and a commercial one is available, then going with a bus allows a fallback position to the commercial card. If, as sometimes happens to the best of projects, the system should be scrapped to make way for a new one, then much of the electronics can be salvaged for use on the new system if both systems were on the same standardized bus. If the hardware is very simple, as it is in the non-bus APT system described later, evolution can occur by designing and building a new interface card.

Just as the interchangeability of cards on a bus increases adaptability, so having interchangeable program "software modules" helps the adaptability of the software. When things don't work, only modules, not entire programs, have to be scrapped. Contrary to popular belief, modularized programs do not take significantly more code or memory, and execute almost as quickly as in-line code. Even if this were not true, memory is dirt cheap these days, and by using smart interface cards, there is no requirement for raw speed.

At first it might seem that having smart cards with unalterable programs stored in their ROMs would reduce adaptability, but this is not the case. The fixed tasks performed by these smart cards are not likely to be in need of change, but by relieving the main CPU of the time-critical tasks and allowing a high level language to be used, changes in the overall control program can be made much more easily, because the structure of these programs is greatly simplified by the use of these cards. High level language programs, particularly if they are well documented, are changed with relative ease compared to a program written in assembler, or a language designed to implement lower level hardware dependent primitives, such as FORTH.

Careful modularization of the hardware and cables along functional lines (just as the software was modularized) improves adaptability. If a function has to be changed, it does not require a rework of all the electronics, half the cables, and the entire back panel--just a single card, one cable, and a modular strip on the back panel. Real thought needs to be given to dividing a system along functional lines and minimizing interactions. Having slightly more hardware or extra cables is sometimes worthwhile if this cleans up things along functional lines. Building in spare capability is usually worthwhile. Extra wires in a cable and extra pins on a connector cost very little and may save making up an entirely new cable later. Terminal strips a little larger than needed often are helpful. Extra reserve in power supplies will almost certainly be used as time goes by. Extra slots in the card cage are always useful.

H. RELIABILITY

Many astronomers rightly fear that a computerized control system will fail on them at some crucial moment. Such fears are often based on past experience with early computer systems which indeed were unreliable. However, if careful attention is paid to reliability throughout the design and construction process, a highly reliable system will most likely result. It might be noted that an increasing number of industrial processes are being computerized without reliability difficulties.

For complex systems, reliability often is enhanced by putting the electronics on a standardized bus. Having all the electronics plug into a single card cage minimizes the number of interconnections to be made, and reduces the susceptability of the system to electromagnetic interference (EMI). The importance of reducing interconnections and EMI can not be stressed too strongly, and a sad but true story is appropriate here.

The first computer at the Fairborn Observatory (east) was a Radio Shack TRS-80 Model I. It served faithfully for a number of years as a data-logger for the photometer, and as an analysis machine. It was then applied with confidence to controlling the 0.4-meter telescope. While it worked in this capacity, it did not work well. Random "crashes" of the system were frequent and extremely annoying, and there were several connector and cable problems. Because of these problems, this approach was abandoned in favor of the STD bus. The STD system which replaced the TRS-80 never "crashed", and its reliability was outstanding.

Going to a bus will not, by itself, eliminate all connector and EMI problems. Good grounding and shielding practices must be followed throughout the entire system. It costs very little extra to run shielded instead of unshielded cables out to the telescope, and the card cage, power supplies, etc., should all be well shielded. All equipment and circuits should be properly grounded, and a single ground point should be used to avoid "ground loops". Connectors should be of high quality, clamp on the cable, and lock to the device being connected, so connections will not jiggle loose. Their contacts should be gold plated to reduce resistance, and should use wiping action to assure a positive connection. The screw-on type A/N connectors are best, although quite expensive. However, reliability is usually a good investment.

An even better approach is to eliminate as many connectors from the system as is practical. In the second generation Fairborn Observatory APT control system, all the electronics except the computer and power supply have been placed on a single card to eliminate connectors. This not only saves money, but it significantly improves reliability.

The quality and reliability of computers and boards varies widely. For some customers, low cost is important, and for them, cheap board materials are used, holes are not plated through, and chips are marginal seconds. This may be perfectly acceptable to someone playing games on his personal computer, where a failure would only be slightly annoying. However, a failure in a telescope control system can not only result in lost observations, but a failure can place the telescope drive in a mode which causes damage to the telescope itself. Therefore, only the highest quality glass epoxy boards, top quality chips, and the best construction techniques are acceptable. Commercial cards are usually burned-in and environmentally stressed before they are sold. It is rare when a card doesn't work the first time it is turned on and keep on working without any failure for years. Reliability costs more to begin with, but it is worth it in the long run.

Even when using well designed equipment of high quality construction, it is prudent not to expose electronics to environmental stresses unnecessarily. Temperature cycling, moisture condensation, overheating, and dust all take their toll and reduce reliability. The floor of an observatory is a particularly inhospitable environment for electronics, and it pays to keep all or almost all of the electronics in an environmentally controlled room. Astronomers also like to operate the telescope in shirt-sleeve comfort! If this is not possible, thought should be given to a small enclosure with a thermostatically controlled heater to provide some protection to the electronics.

However, in spite of all precautions, an occasional failure will occur in almost any complex system. When this happens, maintainability becomes important.

I. MAINTAINABILITY

Systems always fail at the worst possible moment. They fail when they are needed most and when it would be most difficult to fix them. The good design inherent in a control system (or lack of it) shows up when something fails at 2 AM in the middle of a "chance in a lifetime" observing run. Documentation is the most critical aspect of maintenance. If the system is well documented, then troubles can be found in a logical, organized, and expeditious manner. Complete documentation is important. Every card, cable, and wire should be documented--right down to the color of the wires in the cables. The documentation should all be assembled together in one place--preferably in a single bound volume so that nothing gets lost. There should be at least two copies of the documentation. As mentioned earlier, one of the advantages of buying off-the-shelf commercial cards is that most of them are thoroughly documented. Not only do commercial cards rarely fail, but they are relatively easy to fix when they do.

Hand-in-hand with documentation goes labeling. Connectors, terminal strips, boards, etc. all need to be labeled, and the labels should agree with the identification on the documentation. When in doubt, label.

Ease of access is important for maintenance. If things are hard to get at, not only are they hard to fix, but other things can be damaged in the process, not to mention frayed nerves and strained eyes. Large removable covers are helpful.

Wire-wrapped cards are generally more difficult to repair than printed circuit (pc) cards. In repairing a wire-wrap card, it is not unusual to induce another fault, and in trying to repair it, sometimes yet another.

The special little cables, test devices, etc., used to make software development easier are also useful in isolating troubles. A well-rounded selection of modern electronic test equipment is really a necessity. If possible, this should include a good logic probe, digital voltmeter, and wide-band oscilloscope.

Isolating a fault and bringing a system back up on line is much easier if spare boards are available, or even better yet, a completely spare system. Spare pieces and parts should also be kept. Keeping a spare system and spare boards is not cheap, but the cost can be reduced if emphasis is placed on commonality and standardization.

J. CONCLUSIONS

Many different and sometimes conflicting issues must be resolved during the design and development of any complex real-time control system. There are many ways to go wrong, but only a few ways to go right. Of course most observatories have small budgets and there is constant financial pressure to cut corners. However, the greatest expense is having to completely re-do a system, as a system that really never works quite right just is not useful in the demanding world of observational astronomy. Throughout the history of systems development, there is never enough time to do the job right the first time, and plenty of time to do it over. Don't make this classic mistake!

Evolutionary development of a bus-based system using proven commercial cards can result in a highly reliable, easily maintained, and easily modified system. If great design care is taken on a "hardware simple" system, such as the Fairborn Observatory APT's, excellent results can be achieved with just a single board computer and a few fairly simple custom boards.

A. INTRODUCTION

In astronomical research, small telescopes are often used for photoelectric photometry. The wide bandwidth of UBV(RI) photometry, the high quantum efficiency of the photodetectors, and the "zero dimensional" nature of the data all tend to make the best use of the meager photons available to smaller telescopes. While photometry of unique one-time events such as occultations provides valuable data, much small telescope photometry is of variable stars. Such photometry is generally a highly structured and repetitive task that lends itself well to automation.

It is convenient to break variable stars into two broad classes: short period variables and long period variables. Short period variables are those stars that require observation continuously for many hours or all night long, while long period variables only need to be observed once (one set of readings) per night. Typically, one would observe a single short period variable all night long, while one would observe many long period variables in a single night and repeat these observations for many nights. As these two situations are quite different, it would be expected that different systems might evolve to meet each situation.

Short period variables, especially eclipsing binary stars with short duration eclipses, have been favorite photometric objects since the beginning of photometry. The first small automated telescope system to observe them was designed and built by Skillman (1981). This system, which is controlled by Apple and KIM microcomputers, has been in operation for several years now. However, in this chapter we will not be concerned with systems for observing single short period variables hours on end, but will, instead, concentrate our attention on systems designed for observing many long period variable stars each night.

The small automatic photoelectric telescope (APT) developed by Code and others at the University of Wisconsin in the mid-1960's (McNall et al., 1968) was fully capable of observing long period variable stars, although it was not applied to this task to any extent. Rather, it was used to observe a number of standard stars spread across the sky for purposes of determining nightly extinction coefficients. This 8-inch telescope was controlled by a PDP-8 minicomputer, and its photometer was at prime focus. It was housed in a small roll-off roof building, and its operation was fully automated. This promising start on this type of APT was not developed further until Boyd developed a small APT specifically for observing long period variable stars. A number of large telescopes, however, are quite capable of automatic photometry of long period variable stars, but typically are not used for this purpose. One of the most interesting of these is the 1.2 meter Cloudcroft telescope described by Worden et al. (1981). This telescope was controlled by an IBM 1800 computer, and was used in a successful, but short-lived, program to determine the variability of solar-type stars (Radick et al., 1982) until recently, when the telescope was dismantled.

Perhaps the main reason for fully automating a small photoelectric telescope is the efficiency with which such an instrument can observe long period variable stars. Such an automated system is much faster in acquiring and centering stars than any human can be, and an automated system never needs a break, nor does it quit early to catch some sleep. Consider the following comparison between an automated and a manual approach.

A set of data consists of variable, comparison, check, and sky readings. The variable is measured three times, the comparison four times, and the check and sky are each measured twice, all in three colors, for a total of 33 measurements. With 10-second integrations, this can be accomplished in about 7 minutes by an automated system, including the acquisition time. This gives about 10 sets per hour, 6 hours per night, for 100 nights per year (limited by clouds and moon), for a total of 6,000 sets per year. If one is very fast, one might manually do 4 stars per hour (same measurements as above) for 3 hours per night, and (allowing for meetings, vacations, etc.) 50 nights per year, for a total of 600 sets per year. The estimated automatic-to-manual ratio of stars observed is 10:1. This is a sufficiently large ratio to motivate the development of a relatively low cost, fully automatic system based on microcomputer control. If non-automatic systems are manned continuously and have capabilities for fast and accurate computerized setting on stars, then they can give nearly the same observational coverage per night as a fully automatic system.

This chapter discusses some of the considerations involved in designing a small APT for observing a large number of different stars per night. It is not intended to be either exhaustive in enumerating design alternatives or to give specific recommendations. Rather the intent is to introduce the subject and recount some thoughts, experiences, and ideas picked up from others working in similar areas. The hardware design of an actual system was addressed in Chapter 10, while APT software is discussed in Chapter 13. For additional details, the reader is directed to Boyd and Genet (1983), which describes the first Fairborn Observatory APT during its development; to Boyd, Genet, and Hall (1984a), which announced the first fully automatic operation of the Fairborn Observatory APT, along with equipment details and the first observational results; to Boyd, Genet, and Hall (1984b), which describes operational experience and the second generation system; and Boyd, Genet, and Hall (1984c), that describes the second generation system in some detail.

B. ASTRONOMICAL CONSIDERATIONS

For an APT to realize its full potential, it must be able to observe enough different variable stars each night to keep it occupied. If the telescope is too small, it could run out of accessible objects and have to spend time waiting between observations or making less useful repeat observations. If it were too large, it would become too expensive for the typical small college or advanced amateur to afford. The actual size needed to keep such a telescope fully occupied is a complex function of sky brightness, pointing accuracy, slewing rates, etc. Since a star is observed many more times during the process of acquisition and centering than it is for the actual measurements, the acquisition observations must be of short duration. Thus the need for a measurable photon level during the acquisition process actually places the lower limit on the acceptable telescope size. In general, an 8-

inch aperture telescope can be kept fully occupied, and a 16-inch telescope would have a tremendous selection of stars.

Another important consideration is the accuracy of the observations. Small observatories tend not to be located at ideal mountain top sites, and short term variability of atmospheric extinction can be a serious problem. The speed at which an automated system can move between and acquire stars can increase the accuracy of differential measurements since there would be less time spent between measurements, minimizing changes in atmospheric extinction. The same thought applies to variations in the sky readings. APT experience to date suggests that while the results are as accurate as conventional photometry, they are not noticeably more accurate. Generally, the atmospheric scintillation, and not the sensitivity of the photometer, places the lower limit on integration times for an automated system. A larger telescope helps this somewhat, but the difference between an 8-inch and 16-inch instrument might only reduce the required integration time by half. It is not as easy to operate an automated system on stars which are at the threshold of detection, because of the problems with locating and centering the stars. A highly accurate offset from a nearby brighter object may be the most practical approach to such faint objects.

Finally, for long period variables it is possible to arrange the observing schedule so that most observations are made close to the meridian, thus providing the smallest air mass. While this can improve accuracy, it would reduce the year around coverage of the variables. Coverage is maximized by observing stars in the west at the beginning of the night and in the east at the end of the night. These are but two of many strategies that can be used in APT's.

C. THE TELESCOPE

Photometers have been successfully placed at prime, Newtonian, and Cassegrain foci of the APT's mentioned earlier. All things being equal, photometers are easier to design for higher f-ratios. The more timid observer may still wish to be able to look through an eyepiece to see if the automatic system can actually obtain the correct star and accurately center it, but this feature can add to the complexity of the system, and may be of little use once the telescope is operational. A large finder scope is more useful in watching the actual acquisition process.

The software development task on a small APT is generally much greater than the hardware development task. Because of this, the best approach to the hardware is one that simplifies the software. There are two simplifications that should be considered. One is to use a well aligned equatorial mount. The other is to provide the stellar rate motion separately in a way that is transparent to the software. The software task then becomes one of acquiring and measuring stars in a sky that does not rotate.

When an astronomer is at the telescope eyepiece, it is possible to take a few liberties with polar alignment, the orthogonality of the Dec axis to the RA axis, and the alignment of the optical axis with the mechanical axis. A computer is not as forgiving! It is wise to make provision for the fine adjustment of the polar axis in both altitude and azimuth, and also for fine adjustment of Dec axis orthogonality and optical axis alignment. Close attention to these points will help keep the acquisition searches as small as possible.

An APT is unusual in that it makes a large number of quick, short movements during the course of a night's work. During a full night of observing, one of the Fairborn Observatory APT's makes about 25,000 separate movements--most of these in acquiring and centering stars after a slew. All of this starting and stopping places special demands upon the telescope design. It is much easier to move the telescope around quickly if it has a low moment of inertia. A symmetrical fork or yoke design is thus most desirable. Compact and light weight Schmidt-Cassegrain optical assemblies are well suited to this application. As image quality is of little importance, the thin, short focus "Dobsonian" mirrors are also suitable, but the "Dobsonian" mount would be difficult to use. Closed tubes are desirable to keep out stray light, which is present at every observing site when the moon is up. The telescope and photometer should be baffled so that the only light path is through the primary mirror.

All of the short, fast movements tend to induce vibrations in the telescope, and observations can not be made until these vibrations die down. The vibrations are minimized and die down most quickly when the natural resonant frequency of the telescope structure is high. A very stiff, relatively light weight telescope structure tends to have a high natural frequency. Also, the necessity to make many quick, short movements requires drives with a minimum of backlash.

D. MOUNT AND DRIVES

Most large telescopes have separate slewing and tracking motors. The requirements for an APT are different, however, from other telescopes. In a manually operated photographic telescope, for instance, high tracking accuracy for long periods of time is a prime requirement, and rapid slewing rates are desirable for operator convenience. Large telescopes are sometimes equipped with shaft encoders to determine the position of the telescope, but these are unnecessary for the APT. The sudden stopping and starting place special demands on the design of an APT, but the ability to measure the light from a star reduces the requirements for the drives in some respects. Since the APT stays on one star for less than a minute, and since it can update the position of the telescope relative to the sky each time a star is centered, high tracking accuracy is not required. More vibration of the telescope is acceptable in photometry than with photography or visual observation, and this allows the size of the steps of the tracking motors to be greater (even as high as 2" per step). The step size in seconds of arc is not necessarily the same as the angle of movement caused by vibration, which may be greater or smaller than the step size, depending on the mechanics of the mount and telescope.

There are really three different accuracies involved in positioning APT's. First there is the open-loop accuracy needed after making a long move from one side of the sky to the other. Such a long move can be followed immediately by a spiral search for the first "navigation" star. A navigation star can be purposely chosen so that it can not be confused with any nearby stars--thus avoiding the problem of having the system lock onto the wrong star. However, there are time limits to how big a patch of the sky it is practical to search for the navigation star. The area to be searched (and the time taken to search it) is the square of the absolute pointing error, so the accuracy of the open loop "long move" must be reasonable. Our experience so far indicates that 12' accuracy is sufficient. This could be significantly relaxed if long moves across the sky were very rare, or if inefficiency due to long search times were acceptable. To achieve 12' accuracy, it is helpful to precess from year 2000 coordinates to the current epoch, but no other corrections are needed. Open-loop accuracies over long moves of better than 12' reduce the search time, but not enough to make a noticeable effect on the overall efficiency of the APT system.

The second accuracy of importance is that of a local move from one star to another after the navigation star has been initially acquired. We require this to be 3', but have found that it is generally much better--typically about 30". When slewing between stars in the same area, the APT's usually put the stars within the diaphragm on the slew, and if not, almost always they are found in the first loop of the square spiral search pattern.

The third accuracy is that of centering a star within the diaphragm. This is a tradeoff between centering accuracy and time taken to center. At present, we only require an accuracy of about 20% of the diaphragm diameter--about 6" to 12".

When using stepping motors, it is desirable to accelerate and decelerate smoothly (ramp) at the ends of each movement. The system built by Boyd initially ran with no ramping, but its efficiency was greatly improved when it was modified to include this feature.

E. CONTROL SYSTEM

The control systems on large conventional telescopes can be expensive because the pointing accuracy without reference to any stars in the sky may have to be only a few arc seconds. This often requires that encoders be mounted on each axis, and that corrections be made for precession, atmospheric refraction, flexure of the telescope, and other minor errors. Because of the ability to search for, acquire, and center a star, the APT can dispense with all of these corrections, except precession. There is no need for encoders--the loop, so to speak, is closed on the stars themselves.

In a conventional telescope control system, the need to calculate all the many systematic error corrections repeatedly keeps the computer busy, and can require a very capable microcomputer, such as the LSI-11 discussed in Section VI. Also, it is necessary to have some sort of hardware interrupt scheme, and at least some of the programming needs to be done in assembler language or some similar fast language, such as FORTH. With a math coprocessor or a fast 16-bit machine, it may be possible to make all required calculations in FORTRAN or some other high level language. However, by dispensing with the need for calculating extensive corrections, the APT can easily make do with a modest 8-bit microcomputer, and if appropriate steps are taken (pardon the pun), it can directly control hardware stepper motor drivers also. Most of the programming can be accomplished in a high level language such as BASIC, FORTRAN, or Pascal without the use of interrupts. The use of a high-level language without interrupts can considerably reduce the software programming effort.

F. BACKGROUND TO THE APT PROJECT

A discussion of APT's would be incomplete without discussing how Louis Boyd came to develop the first Fairborn Observatory APT, and some specifics of his development program. With an interest in both astronomy and electronics, Louis (who is an electrical engineer) decided to build a radio telescope. In 1972, while looking for parts for his radio telescope in a surplus store, he ran into a fellow "scrounger", Richard Lines. Richard and his wife Helen had a very nice optical observatory in Mayer, Arizona, and soon Louis was helping them on the electronics for a photoelectric photometer. The photoelectric observations at the Mayer Observatory were always made with Richard and Helen working as a team. Louis watched a number of these observational sessions, and began thinking about how stepper motors and a microcomputer might be used to move a telescope between stars -- perhaps even automatically centering them. Helen Lines, after hearing talk about a computerized photoelectric telescope for a while, finally told Louis to "stop talking about one and build one!" This was the beginning of the APT project. The radio telescope was permanently set aside.

The APT project began in earnest in early 1979. First on the agenda was the physical construction of the telescope itself. A modified fork mount was selected for the telescope. The modification consisted of extending the fork beyond the north polar axis bearing to a second cross-piece at the south polar axis bearing. This added stiffness to the fork, but at the expense of limiting the angular motion of the telescope in RA. While this worked well on Boyd's APT, he does not recommend it for other APT telescopes for several reasons. First, a sufficiently rigid fork is possible without the extension (the DFM Engineering mount discussed in Chapter 8 is an example). Second, a larger sky coverage would be helpful in some cases. Finally, if 180 degrees, or even better, a full 360 degrees of rotation can be made, the alignment process is simplified. Although it takes more space than a fork, a yoke mount has many of the properties desired in an APT, and is worthy of consideration.

Besides the mount configuration itself, another important decision was the mechanical drive system. Precision gears are quite expensive, and without the complications of preloading, they are subject to backlash. Boyd chose to use a chain and disk drive. The disks (in both RA and Dec) are 32 inches in diameter, 3/16ths of an inch thick, and are made of aluminum. They were purchased on the surplus market for $5 each, and were suspected of being part of a Honeywell computer memory system in a previous life. While this drive has worked well on the APT, Boyd points out that mixing different types of materials in the mechanical parts of the telescope, such as aluminum disks and steel chains, leads to differential temperature expansions, and these in turn can necessitate more frequent tension adjustments than might otherwise be the case. Sticking with steel throughout the mount has much to commend it. Louis Boyd and his original APT are shown in Figure 12-1.

The photometer itself provided some difficulties that required time consuming redesign. The first photometer utilized thermoelectric cooling, but in the end this proved not only to be unnecessary, but undesirable. As the sky count was always very much greater than the uncooled dark count (for a well selected PMT), cooling really did nothing to improve the signal-to-noise ratio, and only complicated the system. Another complication that proved not to be needed was selectable

12-1 The Boyd Automatic Photoelectric Telescope

diaphragm sizes. Once the system is operational, there is no need to change the diaphragm size, and during development, changes can be made without the complication of some automatic changing mechanism. In the original photometer, DC motors were used to position the filter and diaphragm wheels, and potentiometers were used to sense position. However, this complicated both hardware and software, and placing the filter wheel directly on the shaft of a stepper motor proved to work well.

Although photon counting was used by Boyd in his APT system, he feels that a DC amplifier and V/F converter would work just as well, and would be simpler and less troublesome. Also, the sensitivity of photodiodes has improved to the point where an Optec photometer is being used on the Fairborn Observatory East APT systems.

Not unexpectedly, the software portion of the project was the most difficult task. Software development was started using BASIC. However, in a few areas, it proved to be too slow. With the advantage of hindsight, Boyd feels that the best thing to have done at this point would have been simply to develop a few machine language subroutines for those portions of the BASIC program that required a speed capability beyond BASIC. This is what was done in the second generation system described here and used at the Fairborn Observatory East. However, this was not done initially, and instead, the software was reprogrammed in FORTH. This did provide the necessary speed, and a simplified version of the eventual software that could handle a single group of stars (variable, comparison, and check) was programmed in FORTH and became operational. However, the use of FORTH for the extensive disk I/O operations and floating point analysis required to handle many groups of stars, their selection, and the recording and analysis of the data proved to be difficult. In the end, the Microware OS9 operating system and Basic09 language were used.

The problem of getting an APT to track a single group of stars (variable, comparison, and check) for many hours is much simpler than the problem of getting an APT to observe many groups of stars in an appropriate order, recording and analyzing the data, etc. One needs to know, for instance, where the sun and the moon are located, so as to start and terminate an observing run when the sun is safely out of sight, and to avoid trying to observe stars too near the moon (this can be hard on PMT systems). It was these problems that occupied the last year of the development program. In October, 1983, all the pieces finally came together, and the Boyd APT started observing multiple groups of stars. Genet had been following the development of the system rather closely, and was kindly invited to observe the system on its first full night of totally automatic operation in early November, 1983.

In their present configuration, the Fairborn Observatory APT's can each handle about 75 groups of stars per night (225 different stars), with a record of 93 groups set one winter night. These observations are made at the rate of about one group per 7 minutes, with 33 10-second observations being made per group. Thus about 330 of every 420 seconds (7 x 60), or 67% of the time within a group is spent on the actual measurements, and only 33% of the time is spent on moving between stars, acquiring stars, centering stars, and changing filters. Moving between groups typically takes between one-half and two minutes. While the efficiency of the systems could be improved even further, such improvements would be marginal.

G. THE FUTURE OF APT'S

Now that the project to develop ground-based microcomputer controlled APT's that can handle multiple groups of stars has been successfully completed, it is time to consider what the future might hold for such devices. Not only do astronomical applications need to be considered, but both hardware and software adaptations to make widespread use possible are appropriate topics.

APT's of the type developed by Boyd are adept at observing both short period variable stars, where the same group is tracked for many hours, and at observing longer period variable stars, where many different groups are observed per night. Using APT's to observe short period variables is similar to the existing non-automatic observations, and similar objects can be expected to be observed. These include such types of stars as eclipsing binaries.

The ability to observe a large number of stars in a single night with a number of small APT's in the hands of advanced amateurs and colleges is new. How best to utilize this new resource is worthy of thought by professional astronomers.

The use of APT's need not, of course, be restricted to UBV photometry. Low cost photometers (e.g., Optec) already exist that also give coverage in the R and I bands, and development of an Optec photometer that will also cover the J, H, and K bands (out to 2.4 microns) is under way. Near-IR photometry is particularly useful on late-type objects. While very bright in the I through K bands, these objects can be very dim in the V band, and therefore are difficult to see with the human eye. In non-automated systems, being able to see the object so that it can be properly identified and centered is important. However, with a near-IR APT system, acquisition and centering of the objects can be done in any band sensed by the photometer, and if the objects are brightest in K-band, then this band can be used. That objects might be completely unobservable in the telescope with a human eye is of no consequence to an APT. Similarly, as the daytime sky is reasonably dark at K-band, there is no reason that an APT could not make K-band observations of brighter variable stars day and night--any time the weather is clear.

Having described the rationale for APT's and provided an overview of their operation, we turn our attention in the next chapter to the details of the software which enables APT operation to be fully automatic.

A. INTRODUCTION

In most real-time control systems, the easy part is the hardware. The hard part is the software. In designing a system and making the various tradeoffs between hardware and software, no effort should be spared to make things as easy to program as possible. The most dangerous words of all are "Oh, we can take care of that in software". This is particularly true if one is using software to correct a poor mount or a drive with backlash or excessive periodic error. The software to drive an APT could be written in almost any language, and could be made to work on almost any computer, but if consideration is given at the start of the project as to the exact requirements and objectives, a lot of backtracking can be prevented. Sometimes a few external chips can eliminate a large portion of the software, saving large amounts of programming time.

In the case of the Fairborn Observatory Automatic Photoelectric Telescopes (APT's) the software task was greatly eased by (1) using an equatorial mount, (2) using a single motor per axis without clutches, gear shifts, tangent arms, etc., (3) a tight drive system without backlash, (4) not using shaft angle encoders or any other form of position feedback, (5) supplying the stellar drive rate via hardware, not software, (6) programming in a high level language (with only one exception), (7) using an "assembly" of independent program modules, and (8) having a control program that is "serial", i.e., it has no interrupts. Without these simplifications, the job would have been much greater, although the current system did require considerable work.

Even under the best of circumstances the software development will not be an easy task. A decision must be made early in the system design as to how much data reduction will be done in real-time, and exactly how automated the system will be. It is also important to define the file structures for the input and output files and internal data structures before coding begins. The software task for an APT is sufficiently immense that thought needs to be given to breaking it into separate programs. There are undoubtedly numerous ways that this could be accomplished.

One approach to APT software that works is covered in some detail in this chapter. Before the various algorithms used to control the APT are discussed in detail, it would be appropriate to first consider, in a summary fashion, a few of these.

Slewing can be handled by making each move in two segments, the first a diagonal move with both motors running for an equal number of steps; the second move with one motor running for the remaining number of steps. This is shown diagrammatically in Figure 13-1.

Searching can be handled by a square spiral pattern, in which the size of each side is incremented by a constant on opposite corners. The constant should produce a motion about half the size of the diaphragm. This insures sufficient overlap on each loop so that a star will not be missed in the presence of vibration or small

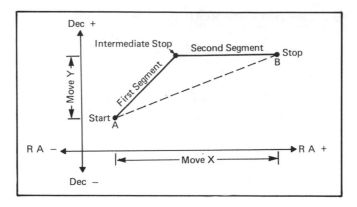

13-1 The Automatic Photoelectric Telescope MOVE Command

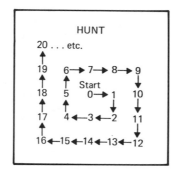

13-2 The Automatic Photoelectric Telescope HUNT Command

errors in the drive. The pattern is shown in Figure 13-2. At each numbered position a reading would be taken for about 0.2 seconds to determine if a star is present. In the Fairborn Observatory systems, the star is always found within twelve loops, but provision must be made to exit after a reasonable search if the star isn't found. Clouds cause that!

Centering is no more difficult. The most obvious method is to make a cross-shaped pattern to find the edges of the diaphragm. This method is slow and vibration of the telescope can cause false centering. A much simpler way is to move the telescope to four corners, each just inside the edge of the diaphragm by a small amount when moving from a properly centered star. This is shown in Figure 13-3. A reading is taken at each position (1-4) and the following logic applied. If all spots indicate a star is present, the star is centered. If one spot is outside the diaphragm, make a small move toward that direction in both axes. If two adjacent spots are outside of the diaphragm (e.g., 1 and 2), then move on one axis a small amount. If three spots are outside the diaphragm, move away from the one which is in the diaphragm. In each of these cases, repeat the process until all four corners show the star inside the diaphragm. If an invalid combination occurs, such as showing spots 2 and 4 out with 1 and 3 in, the process should be repeated. Scintillation can cause this.

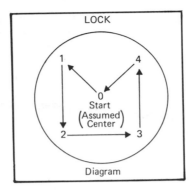

13-3 The Automatic Photoelectric Telescope LOCK Command

When making the scientific measurements on the centered stars, it is suggested that a fixed, short integration time be used on all readings (perhaps 0.1 seconds) and then as many of them be made as required for the actual measurement integration times. A number of short integrations will allow a sum-of-squares computation to be made while making the readings (if desired) for a statistical evaluation of the quality of each reading. If the reading proves to be poor statistically, it may be repeated, reducing the errors caused by scintillation. The other advantage is that the hardware counter need only count the maximum number of pulses from the PMT or V/F converter in the shorter period. This may simplify the hardware.

When searching or centering, it is important to select a decision threshold (star present or not) which will be reliable. That threshold must be greater than the sky readings. It has been found that the most reliable threshold will be slightly less than half the difference between the expected values of the star and sky readings. Scintillation produces greater variation in the value of the star reading than in the sky reading. A threshold must be selected which will not lock onto a fainter nearby star, or miss the desired star entirely. This problem is the major factor limiting the faintest star which may be measured automatically. In practice, a star which gives only four times the sky reading can be easily acquired. Usually, searching and centering should be done with a filter in place so that the threshold may be set as a function of some known magnitude (such as "V" in the UBV system).

Storage of data files can be on any medium, but a floppy disk might be most convenient. Manual entry of data from printed output is much more error prone than direct machine-to-machine interchange. The only files which are likely to require exchange are the final reduced data, and possibly the input program.

There are many ways in which an APT could be programmed. In any program, some of the portions will be hardware dependent, but in terms of functions to be performed, there is likely to be considerable similarity between programs. In discussing microcomputer APT's, going over each line of code in some specific program would only be useful to someone about to implement the same hardware and software. The approach taken is to discuss the control program in somewhat general terms in this chapter, and then give the specific algorithms used in Appendix K and the software listings in Appendix L.

The language used in the control programs is Basic09, which runs under the OS9 operating system. Both the language and operating system are by Microware. This is available for 6809-based microcomputers, but not for other microprocessors. With Basic09, various portions of the total system can be written and debugged as separate programs. When other programs are needed, the OS9 operating system automatically obtains them from memory or pulls them off disk as required. Some thought needs to be given to what should or should not be put into separate programs. Basically, those functions which must always be performed together synchronously as a group are usually best put together in the same program. Conversely, those functions that are usually not performed together, or might not be performed together in future situations, should probably be in separate programs. An overall supervisory program can be used to tie the subprograms together, although this program is likely to be structured in a manner which depends on the type of observing program being pursued.

The supervisory program being used at the Fairborn Observatory is called "MAIN", and it is a specialized supervisor to the extent that it is intended to do UBV or VRI photometry of medium to long period variable stars. It is not difficult to modify it to do continuous photometry of a short period variable star, as this is a much easier case. Many other modifications for different observing programs are possible. For instance, an asteroid photometry MAIN program could calculate the position of each asteroid at the time of observation and automatically select the nearest suitable comparison star. A long period quasar photometry program would be identical to the long period variable star case except that the objects would be fainter and the telescope larger.

MAIN will first be summarized, then discussed in some detail. Then directly required procedures and files, such as HUNT, which acquires the stars, LOCK, which centers the stars, and MOVE, which moves the telescope, will be discussed. Finally, auxilliary procedures such as SET, which allows the clock to be reset, will be discussed.

B. APT SOFTWARE FUNCTIONS AND SUBPROGRAMS

1. **MAIN** has been divided into ten areas for convenience. Each of these will be mentioned briefly below, and then will be discussed in more detail in the following sections.

a. **BUILD RISE/SET TIME TABLE.** An efficient APT stores all the objects to be viewed year around. On any given night, only some of the objects will be observable at some time during the night. Which ones these are can be determined by calculating the rise and set time of all objects for the night in question.

b. **OPEN OUTPUT FILE.** An output file to store the photometric data gathered during the night is opened. About 14 files from two weeks of observing will fit on one 5-inch diskette. Later, data on a given star is consolidated with data on the same star from other nights and other weeks.

c. **INITIALIZE TELESCOPE.** The telescope and photometer are readied for the night's observations. This includes assuring standard starting positions for the telescope itself, and initializing the filter wheel in the photometer.

d. **DETERMINE WHICH GROUP TO OBSERVE.** By finding which group (variable, check, and comparison stars and sky background) will "set" (leave the photometric observing area of the telescope) first, this group can be observed first.

For longest year around coverage, the westernmost groups need to be observed first in the evening, and the easternmost observed last.

e. SKIP IF NEAR THE MOON. Photometry too near the moon can reduce accuracy, and for PMT-based systems, it can damage the PMT.

f. MOVE TO GROUP/STAR. In this portion of MAIN, the telescope is moved to the group in question.

g. HUNT AND LOCK. Before the stars can be measured, they must be acquired (in a spiral search pattern), and centered in the diaphragm.

h. MAKE PHOTOELECTRIC MEASUREMENTS. This is what it is all about. Measurements are made in the U, B, and V passbands (or in the VRI system). Results are stored in a temporary buffer.

i. STORE MEASUREMENTS. The measurements are placed in the output file. This is only done after all the measurements in a group have been completed, and then it is done all at once. This avoids unnecessary operation of the output disk drive and the recording of incomplete groups.

k. EVALUATE GROUP DATA. When measurements of a group are attempted, there are three possible outcomes: (1) the system will be unable to complete measurements on the group due to clouds or other reasons; (2) the measurements will be completed, but are of unacceptable quality -- perhaps suggesting a remeasure; and (3) the measurements are of an acceptable quality and the next group should be attempted. An evaluation of the group data can distinguish between the latter two cases.

Each of these ten portions of the main control program will now be considered in more detail.

2. Build Rise/Set Time Table. There are two basic approaches to determining what objects an APT should observe during some particular night, and what sequence it should observe them in. One is for some human to choose the objects and their observational order in advance. The other is to let the APT do this itself. The systems at the Fairborn Observatory use the latter approach. From a set of groups for year round observing, the systems automatically determine which are observable on a given night. The order in which they are observed is determined as the observations are made, and is dependent on the outcome of the observations as they progress.

This approach has the advantage of eliminating the nightly human input that would otherwise be required. If there are no equipment failures, the system will continue to observe the program objects until the output disk or tape is filled, or a new program list is given to the APT. This approach also has the advantage of allowing the system to adjust to the circumstances of the night's viewing as the night progresses. The disadvantages of this approach include complications in the software to implement such "smarts", and a selection process less sophisticated than that of some humans (but very consistent!).

The first task in this routine is to build a table of the rise and set times for all the groups in the total observing program. These are not the rise and set times of the groups at the physical horizons in the east or west, but are the times for a predefined area of the sky above the telescope that is deemed to be physically and photometrically observable. In the case of the Fairborn Observatory APT in Arizona, this is defined as all of the sky that is within 45 degrees of the zenith AND that is also within four hours (60 degrees) of the meridian. The 45 degree "cone" assures that high air masses that could reduce photometric accuracy are avoided, and also avoids letting the walls of the observatory (a converted garden shed) block off any of the light. The hour angle restriction is due to the telescope mount--a modified fork. The fork extension limits telescope motion in RA. In Ohio, the APT's are operated essentially as meridian telescopes. All objects are observed within one hour (RA) of the meridian. This increases photometric accuracy (at the expense of some seasonal coverage) and allows narrow "hatches" in the roofs to be used. Not only are these narrow hatches easily motorized, but rolls of very thin mylar can be used to protect the telescopes and equipment from the often cold Ohio climate.

To determine when a group will rise and set in the observing window for that night, an angle is calculated that is the plus or minus angle from the group's RA where the group will rise or set in the observing window. The group's RA is taken as that of the centroid between the variable and comparison stars, which is also usually the sky measurement position. This angle is a function only of the observing window, the Declination of the group, and the latitude of the observatory.

A table is then formed that contains the group name (usually named after the variable star in the group, such as "RS CVN"), the rise and set times, and an indicator of how many times the group has been observed. This indicator is called "STATUS" (for group status) and it is initially set to zero for each new group.

In setting the edges of the observing window, it should be kept in mind that the groups have some size themselves, and that they will move some amount in the time it takes to make the measurements. Typically, one or two degrees might be added to allow for group size, and seven minutes' time (2 degrees in angle) to allow for the time it takes to measure the group. In the case of some fixed portion of the window (such as the physical fork limit mentioned earlier), it might pay to add a small safety margin--perhaps a couple of degrees. This could avoid hitting any of the limit switches.

3. Open Output File. Before any actual measurements are made, a name needs to be determined for the output file, the output file opened and tested to see that it is there, and then closed again until the first data are ready. The opening and closing of data files is unique to particular equipment and operating systems, and will not be discussed further. The determination of the name of a file is of interest, however.

In an automatic system, the name of the output file needs to be determined by the system itself. It helps later data consolidation efforts if the file name is unique to the night of the observations. Then if for some reason observations should be interrupted and the system restarted in the same night, a new file name will not be created, but any further data will be added to the previously opened file. As the system contains a clock/calendar that runs at all times (with battery backup), the modified Julian date (MJD) can be used for a unique name each night. A typical file name is "MJD45641".

4. Initialize Telescope. To operate properly, the telescope pointing angle and the time of day must both be known accurately. As the telescope control system does not include angle encoders of any kind, it must be provided with some other way of determining at least one position each in RA and Dec. If a human operator were allowed to position the telescope initially on a star of known position, then this could be used as the starting point. While easily done, this does not allow fully automatic operation. The approach used in the Fairborn Observatory APT's utilizes the RA and Dec limit switches to place the telescope in a known initial position.

The limit "switches" are actually knife edges placed on the RA and Dec drive wheels that pass through LED/photodiode pairs. A Schmitt trigger is used (three in series) to sharpen the switchpoint. How the telescopes are brought precisely to the start position on the southeast limit switches is discussed later in STARTSCOPE.

5. Determine Which Group to Observe. This portion of MAIN determines what group should be observed next. Once this determination is made, a check is made to see if the group is too near the moon. If not, then the telescope moves to the selected group.

Before considering the complexities of automatically determining which group to observe next, it is worthwhile to consider why the observing order can't be fixed for all time, or at least computed once and for all at the beginning of the evening. In theory, perhaps the order could be determined once and for all, and the starting point on the list just adjusted for the starting date and time of the system. However, this would assume that the time taken to observe each group of stars is known. In fact, this time is not actually known in advance. A gust of wind could throw the telescope off a bit during a long move, and this could result in a longer spiral search time. Also, if the order of observations were fixed, then if observations on one group were not really satisfactory, they could not be repeated, and the system would have to blindly plunge on to the next group so as not to upset the pre-arranged timetable. Finally, the sky simply is not clear all night every night, and the system can get delayed by poor weather, and this alone makes a fixed schedule far from optimum or even workable. By building "smarts" into the control program, it can respond dynamically to the actual situation as it happens.

Before selecting the next group to observe, a parameter known as (the system) DEPTH is always set to zero. DEPTH is an overall property of the system for any given night, and when it is increased, the system will start repeated measures of groups. The number of times any individual group has been measured is its (group) STATUS. Initially, the system DEPTH and the STATUS of each group are both 0. As measurements are made, the STATUS of many groups will change from 0 to 1. If a small cloud were over some group, the attempt to measure it would not be successful, and its STATUS would remain at zero. If there were a large number of groups in the overall observing program, then on a given night, there might not be enough time to observe all the groups that were observable, so the DEPTH of the system would not go from 0 to 1 in this case. On the other hand, if the number of groups in the observing program were small, the system would run through them all quickly. Rather than do nothing for the rest of the night, the system would start repeat measurements. Two points need to be made.

First, the system assumes that at least one group is observable at all times. If this is not the case, then the system will get in a loop where the DEPTH is continuously increased. With a 90 degree window (2 x 45 degrees), four evenly spaced objects would meet this criterion, but more typically there would be a minimum of a dozen groups. This point should be kept in mind when system tests are first initiated.

Second, why is the system DEPTH always reset to zero each time the system goes through the process of deciding what group should be observed next? Would it not be more logical for it to progress from 0 to 1 to 2, etc.? If a group is not successfully observed, then its STATUS will remain at zero. Some temporary problem (such as a cloud) may have gone away, and by always resetting the DEPTH to zero, an attempt will be made to observe this group again before other groups are observed twice.

The reason that at a given system DEPTH the group that will set first is observed next is that this will maximize the seasonal coverage of groups. As soon as it is dark enough, the group that will set first is observed first, and the system works eastward. At the end of the night's observing session, some new group, not previously observed, will just make it past the eastern edge of the observing window for the first time. As this group will not have been observed before, its group STATUS will be zero. It will then take precedence over any group that has been observed at least once.

With a moderate number of program groups (50 to 100), the system will start with groups at the western edge of the window and work its way east. This will take most of the night. When it finally arrives at the eastern edge of the window, it will then slew to the west and start working groups in the west a second time. However, any time a new group comes up in the east, it will stop working the groups in the west, run over and work the new group in the east (for the first time), and then return to working other groups for the second time.

There is nothing sacred about the above logic. It works well for the purpose intended, but many refinements in the logic are possible in this application. For other applications, an entirely different logic may have to be used.

Having tentatively selected the next group to observe, some other checks need to be made before the telescope is moved to the group and observations actually made.

6. Check Moon. Before the telescope actually moves to a group to make observations, a check is made to see that the group is at least an appropriate distance from the moon. To do this, program LUNAR is run. Given any date and time--the modified Julian date including the fractional (time) portion--LUNAR returns the current RA and Dec of the moon. The distance from the moon in RA and Dec can then be calculated simply by taking the absolute differences between the position of the group and that of the moon in RA and Dec.

If the group is closer than 10 degrees to the moon, the STATUS of the group is incremented by one. This gives this group another chance later on (the moon moves about 15 degrees in 24 hours). After incrementing this group's STATUS, the DEPTH is reset to zero, and a fresh start is made to find the next group to observe.

If the group is not too near the moon, then--at long last--it is almost time to move to a group. On the way to doing this, the system identifies the group it is about to move to, and displays its coordinates.

7. Move to Group. To prepare for the first set of measurements, not only does the telescope have to move (using program TRAVEL) to the area of the next group, but any flip mirrors, and filter and diaphragm wheels must all be properly positioned. Also, some rough idea of sky brightness is needed before a search is even started for the first star so that an appropriate brightness threshold can be set. If for some reason the sky is abnormally bright (dangerously so), then shutdown would be appropriate. TRAVEL computes the number of steps to make along each axis, and the length of each of the two submoves which comprise the total move. The move is executed, which should bring the telescope close to the navigation star.

The telescope is now in the area of the group, and it has been readied for making measurements. However, as the first star has not yet been centered, the telescope does not know exactly where it is. Also, it does not have a good idea of how bright the sky is, and this information is needed to make the search for the first star. With a very uncertain position, a sky brightness measurement could very well contain a star, certainly fouling everything up. This ambiguous situation is resolved by using procedure THRESH to make five background measurements in a "square with center" pattern. The lowest of the five readings is then used as one of the inputs to determining the threshold level for acquiring the first star. Note that this sky measurement is not a data point. This first measurement is only used to set the threshold for the HUNT routine.

Finally, it might be noted that after a long move between groups (which can be clear across the sky), the accuracy of the system on arrival will not be as great as it will be within a group once the first star has been acquired and centered. To account for this difference, the acquisition search (in the HUNT routine) for the first star in a group is widened by allowing the square spiral pattern search to make more "loops", and thus cover a wider patch of the sky. The number of allowable loops (variable LOOPS) before the system decides it can not find the first star of a group is set to 25. Within a group, it is reduced to 5. Also, the first star is the check star, and it is purposely chosen to be in an easily acquired brightness range, and to be free of nearby stars of similar brightness that might confuse the system. Thus the check star is also the "navigation" star.

Now the system is ready to acquire, center, and measure the first star. It will then move on to measure the other stars and the sky, until all the necessary measurements (in all three colors) have been made on the group.

MEASUREMENT SEQUENCE	U	B	V
0. Check Star			
1. Sky			
2. Comp Star			
3. Variable Star			
4. Comp Star			
5. Variable Star			
6. Comp Star			
7. Variable Star			
8. Comp Star			
9. Sky			
10. Check Star			

13-4 The APT Measurement Sequence

8. Hunt and Lock. The hunt and lock procedure described in this section, and the UBV (or VRI) measurements described in the next section are usually performed 11 times in each group. The order of measurements is given in Figure 13-4.

As is customary in photometry, a "balanced" measurement sequence is used, so the check star is measured first and last. The sky is then measured at a position halfway between the variable and comparison stars. If this position is too near a known background star, the "group coordinates" (this halfway position) are shifted slightly to avoid this problem. Traditionally, the sky measurements are made immediately following each star measurement and next to the star, but as this system never works at high air masses (low elevations) or in close proximity to the moon, and the variable and comparison stars are always within a degree or two of each other, the two sky measurements (1 and 9) at the group center are more than adequate. Also, as is customary, each variable star measurement is "bracketed" by comparison star measurements. Thus for the three variable star measurements made, four comparison star measurements are required.

The RA and Dec of the appropriate star or sky are obtained (in epoch 2000 coordinates), and precessed to current coordinates with procedure PRECESS. Procedure TRAVEL is then used to move the telescope to the precessed coordinates (always a short move because it is within the group). If the sky is being measured, then the HUNT and LOCK routines are not needed, and the UBV measurements are made.

Now the system is ready to start the square spiral search for the star. If the star is not found within the allowable number of loops, the HUNT program automatically returns the telescope to the center of the spiral search pattern, and the STATUS of this group is incremented by one. A system level parameter called GROUPABORT is set to TRUE, and another system level parameter called ABORT is incremented by one (ABORT is initially set to zero at system startup). A check is then made to see if ABORT is 2. If it is, then STOPSCOPE is called immediately, and the system is shut down. If ABORT does not equal 2, it must then equal 1, and the systems picks a new group. If this new group, or any subsequent group aborts, then ABORT is 2, and the system will shut down. In short, one abort will not cause a shutdown, but two will.

If the star is found, then the star is centered with procedure LOCK. LOCK is an iterative procedure that makes only four measurements per iteration. The measurements are made at the four corners of a square that fits within the diagram. At each corner, a determination is made whether or not a star is there, based on whether it exceeds a threshold or not. Thus the result of each corner measurement is a true or false indicator of whether or not a star was in that corner. There are 16 possible combinations of these indicators. Of these 16, there are 14 combinations from which the logic can determine the direction the telescope should take to center the star. Two of the 16 combinations are not logical (with a single star in the field), and call for a remeasurement. One example of a proper combination would be one in which the two measurements on the right of the square showed stars, and the two on the left did not. This would suggest that moving the telescope to the right will move the star closer to the center of the diaphragm. An example of an improper combination is where one pair of diagonal corners contains stars, while the other pair does not. When the star appears in all four corners, it must be reasonably centered, and LOCK is deemed successful. If the lock is unsuccessful, then HUNT is tried again.

In searching for a star it is necessary to set a threshold such that counts below the threshold will indicate the star is not in the photometer diaphragm, while counts above the threshold indicate it is. The threshold is set by MAIN by first measuring the sky background (see the description of THRESH later in this chapter), then computing an equation to predict the expected count. The threshold is then appropriately set between the sky background and the expected count.

In the second generation software used in Ohio, the integration time used in the spiral search pattern is 0.1 seconds during the first search in a group for the check star. This star is always chosen to be reasonably bright, and is sought after a long move from another group when position uncertainty is highest. Once the check star is found, then searches within a group are made with one second integration periods. This allows much fainter comparison and variable stars to be found than would otherwise be the case. As the pointing accuracy of the moves is high within a group, there is little searching, and the longer integration times do not hurt the total system efficiency.

Assuming a successful lock, the system--at last!--is ready to make the actual measurements.

9. Make Photoelectric Measurements. The measurements are made by sequencing the filters and making timed integrations of 10 seconds each. These counts, along with the time and zenith angle, are placed in an output buffer as each star is measured. For the middle variable star measurement, the heliocentric correction for the group is calculated.

10. Store Measurements. Once the measurements of a group have been successfully completed, it is necessary to store them on the output file. On the way to locating the end of the output file, the name of each group observed so far is displayed on the monitor. Once the end of file is reached, the output data are transferred from the output buffer to the output disk, and the file is closed. Finally, the system DEPTH is reset to zero so that when the system starts to select a new group to observe, those observed the least (or not at all) will be given priority. However, before a group is chosen, an evaluation is made of the quality of the present group's data, and if it is not acceptable, it is measured again.

11. Evaluate Group Data. There are 11 x 3, or 33 10-second integrations in a group. There are 12 different "types" of measurements, namely U, B, and V measurements for the check, variable, and comparison stars, and the sky. Two types have two measurements in each color (check star and the sky), one type has three measurements in each color (the variable star), and the last type has four measurements per color (comparison star). A statistical test is made of all nine star ratios (three colors on three types of stars). No test is made on the sky ratios, although they are reported on the monitor for the interest of any human who might happen to be watching. If the test is passed, then the STATUS of the group is incremented by one and the selection of a new group to observe is initiated. If the test fails, then the group will be reobserved until it passes the test, the group sets in the west, morning arrives, or there is a "system abort". A system abort condition occurs when not all the stars in a group can be observed. If the test eventually passes or the group sets in the western edge of the observing window, then the system selects the next group to observe.

It may seem futile to keep repeating observations as a group moves towards the west. Experience suggests, however, that any continuing problem not severe enough to cause an abort (i.e., the system is successfully finding all the stars in the group), yet serious enough to produce observations of poor quality could be a transient problem. Observations on the group might as well continue in hopes that it will clear up during the evening. This has happened a number of times.

The evaluation test is used in the system in Arizona. In Ohio, which has "quasi-meridian" telescopes, the evaluation is made, but not acted on. We prefer to keep on observing until all the stars in our "slot" centered on the meridian are observed, then just repeat the observations a second time if new stars have not popped into the slot from the east.

On a decent night, the system will not abort, nor will it get hung up on a group that fails repeatedly to pass the quality test. Instead, it will make it all the way through the program list of stars observable on that night. Dawn will find it working its way through the list for a second time, picking off the high-priority unobserved groups in the east as soon as they enter the observing window. With dawn, the computed angle is greater than the twilight angle, and the system executes procedure SHUTDOWN.

C. APT SUPPORTING PROCEDURES AND FILES

In the previous section, the overall control program (MAIN) for the APT's at the Fairborn Observatory was described. It uses 21 different supporting procedures to carry out its various tasks. Each supporting procedure is discussed briefly below (in alphabetical order).

1. **COEFFICIENTS** is a file that contains the assumed "average" extinction coefficients. These are used to set the THRESHOLD value in MAIN, and in data reduction in TRANSFORM.

2. **DIGITAL** is a short procedure which reads the manual control paddle. In automatic operation with MAIN, the paddle input ports can be checked occasionally for human override requests.

3. **HELIO** provides the correction from geocentric to heliocentric times. It is customary in variable star research to record observational times as if they were viewed from the Sun.

4. **HUNT** executes a square spiral search routine to locate a star. The input parameters to the procedure are RADIUS, LOOPS, THRESHOLD, and DURATION. RADIUS is the radius of the diaphragm in steps. By making this a variable, different diaphragm sizes can be used. Although for any given group, the current system uses the same size diaphragm for both acquisition and observation, it is conceivable that some systems might use a larger diaphragm for acquisition, and a smaller one for the actual observations. LOOPS is the maximum number of square spiral loops that will be taken before the system aborts. Up to 25 loops are used to acquire the first star in the group, then this is reduced to 5 for all subsequent stars in the group. THRESHOLD is the number of counts above which it will be concluded that the correct star is in the diaphragm, and below which it will be concluded that there is no star (or a fainter "wrong" star) in the diaphragm. A more complex scheme could be used, such as requiring the star to be within a window defined by both maximum and minimum values, or measuring it in several colors and evaluating the color index, etc. However, at least with smaller telescopes which are, by necessity, observing brighter stars, this simple scheme works very well. DURATION sets the number of 0.1-second integrations used at each stop in the square spiral search pattern. Typically, DURATION is 1 during the first search for the bright navigation star, and a larger value for fainter "within group" stars.

HUNT makes a series of short integrations. After making the first one, it moves one radius to the west and makes another integration. It then moves one radius to the south and makes the third integration. It then goes to the east two radii. Then two radii to the north, completing the first loop. Every second "side" of a loop includes one more move, and the repetitive nature of the moves makes for simple logic.

After each integration, the COUNT from the photometer is compared with the THRESHOLD. If the COUNT is larger than the THRESHOLD, then the telescope stops at this position, and THRESHOLD is set equal to COUNT. If THRESHOLD is equal to or less than COUNT, then the system moves to the next position in the spiral. If the system makes the maximum number of loops, it will stop northeast of the original position (at the end of the last loop), and on deciding an error condition exists, it will return the telescope in a southwest direction to its starting point in the center of the spiral. In this case, the THRESHOLD value will be unchanged from its original value, signalling that the search was not successful.

5. LOCK was described earlier in some detail, and is an iterative procedure used to center a star in the photometer's diaphragm.

6. LUNAR. To avoid doing photometry too close to the moon, program MAIN calls LUNAR to check if a group is near the moon before commencing measurements. The modified Julian date (including the fractional "time" portion) is provided as an input, and the position of the moon is returned.

There are some very accurate algorithms for determining the position of the moon in the "Almanac for Computers", but these are total overkill for this application. Instead, the approach suggested by Burgess (1982) was used as the basis for the algorithm. It is not very accurate, but is usually within a degree, which is close enough for an APT trying to stay at least 10 degrees away from the moon.

7. MEAS sets the integration time for the photon counting system, and returns the final count. The parameter TIME sets the integration time in tenths of a second. If TIME = 100, then the integration time is 10.0 seconds. The final count is normalized to counts per tenth of a second, and is returned as COUNT. As the specifics are hardware dependent, no further details will be given here.

8. MOVE receives the input parameters MOVEX and MOVEY, and moves the telescope these amounts. MOVE has no outputs. MOVEX and MOVEY are signed real numbers. An integer could not be used, as some moves are greater than the limit of 32767 imposed by Basic09 on signed integers. The sign convention is +X is west, and +Y is north.

MOVE breaks the total move into one or two separate moves. The one move case is where either MOVEX or MOVEY equals zero, but not both. In this case, the telescope would make a straight move in either RA (alone) or Dec (alone). The two move case is the general case where motion is required in both RA and Dec. MOVE takes the total move (in the general two move case) and breaks it down into two separate moves. The first move requires motion in both RA and Dec of an exactly equal number of steps. The length of this move is set by the shortest of the RA or Dec motions, and with both motors operating at once, this motion is always in one of the four "45 degree" directions. As full slewing velocity is much higher than the steppers can obtain from a dead start, ramping up and then ramping back down is required. The second part of the total move takes place in either RA or Dec, but not both, and it completes the "unexecuted" steps remaining in the longer of the two-axis moves.

9. PRECESS takes year 2000.0 coordinates in radians and precesses them to the current Julian date. The simple algorithm that is used does not account for changes in the precession rate.

10. PTCLK reads the clock/calendar chip on the PT69 computer used in the Fairborn Observatory systems. It is hardware dependent.

11. RAMP is a short assembly language routine that generates the pulses for the stepper electronics. RAMP starts at an initial slow step rate, ramps up to a maximum step rate (slew), and then ramps down to a complete stop.

12. SHOCO. To SHOw COordinates, procedure SHOCO takes RA and Dec as inputs (in radians, epoch 2000.0), and converts them to degrees or hours, minutes, and seconds, and displays them in the format "XX HH MM SS". SHOCO is used by JOY4 and MAIN.

13. SOLAR, given the modified Julian date, including the fractional time, calculates the RA and Dec of the sun and returns them as variables RA and DEC. The algorithm is based on material in the U.S. Naval Observatory "Almanac for Computers".

14. STARFILE is a file that contains data on all the groups in a given observing program. Data includes the number and name of each group, the identification number (HD) of each star in the group, as well as its coordinates and magnitude. Additional information, such as the sky measurement position, is given.

15. STARTSCOPE. As its name implies, this procedure is used to initialize and start up the telescope. It is a procedure called by the main control program.

First, any diaphragm and filter wheels are placed in their home positions. Next, the telescope is moved exactly to its home position. When it was previously shut down, it was moved to its home position, but rather than trust this (it could have been disturbed), the telescope is backed off the home position 500 steps in both RA and Dec, and then moved back to home. It is then moved off a tiny amount (20 steps) and moved back to home. This latter small movement makes sure that the telescope is travelling very slowly when it hits the limit switches. The limit switches are not physical switches, but are small IR transceivers. A "knife edge" on each main drive disk interrupts the light beam for very precise positioning.

With the telescope in its exact home position, all that is left to do is to read the time, and start the telescope at this exact time. The telescope immediately starts moving to the west, and is now in celestial coordinates.

16. STOPSCOPE. When the MAIN program determines that morning has arrived, or that clouds have moved in and two groups have been missed (ABORT = 2), it calls STOPSCOPE to shut down the telescope. This is done by moving any opaque filter or diaphragm into the light path, and moving any flip mirror to the viewing position (to keep light out of the photometer during the day).

The sidereal rate clock is then turned off (and the telescope stops tracking). A move well beyond the limit switches is commanded, and when the scope runs into the RA and Dec limit switches, it has arrived at the home position. The program then closes the roof and does other shutdown functions as appropriate.

17. SUNANGLE calculates the zenith angle of the sun. First it runs TIME to get the current time, Julian date, and local mean sidereal time. It then uses the Julian date as an input to SOLAR, which returns the RA and Dec of the sun. The RA and Dec of the sun and the local mean sidereal time are then used as inputs to program ZENITH to determine the zenith angle of the sun. This is used to determine when it will become dark in the evening or light in the morning. This allows automatic startup and shutdown, respectively.

18. THRESH is used when the system first moves to a new group   to determine the background count that is used to acquire the first (check) star. Five measurements in a "square with center" pattern are made, and the lowest count is retained as the result.

19. TIME. Given the time and date as inputs (using PTCLK), and knowing the local longitude, procedure TIME determines the modified Julian date (MJUL) and the local mean sidereal time (LMST).

20. TRAVEL determines the total number of steps in X (RA) and Y (Dec) to go from the current position to a new position. To make the actual move, TRAVEL calls MOVE, and MOVE splits the total travel into two discrete moves, then executes the moves. The inputs to TRAVEL are the current position and the desired new positions.

In general, a move from point A to point B is made in two distinct submoves. In the first segment of the move, both RA and Dec steppers are stepped together in exact synchronization (driven from the same pulse source). As each motor can be moved independently in a positive or negative direction, there are four different directions the telescope can go (++, +-, -+, and --). These four directions are at "45 degrees" to the "ordinal" RA and Dec directions. At the end of the first segment of travel, the telescope is ramped down to a complete stop, and then the second segment of travel is initiated. It is also only in one of four directions, but these directions are along the regular RA and Dec directions. In the second segment of travel, only one stepper motor on a single axis is active.

21. ZENITH. Given the time and the position of any object and that of the observatory, this subprogram calculates the object's zenith angle.

D. AUXILLIARY PROCEDURES

Auxiliary procedures are those not required in the actual operation of the system, but which are needed either to enter program objects, bring data out of the system in raw or reduced form, or, initially, to align and check out the system. In some systems, it may be desirable to keep all the auxilliary programs separate.

1. BUILDFILE provides for manual entry of the objects to be observed onto the master file. After building the initial file, objects can be changed, corrected, or added. BUILDFILE allows changes to be made on just those entries in need of change.

A number of conventions are used. The name of a group is usually the name of the variable star, such as R Leo, V411 Cyg, etc. The name of a star, including any variable star, is its HD catalog number, including the letters "HD" as a prefix. Magnitudes are the V magnitudes, and in the case of variable stars, the magnitudes are the magnitudes at minimum light, not maximum, as is usually given in star catalogs.

Coordinates are entered using epoch 2000.0 . The input format is somewhat freeform, as program MANCO is used to digest the inputs and convert them to radians. In general, the system displays epoch 2000.0 coordinates in degrees or hours, and stores epoch 2000.0 coordinates as radians, since most microcomputer trigonometric functions require angles in units of radians. In commanding movements of the telescope, the coordinates are precessed as the last action, assuring that precession is not accidentally done twice. The Sky Catalog 2000.0 (Hirshfeld and Sinnott, 1982) has been found to be particularly useful, as it contains all the stars within range of the telescope in 2000.0 coordinates. Other data provided in this catalog, such as the V magnitudes, is also helpful. Use of epoch

2000 coordinates throughout the system aids an observer who is verifying system performance using the new and popular epoch 2000.0 catalogs and atlases.

2. DATREAD displays the contents of raw data files. It adds in the appropriate header information, and formats the data in a convenient manner.

3. JOY4. Manual control of the telescope is needed for making observations, for initial telescope alignment, for letting the neighborhood kids look at the moon, and for "fuzzballing" (unauthorized human use of an APT during prime dark observing time to look at Messier objects).   A menu provides such options as returning to the home position, preselected stars, and interesting objects; pure manual control; and photometric measurement (manual with computer assistance). Current coordinates are displayed after each motion.

4. MANCO stands for "MANual COordinate entry". It allows entry of the coordinates of program objects (epoch 2000.0 coordinates) in a reasonably "freeform" format.

Procedure JOY4 can call MANCO for coordinate entry during manual operation of the telescope. However, the most frequent use of MANCO is by program BUILDFILE when adding new objects to the observing program or modifying existing ones.

5. SET allows the clock to be set against WWV. The clock should be set (and maintain time) to within about one second of time. Note that a one second time error translates into a 15 arc second pointing error in RA. This error is eliminated as soon as the first star is centered, but an error as large as one minute of time is 15 arc minutes of error in RA, and this would cause the system to miss its first acquisition.

6. SHOWTIM allows the current time to be displayed.

7. TILDARK has no input or output parameters. It repeatedly runs program SUNANGLE until it finds that the angle of the sun is greater than the variable TWILIGHT. When this occurs, it runs procedure MAIN. This allows the system to start observations automatically as soon as it gets dark.

8. TRANSFORM takes the raw data and reduces it to differential magnitudes. Currently this is done off line, but it does not take long to reduce the data from a group, and in the future it may be done either immediately after the data from a group has been taken, or at the end of the evening. Currently, REDUCE accounts for extinction with assumed nightly extinction coefficients (using seasonal averages based on past experience), transforms results to the standard UBV system, and adds in the heliocentric correction. The procedure is similar to that given by Hall and Genet (1982).

Future improvements might include automatic calculations of the nightly extinction coefficients based on the comparison and check star observations. As observations are not made at high air masses, comparison stars are always close by, and comparison stars of similar spectral type are usually chosen, the use of assumed extinction coefficients is actually quite accurate. However, as the data to calculate the extinction coefficients would be at hand, it might be interesting to do so.

This completes the discussion of the Fairborn Observatory APT software. Details on specific APT algorithms are given in Appendix K, while complete program listings are provided in Appendix L. We now turn to the Winer Mobile Observatory design for a transportable telescope.

Chapter 14. WMO PORTABLE TELESCOPE DESIGN CONSIDERATIONS

In the two chapters which comprise this section, the plans for a trailer-mounted portable telescope are described. I (Trueblood) have established a private nonprofit observatory named the Irvin Marvin Winer Memorial Mobile Observatory, Inc., or Winer Mobile Observatory (WMO) for short, after a man whom I knew only a relatively short time, but who left a permanent impression. Irv died prematurely in middle age in 1982, so it was felt a lasting tribute was in order.

A. SYSTEM DESIGN METHODOLOGY

Although the design presented in these chapters may be of interest to those contemplating an instrument of similar capabilities, the important issue is not the information, but the process of design and development. The data presented earlier and in the appendices on hardware will soon be outdated, but the lessons I learned over the past 10 years doing systems design and integration will last a lifetime.

For a control system as strightforward as Tomer's, a formal design approach is not necessary. Although Tomer did spend considerable time designing his system, he spent more time building it. The drive train and the software are simple both in concept and execution. The pointing accuracy, though quite useful, is moderate when compared to that found in professional observatories.

The WMO telescope, on the other hand, will be considerably more complex, due to its stringent performance requirements. The task of developing a complex system can quickly become disorganized if it is approached in a haphazard manner. The price of this disorganization is wasted time and money. Small projects can afford small schedule slips and cost overruns. A project of this size quickly generates unacceptable delays and cost overruns if not managed properly.

The history of developing systems employing the kind of high technology used in computerized telescope control systems is chequered. It is the high-speed electronic digital computer that has encouraged the construction of sophisticated, and therefore complicated, systems. In the roughly 25 years in which large computerized systems have been developed, there are few customers who have been satisfied with both the performance and the cost of the system that was delivered.

What is needed is an approach for converting the user goal into a computerized control system. This process can be represented as a hierarchical structure, as follows:

DESIGN   IMPLEMENTATION

User Goal   Tune System
    System Performance Requirements   Command Generation S/W
    Top-Level System Design   Hardware Interface S/W
        Detailed System Design   User Interface Software (S/W)
            System   Install Hardware

Since the leap from the user goal to the system is too large to manage effectively, intermediate levels have been created, each of which represents a definable and controllable step toward the fully working system. The step from one level to the next level involves a well-defined transformation which requires an increase in detail or a change of focus.

The WMO system has been designed from the top down. The system performance requirements state specifically "what" the system will do, not "how" it does it. They are formulated with the user goal and the system operational environment in mind. The top-level system design is the top-level "how" document, and identifies all hardware and software functions without describing them in complete detail. The overall software organization is described, and major software modules are identified and described. The detailed system design is a set of hardware schematics and software module specifications that give sufficient details to build the system.

The system is being implemented from the bottom-up. In the early stages of development, apparent place and mechanical corrections will be ignored. These corrections are simply equations which can easily be programmed in later. The hard work is always the interfaces--to motors, encoders, counters, and the observer. The best approach to developing the system is "build a little, test a little". The typical computer hobbyist (myself included) is used to sitting down at the computer and writing a fairly complex program all at once in an evening, or a few evenings, without doing any testing of the program while it is being developed. That approach is adequate for scientific computations, but it simply will not work when complex software and hardware are involved, and is doomed to failure.

B. THE WMO OBSERVING PROGRAM

The emphasis of the research programs at the WMO is on solar system astronomy--primarily grazing lunar occultations, minor planet occultations, and, possibly in the future, occultations by comets. As lunar laser ranging becomes more routine, the scientific importance of observing grazing lunar occultations diminishes, although these observations are still useful for detecting previously unmapped features on the lunar limb, and for detecting double stars. Photoelectric observations of total lunar occultations are still useful for determining stellar diameters and binary separations, when diffraction patterns can be obtained. Although both lunar and minor planet programs are being pursued at the WMO, the minor planet occultation program will receive greater emphasis in the future.

Both programs require that observations be made at a narrowly defined place and time. The typical grazing lunar occultation observation must be made inside an area that is a few hundred miles long by about a mile or two wide. Similarly, minor planet occultations must be observed inside an area roughly several hundred miles long by about 50--200 miles wide. In most cases, there is not an observatory of even modest size in the path, and on those few occasions when there are, telescope time usually cannot be granted to do these observations, or weather interferes. Timing is critical, so that one cannot wait for the weather to clear to make the observations. All this points to the need for a transportable telescope system.

Minor planet occultations are predicted by Gordon Taylor at the Royal Greenwich Observatory, by Larry Wasserman and Bob Millis at Lowell Observatory, and by David W. Dunham. The main searches for events are made using the AGK3 and SAO catalogs, so stars down to 9th or 10th magnitude are included. Lowell Observatory conducts searches down to 13th to 15th magnitude. When searches are made for visual observers in the United States, events are sought in which the star is no fainter than 10th magnitude, the minor planet is fainter than the star, the predicted drop in magnitude during the event is at least 0.8, and the minor planet has a diameter of at least 100 km. There are some seven or eight such events predicted each year using current prediction techniques. Of these, only two to four are actually observed, because of bad weather, last minute shifts in the predicted paths away from populated areas, or a host of other reasons. For photoelectric observations, events are sought in which both the star and the minor planet may be as faint as 15th magnitude, either one may be the brighter object, and the predicted drop in magnitude may be as little as 5%. There are some two to four dozen such events predicted each year, but at the present time, the lack of good astrometry data limits the number of these for which observations are actually attempted.

Although both professionals and amateurs make minor planet occultation observations with telescopes under 12 inches in aperture, one of the more popular observing configurations used by professionals at Lowell Observatory and the University of Maryland is a C-14 and a high speed photometer capable of 1 millisecond time resolution (see Schnurr and A'Hearn, 1983). Although this makes a very convenient and portable system capable of obtaining publishable results on a wide range of events, it is capable of observing only about half of those events which are predicted each year. Of these, only about half are observable with the full 1 ms time resolution.

The problem is collecting enough photons for high time resolution photometry. A rough rule of thumb is the 6-6-6 rule: a 6-inch telescope receives 10^6 photons per second from a 6th magnitude star. A 12th magnitude star would yield about 1/250 as much light, or 4,000 photons per second. With a 1 ms integration period, one would observe only four photons per integration period. This is indistinguishable from the combination of the dark count of even a good PMT, and statistical fluctuations in photon arrival times. The C-14 would receive 5.4 times as many photons, or 22 per millisecond. If the event produced a two magnitude drop, this would be detectable, but such a large delta magnitude is not typical of most of those events predicted. The integration period could be increased to observe these other events, but the loss in time resolution translates directly into an equal loss in accuracy of measuring the minor planet diameter, which is one of the key goals of a minor planet occultation observing program. Such information, in conjunction with measured albedos, is useful in determining the composition of the minor planets, and both size and composition information are useful in discriminating among the half dozen or so serious contenders for a theory of the formation of the solar system.

Therefore, by using an aperture considerably larger than 14 inches, one can observe a larger number of events each year. If a 30-inch aperture were used, there would be 25 times as many photons as with a 6-inch aperture, or about 100 photons per millisecond. This would be adequate to observe a 0.5 magnitude change in light from a 12th magnitude star, and would be able to detect a similar change in light from a 15th magnitude star with 1/15 th of a second time resolution. Therefore, practically the entire range of predicted events would be available to a portable telescope of such a large aperture. The occultation is not observed to happen between two 1 ms integration periods. Instead, a standard algorithm is used to fit a curve to the data, and to pick an event epoch time based on standard criteria. However, the greater the time resolution, the better this algorithm works. Another reason for using large optics is to reduce scintillation, which can be a significant component of noise in high speed photometer light curves.

The minor planet occultation data will be gathered using a single-channel high speed photometer capable of 1 ms time resolution. Photon counting electronics and a low dark current PMT will be used. As is typical of photometers used in this application, the diaphragm will be only 15" in diameter, and equipped with a Fabry lens to reduce photon count fluctuations due to short term tracking errors (A'Hearn, 1984).

C. SYSTEM OPERATIONAL ENVIRONMENT

The WMO environment during normal observing conditions will usually be very dark, with temperatures ranging from 0 - 90 degrees Fahrenheit and humidity from 0% to the dew point. The observer should not be required to move around a great deal to use the computer, since he may stumble over or into a piece of equipment and injure himself. The equipment handled by the observer (such as a hand paddle) should be able to tolerate the range of temperatures and humidity the observer himself may encounter. Observing will be done at remote locations, so wind and rugged terrain will be more important factors than at a fixed observatory. Adequate power for the telescope, computer, other electronic equipment, and heating (or air conditioning) must be provided.

The time required to set up the system should be minimal, so that useful work can be done as soon as possible after applying power to the telescope drive and the computer. The time required for the observer to enter a command and for the computer to execute it should not be appreciably longer than the time required for the observer to perform the same function manually. This is not necessarily a requirement for most computerized systems, such as the Fairborn Observatory APT, where the goal of unattended operation permits, and design tradeoffs may even force some tasks to take longer under computer control than when performed manually. However, in the WMO system, setup time is an extremely precious resource, so the sole justification for automating any function in this system is to save time. Physical and electronic setup, and alignment of the telescope for accurate tracking should be as rapid as possible, to allow last minute changes of observing site to avoid clouds, and to permit travel mishaps, such as flat tires or getting lost, to be absorbed into the schedule without missing the event.

The observer using a small telescope typically would be alone, so features of the system used for centering objects in the field of view should be available to a user at the eyepiece of the telescope, as well as at the computer console. This is to permit alignment of the telescope with the celestial coordinates by sighting stars of known position.

The amount of direct interaction with the computer, for example, through a CRT terminal with keyboard, should be minimized, and should use keywords and prompting phrases that are in common use among astronomers. At least half of the telescope time will be set aside for guest observers, who typically will not be computer specialists.

D. SYSTEM PERFORMANCE REQUIREMENTS

System performance requirements should be set keeping in mind not only the observing program and the operational environment, but also the user budget. This requires a good knowledge of the available technology and, specifically, a knowledge of how useful the technology is in the operational environment.

The functional requirements and available budget for any project should be the driving forces behind all tradeoffs and design decisions that are made. The requirements should be based on a knowledge of the operational environment and the available technology, but the first version of the functional requirements is written without benefit of a top-level design. As the project proceeds in its development, the functional requirements should be reviewed periodically and updated, if necessary. The following performance requirements for the WMO telescope are listed as an example of how to begin the project:

1. Portability: The WMO telescope will be used primarily for observations that require the telescope to be portable. The vehicle used to transport the telescope should be inexpensive, readily accessible by the observatory staff, and, preferably, be capable of being used for activities other than merely transporting the telescope. In addition, a source of power must be provided for operation in remote locations where local power is not conveniently available.

2. Setup Time: The telescope and control system should require no more than one hour for one person working alone to perform all setup and initialization functions, measured from the moment the telescope arrives at the observing site until the telescope is tracking the target star with the accuracy specified in Requirement 6. These setup and initialization functions include activating the remote power source, locating the telescope on solid ground, levelling the trailer and raising the tires off the ground, physical setup of the telescope tube or truss and drive, aligning the telescope for tracking, readying the photometer, initializing the computer control system, and acquiring the target star. Further reductions in setup time would add greater flexibility to handle last minute site changes and in absorbing travel schedule impacts.

3. Optics: The telescope optics shall be of sufficient aperture as to be capable of recording occultations of magnitude 12 stars by minor planets one-half magnitude fainter. This is to make available more events than are currently attempted by existing portable systems.

4. Telescope Pointing Accuracy: The telescope pointing accuracy should be at least 30" if fewer than four star sightings were made after arriving at the observing site, and better than 15" if at least one dozen star sightings were made after arriving at the observing site. This requirement should be met in all areas of the sky within 30 degrees of the event in RA and 10 degrees in Dec. This requirement reflects the fact that the telescope location changes frequently, and that unforseen last minute events (flat tires, traffic tickets, clouds, last minute astrometry indicating a path shift) may limit the time available for star sightings after arriving at the site. It also reflects the fact that if the target star is very faint, it must still be the correct star, and it can be located almost anywhere in the sky, including very near the horizon. The reason for relaxing the requirement from 15" to 30" is that if the telescope arrives at the observing site with too little time remaining before the event to achieve 15" pointing accuracy, observing conditions may permit the use of a larger photometer diaphragm.

5. Telescope Pointing Time: The slew rate should be as fast as is practical, with a goal of no more than one minute from any part of the sky to any other part of the sky, and no more than 30 seconds if a slew of less than 90 degrees is made. This includes both the time spent at the maximum slew rate and the required ramp times. These figures are somewhat arbitrary, and might be modified as field experience is accumulated, but are based on predicted slew rates of the microstepped drive described in the following chapter. Rapid slewing is required to sight enough stars during the setup period to align the telescope for accurate pointing and tracking.

6. Long Term Tracking Accuracy: After reaching the commanded position, the telescope should track the desired object with a position error accumulation rate of less than 10" per hour if fewer than four star sightings were made after arriving at the observing site, and 5" per hour if at least one dozen star sightings were made after arriving at the observing site. Often one can find the star soon after setting up the telescope, but after the observer has fussed for an hour with a photometer that is indignant about being asked to function normally in 10°F weather in the middle of a corn field, the star has given up and wandered out of the field of view (the photometer diaphragm is 15" in diameter). If all goes well, it could be as long as two or three hours from the time the target is acquired until the main event, which lasts from a few seconds up to about 30 seconds. Observations are often made for 5-20 minutes before and after the event to search for secondary occultations caused by possible satellites of minor planets. Ideally, the observer should have the luxury of centering the object in the diaphragm, then feeling free to devote his attentions to monitoring the data, time receiver, generator voltage, and other field equipment without being required to check the eyepiece every few minutes to determine whether the object is still in the diaphragm. This means that accurate tracking is at least as important as accurate pointing.

7. Short Term Tracking Accuracy: The drive and control system should move the telescope smoothly enough to track the target star without fluctuations greater than 2". Although telescopes at choice observing sites often enjoy 0".3 seeing, and astronomers want short term tracking on most telescopes to be this accurate for photography, the WMO telescope will rarely enjoy such excellent seeing conditions, and will not be used for photography. A Fabry lens in the photometer will prevent any 2" short term errors from affecting the photon counts significantly.

8. Data Input and Control Device: The keyboard or other data input and control device, and other exposed equipment should be able to withstand temperatures ranging from -10 to +90 degrees Fahrenheit, humidity ranging from 0% to the dew point, and windblown dust or sand. This device will be lit with a red lamp or otherwise be easy to use in the dark without imparing the observer's dark adaptation, and will be conveniently positioned within arm's reach of the eyepiece. Manual slew commands will be reviewed by the control system software before they are executed to prevent damage to the telescope and drive, and to give optimum pointing and setting performance.

9. Computer Environment: The control system equipment (the computer) should meet the temperature and humidity requirements described in Requirement 8, or it should be placed in a protective enclosure that can keep the equipment within its operating ranges throughout the specified ranges of temperature and humidity. Disk drives and other equipment sensitive to foreign particles should be protected against wind-driven dust and sand.

10. Commands: Command inputs should be in the form of menu selections or plain language commands whose meanings are known unambiguously throughout the general astronomical research community.

11. Extraneous Light Control: The optical assembly will be constructed to minimize the amount of stray light that reaches the focal plane. This is particularly important in a portable telescope, since the typical occultation observing site is right beside a country road with headlights shining directly on the telescope tube during the most critical part of the observation.

As the system design evolves, these requirements will be updated and additional requirements will be added. This is a normal part of the iterative design and build process.

E. OVERALL SYSTEM DESIGN AND EVOLUTION

After the performance requirements have been formulated, major design decisions need to be made to arrive at a top-level system design. These decisions include the following for the WMO telescope:

1. method of transport
2. telescope mount type
3. optical system
4. drive design approach
5. control system approach

Each of these topics is addressed below.

1. Method of Transport. There are three basic telescope transport options: (1) Make the telescope tear down into pieces which can be packed into a van or small truck. (2) Mount the telescope permanently inside a van or small truck, and have a roll-off roof section or side panels which fold down to reveal the telescope. (3) Mount the telescope permanently on a trailer and tow the trailer to the observing site.

The stringent requirement for rapid setup (2) eliminates the first option. It takes two people almost a half-hour to set up a C-14, which is a relatively light weight and compact 14-inch telescope. To meet the optics requirement (3), a 30-inch telescope is required. If the telescope had to be set up and torn down at each site, the time required to do this would severely limit the flexibility of the system to respond to changes in the weather or delays in transporting the telescope to the observing site.

Of the remaining two options, a truck with a 30-inch telescope mounted permanently inside it would need to be rather large and expensive. Jack screws to lift the vehicle off its tires to prevent vibrations would have to be of the type used on backhoes and firetrucks (big and expensive). Room needed for the control computer and the instruments would further increase the size and expense of the truck needed. Such a vehicle would not fit in my home garage, and the residential covenants prohibit parking large commercial vehicles on the street. If the truck is stored in a commercial garage, it would add to both the operating expenses and the time required to get the telescope on the road. All this rules out mounting the telescope inside a truck, which means the telescope will be mounted on a trailer.

14-1 "Phoenix III" Telescope Trailer--Front View

14-2 "Phoenix III" Telescope Trailer--Side View

2. Telescope Mount Type. The fact that a telescope trailer must be towed to the observing site plays a larger role in the selection of the telescope mount than many people would imagine. While a graduate student at Wesleyan University in 1971-72, I helped Andy Tomer build a trailer mounted telescope. It was similar to a "Porter's Folly" equatorial mount, and consisted of a polar axis disk (three manhole covers welded together and machined true) which separated two structural "cones", as shown in Figures 14-1 and 14-2. The apex of the cone below ("south" of) the disk fit into a thrust bearing at the front of the trailer. Automobile wheel bearings mounted inside assemblies attached to the two triangular plates bore the weight of the telescope at two points on the disk. Therefore, three points (the two disk bearings and the thrust bearing) defined the right ascension axis. The second cone was located above ("north" of) the polar disk. Its apex was the declination bearing housing. Andy used a 12-inch Cassegrain on this mount, but it was sturdy enough to hold a 24-inch telescope. The entire assembly weighed slightly over one ton.

We towed this trailer (nicknamed "Phoenix III" because it contained parts from two previous telescopes) to Nova Scotia behind a rental truck for the July 10, 1972 solar eclipse. Although it had two axles (four tires) to bear the weight, it bounced and slid its way there and back. It did not corner or maneuver well at highway speeds. During this trip, we suffered a blown tire, and later we were forced to stop to weld first a fender and then an axle spring back onto the trailer. After these experiences, we concluded that: (1) the trailer weighed far too much, (2) the center of gravity was too high (which caused the cornering and highway maneuvering problems), and (3) polar alignment was too difficult, as it required considerable effort on the part of two strong people for several minutes to swing all this mass to within several degrees of the pole--finer alignment proved impossible.

Tomer solved these problems in his next version (Phoenix IV) by making the mount and telescope tube fold down into a low-slung box that formed the basis of his trailer. Setup consists of lifting the telescope out of its box using a gearhead motor and a steel cable wrapped around a series of pulleys. It takes about 15 minutes for the motor to lift the mount up until the polar axis angle equals the latitude. The whole equatorial mount is then rotated on an azimuth bearing to complete polar alignment. The new telescope is shown set up and aligned on the pole in Figure 14-3, and nearly ready for towing in Figure 14-4. The counterweight arm shown in Figure 14-4 is removed, and a canvas cover is fastened in place when towing the trailer. Tomer's solution embodied in Phoenix IV works well for his 12-inch Cassegrain, but there are problems when a 30-inch or larger telescope is used. The geometry involved, and state and federal limits on trailer width, are the constraining factors. Furthermore, the motor and pulley system required to lift the larger telescope and its mount would consume a great deal of power, and would not meet the setup time requirement.

Referring to Figure 14-2 again, if the south polar cone were removed from the mount and the disk flipped forward to a horizontal position, the remainder of the mount could be lowered almost two feet. This would remove a fair amount of weight from the mount, and lower the center of gravity considerably. The result would be an altitude-azimuth (alt-az) telescope mount.

Experience in towing both of Tomer's telescope trailers emphasizes the difference in towing characteristics that result from the lighter weight and lower center of gravity of the Phoenix IV trailer. The ease of towing, and the elimination of the need for accurate polar alignment in a short period of time both point toward an alt-az mount as the best way of meeting the portability and rapid setup requirements.

Although an alt-az mount would solve many of problems discussed so far, it introduces new problems to be solved. The most difficult are (1) two axes must be driven at varying rates, instead of driving only one axis at a relatively constant rate, and (2) the image field of an alt-az mount rotates at the rate of change of the

14-3 "Phoenix IV" Telescope Trailer--Polar Aligned

14-4 "Phoenix IV" Telescope Trailer--Stowed for Towing

parallactic angle. These topics were discussed in Chapter 7, in which correction equations were given. A third problem for which there is no correction is that celestial objects cannot be tracked through a cone centered on the local zenith. A powerful microcomputer and the use of the right kind of motors and shaft encoders should solve the first problem, within the performance requirements specified above. Since long-exposure photography is not one of the programs of interest, field rotation is not a problem. (This could be solved, though, using a third motor and rotation axis, and by using the equation developed in Chapter 7.) Experience on previous expeditions indicates that occultation observations are rarely made within 5° of the zenith. The sky area of 5° represents about 1% of the total visible hemisphere.

The optical assembly will consist of a rolled steel tube mounted to the altitude bearings, with structures for holding the primary and secondary mirror assemblies. A solid thin metal tube is more rigid pound for pound than a Serrurier truss, and even though the wind load on a portable telescope located in an open field is significant, a closed tube with proper baffling is much better at handling the stray light problem than is a Serrurier truss. The end of the tube could be closed with a flat glass window if windblown particles pose a problem.

The optical assembly will be supported by a symmetrical fork. This minimizes the moment of inertia about the azimuth axis, keeping the natural frequency of the mount high. It also eliminates counterweights, which are otherwise useless masses that must be towed and maneuvered on the highway. Each fork arm will consist of three lengths of steel tubing welded in a tripod configuration.

Despite the fact that the choice of an alt-az mount complicates the control system software, it results in a compact and lightweight assembly which is easily transported to a remote observing site, and is easily and quickly aligned to the equatorial coordinate system.

3. Optical System. The primary requirement affecting the optical system design is portability. This means the primary-secondary mirror separation must be minimized. This means the primary should be at least f/4, if not faster. A Newtonian system has two disadvantages. First, it places the photometer out at the secondary mirror assembly. This increases the moment of inertia, which decreases the natural frequency of the tube assembly, and greatly increases the overall mass of the system needed to support the bulkier tube. Second, a Newtonian optical system places the eyepiece in awkward positions for sighting stars during the alignment process. The photometer will weigh about 2-3 pounds, so the added moment of inertia is not significant. However, there is rarely good footing at remote locations to place a tall ladder for viewing through large Newtonian optics. If a high power refractor is attached to the main tube for sighting bright stars, it will take too long to realign it with the main scope after a trip over rough roads. Differential flexure, temperature effects, and other problems will also introduce significant errors into the encoder calibration constants. To obtain easy visual use for telescope alignment, Cassegrain optics will be employed. A Richey-Chretien design is not necessary because a wide, flat field is not required, and it is undesirable because of the additional expense involved in figuring the optics.

Ordinarily, a Cassegrain optical system requires about 18 to 24 inches additional space behind the primary mirror for attaching the photometer or other data collection equipment. The problem with this approach is that it makes the fork arms longer to provide clearance for the photometer when observing near the zenith. This increases both the total weight and the moment of inertia of the mount. To avoid this problem, a flat third mirror 45 degrees to the optical axis will be used in a Nasmyth configuration to direct the beam through the altitude bearing to a focus near an equipment mounting plate attached to the fork arm. The flat will be mounted on a rotating platform so that an eyepiece assembly can be mounted on one fork arm, and the photometer can be mounted on the other fork arm. This allows the telescope to be aligned to the equatorial coordinate system using the eyepiece and a hand paddle to sight stars without the need for changing equipment assemblies. It also permits using a photometer with no built-in eyepiece or flip mirror. If present, these features would add weight to the instrument, consume valuable setup time, and unnecessarily complicate the photometer computer interface hardware and the control software.

In sum, the use of large aperture folded Cassegrain optics is a result of the requirements for both portability and the ability to observe the events which are the focus of the observing program.

4. Drive Design Approach. The basic options for driving a telescope were reviewed in Chapter 3. These are the worm gear, band, chain, and friction drives. As mentioned in that chapter, a worm gear drive requires special consideration from the control system. The chain drive suffers from excessive periodic error, and both the chain and band drives do not have the stiffness required to keep the natural frequency of the drive high. The disk and roller friction drive does not have these drawbacks, and offers the kind of performance needed in an alt-az mount, with its wide variation of drive rates.

5. Control System Approach. A classical closed loop servo was chosen as the control system approach for the following reasons:

1. The observing program will not always include photometry, so a photometer will not always be mounted at the Nasmyth focus to be used to acquire and center targets. When doing photometry of comets, the targets may be so faint that long integration periods are required just to get a reading, which would slow down the entire process of target acquisition. Furthermore, it is difficult to offset from bright stars to acquire faint moving targets, particularly when working under severe time constraints and field conditions.

2. Faint targets must be acquired in dense fields of stars of similar brightness. The Fairborn Observatory APT's can tolerate moderate pointing errors because there are relatively fewer stars of the brightness of the target stars used in the APT research program. It is as yet an open issue how faint stars can be and still be accurately acquired using APT techniques. High accuracy position feedback can only be provided by shaft encoders.

3. Targets must be tracked accurately during an hour or more with an alt-az mount. A powerful 16-bit microcomputer and real time operating system are available, and I have over 10 years' experience in programming real time control systems, so insoluble software problems are not anticipated.

The control computer will be capable of computing all the correction equations discussed in Chapters 6 and 7 rapidly enough to ensure accurate pointing and smooth tracking. A great deal of software must be developed, not only to handle these basic servo functions, but also to aid setup and alignment (measuring the encoder zero offsets and azimuth tilt angles) and to reduce the time these activities require. If actual field experience indicates it is needed, an extended Kalman filter can be added in the future to improve the pointing and tracking by adjusting the calibration constants in real time in response to manually entered pointing corrections.

6. Top Level System Design. The basic system concept is to build a single-axle trailer holding an alt-az mounted telescope of 30-inch aperture. The trailer will be towed by a van which contains the control computer and related equipment, a gasoline powered generator, power conditioning equipment, and extra heating and cooling capability to protect all this equipment and the observer from the elements.

The trailer will be built around a 45-inch diameter bearing that will serve as the azimuth axis plane. Most states permit trailers to be a maximum of 8 feet wide. Before construction of the telescope begins, a 1:4 scale balsa wood model will be built to ensure that geometrically related designs are correct. Next, a test bed system will be developed using an existing 12-inch telescope. The first stage of the test bed trailer is shown in Figure 14-5. This will be used to verify the basic system concept, and to gain field experience using the telescope in the WMO observing program. During this phase, the operational procedures, especially those of setting up at a remote site, will be refined. The final step will be to procure and install the 30-inch optics. The larger telescope will require re-tuning the drive and control system, but by the time this phase is reached, experience at these tasks will have already been gained with the smaller telescope. If the concept does not prove out, the investment in large optics will not have been wasted.

14-5 Test Bed Trailer During Construction

Inside the van, the computer rack, generator, power conditioning and monitoring equipment, and other equipment will be mounted on pallets that slide in and out of the van easily, and lock to the floor of the van with quarter-turn fasteners. This will permit the van to be used for more mundane purposes, and will permit the computer to be used at home outside the van. The generator will be mounted inside a large plywood box with fiberglass sound baffling to keep the noise down to a tolerable level. The end of the box will be fitted with a duct to vent generator exhaust gasses outside the van.

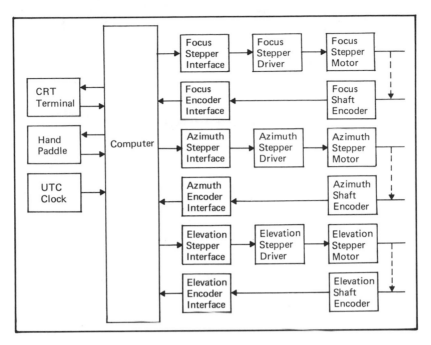

14-6 System Block Diagram

Figure 14-6 shows the system hardware configuration. The same computer is used to control both axes. The UTC clock is used to compute sidereal time, which is used, in turn, to compute hour angle from right ascension. In addition to the components of the operational system shown in Figure 14-6, thought must be given to those items needed to build and maintain the system. Since some custom digital design and fabrication will be done, an oscilloscope, digital test probe, VOM, wire wrap tool or a printed circuit board etching kit, and other such equipment will have to be procured. Tools for software development will also be purchased as part of the computer system procurement, including a video terminal for editing source code and data files, and a hard copy device for printing compiler and assembler listings in a reasonable amount of time. Since a video terminal is needed for software development, it is reasonable to use it for interactive entry of commands to the computer to perform positioning and instrumentation control functions.

This completes the discussion of the top level design. Before a detailed design can be set forth, fundamental questions must be answered. Included are the issues of how to calibrate the encoders, processor performance, and selection of the control computer. These issues are addressed in the remainder of this chapter.

F. POSITION ENCODER CALIBRATION

Before the telescope can be used with closed loop servo control, the shaft angle encoders must be calibrated. This consists of developing an algorithm for converting the raw encoder readings into the true position of the telescope optical

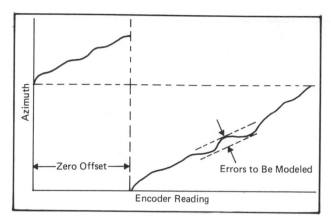

14-7 Encoder Calibration Curve

axis. Figure 14-7 depicts the calibration curve that results from this process. One method of calibration is to sight a large number of stars and compute the coefficients of a simple polynomial power series, which is then used as the model to transform raw encoder readings into alt-az coordinates. An example is position = Sum $_{i=0,n}$ [A_i (raw encoder reading)^i] where the A_i are determined by recording the encoder readings for n stars, then performing a least-squares fit. This method has several problems:

1. n can be quite large to obtain a good fit, especially if the function has several high frequency components.

2. There is no control over the excursions of the algorithm from the true functional relationship (i.e., the peak-to-peak variations may be large).

3. This method will not work for portable telescopes, which must determine new zero offsets and polar axis or azimuth axis tilts for each new observing location, without necessarily re-determining those errors which are intrinsic to the telescope itself.

4. There is no way to monitor how telescope characteristics change over time. This monitoring can reveal design or construction weaknesses, or a need for maintenance or repair before the problem causes a loss of observing time. This is particularly important with portable telescopes that are subjected to the shocks and vibration of transport over rough roads.

Objections 1 and 2 can be met in part by using a Chebyshev polynomial instead of a simple power series. According to Arfken (1970, p.629), such an approach limits the maximum error (but does not minimize the RMS error), and prevents the errors from bunching up at the ends by distributing them equally throughout the range of the function. It also gives the desired accuracy in the fewest number of terms. Spherical harmonics would also provide a more efficient fit than a simple power series.

However, objections 3 and 4 still stand. Because it is quite useful to know how telescope characteristics change with time, the method of power series or Chebyshev series will not be discussed further.

Another approach is to model the errors. The equations for modelling sources of systematic error were presented Chapters 6 and 7. These equations use coordinate variables, such as (h,δ) or (Z,A), and constants embodying characteristics of the telescope itself. The model should include all sources of error between the encoder and the optical axis. Errors between the motor and the encoder need not be modelled, since such errors do not affect the functional relationship between optical axis position and raw encoder reading. The model constants are as follows:

Equatorial Mount	Alt-Az Mount
1a. Arc seconds per encoder count (2 axes)	1b. Same (2 or 3 axes)
2a. Zero offset (2 axes)	2b. Same (2 axes)
3a. M_{el} (polar axis alignment)	3b. Angle a (azimuth tilt)
4a. M_{az} (polar axis alignment)	4b. Angle b (azimuth tilt)
5a. p (non-perpendicular axes)	5b. Same
6a. C_{ns}, C_{ew} (collimation)	6b. Same
7a. $F(90°)$ (tube flex - 2 axes)	7b. Same (2 axes)
8a. Mount flexure	8b. Ignored -- constant
9a. t' (servo lag - 2 axes)	9b. Same (2 axes)

The constants are determined ahead of time by sighting stars and inferring the constants, or by modelling (e.g., Serrurier truss flexure, using a mechanical structure modelling program). Constants related to position and orientation cannot be measured or modelled ahead of time for portable telescopes. Direct measurements of these constants are greatly preferred over modelling. During observations of regular program objects, these pre-determined constants are used in the equations given in Chapter 7 to perform the conversion from raw encoder units to optical axis position. The calibration problem is reduced to the problem of obtaining optimal estimates of the error model constants from measurements.

The easiest method of measuring these constants is to construct observations which tend to isolate one error source from all others. Procedures for doing this were discussed in Chapter 7. After the error constants have been measured, their initial measured values can be averaged to eliminate much of the random error. This simple procedure is the recommended approach for most applications.

However, for those requiring even greater accuracy who are familiar with the mathematics involved, the error constant values can be refined further by using modern digital filter optimal estimation techniques. A large number of stars (e.g. 200) can be sighted and the data used by statistical procedures to obtain very good estimates of the error constants. Not only does this yield improved accuracy when computing telescope pointing corrections, but it also permits calibration observations to be made all over the sky, not at particular places determined by a procedure to measure one error constant. The observation equations in the digital filter relate the observables (A and Z) to the estimated parameters (the error constants) in a way that permits separating the various error effects, which are not intrinsically orthogonal. Such methods require that the observation equations be overdetermined, that is, there are many more star sightings made than there are constants to be estimated.

The familiar technique of using weighted least squares is not the most computationally efficient, since the covariance matrix used for weighting the observations must be inverted, and matrix inversion cannot be performed rapidly.

The Kalman filter technique also yields optimal estimates of the error constants, but without inverting a matrix. The extended Kalman filter can be used during normal observing sessions, rather than being confined to special encoder calibration sessions. Observer inputs to correct telescope pointing errors during tracking can also be used by an extended Kalman filter to improve the values of error constants in near real-time, so that tracking accuracy can be improved while observations are being made. This, too, is of considerable help when using a portable telescope, since the location and alignment of the telescope (and other inputs to error constants) can change considerably from one night to the next. For those with advanced skills in mathematics, these digital filtering techniques are discussed in greater detail in Appendix N.

G. COMPUTATIONAL REQUIREMENTS

One of the criteria for selecting a control computer is the speed of its processor. Only when the rate at which calculations must be performed is known can the computational requirements be understood. In a digital servo, the rate of performing calculations is determined in part by the rate at which errors accumulate during the interval between the calculations. Since all the equations of interest in Chapters 6 and 7 depend on hour angle, elevation, or some other parameter that changes with time, they can be used to assess error growth rates.

Each error correction can be assigned an error budget. The simplest way of doing this is to give each error source an equal weight, and then determine the error budget for each calculation based on the total number of corrections to be made. There were eight significant calculations discussed in Chapter 6 in the reduction from mean to apparent place, and nine calculations in Chapter 7 for mechanical and alignment corrections. If it is assumed these 17 random errors accumulate in quadrature, then the root sum square total error budget should equal the short term tracking error maximum of 2". This means that

$$\sqrt{17 \ e^2} = 2"$$

for equally weighted errors, which gives individual error budgets of roughly e = 0".5. In the following sections, the time to perform each error correction calculation once and the number of times per second it must be performed to stay within the error budget are found. The calculation times are given in terms of the computational units developed in Chapter 6. This system of computational units assigns 1 unit for additions and subtractions, 2 for multiplications, 3 for divisions, and 22 units for trigonometric and other complex functions. The results are then summarized at the end of the section.

1a. Apparent Place Correction–General Precession. In Chapter 6, precession was found to require 272 computational units. To determine the allowable time interval between precession calculations, the rate at which errors grow must be determined. The time dependence of precession enters into the evaluation of ζ_o, z, and Θ in [6.3] through T and t. These quantities are in units of 100 years, and produce changes on the order of 50" per century, so that during the course of observing the same object over an evening, precession will change insignificantly. Therefore, precession needs to be computed only once an evening per object, when the object's catalog coordinates are entered.

1b. Apparent Place Correction-Nutation requires 524 units to perform the necessary calculations. The periodic term which dominates nutation is Ω, through the sin Ω term of $\Delta\psi$ and the cos Ω term of $\Delta\varepsilon$. The periodic term of Ω that changes most rapidly is $(5r + 482890''.539)$ T where T is in Julian centuries. During a single evening of observing, T will change by roughly 0.00001, so Ω will change by roughly $0°.018$. This term goes through 360 degrees once in approximately 19 years. Depending on the value of Ω in that 19 year cycle, a change of Ω by $0°.018$ could affect the sin Ω or cos Ω by at most 0.0003 (sin $0°.018$ - sin 0). This factor times $17''.2289$ is only $0''.005$, so there is no need to compute nutation more than once per evening per object.

1c. Apparent Place Correction-Aberration. The following table summarizes the operations involved in computing annual aberration and applying it to (α,δ) using equations [6.14]:

Item	Equation	Trig	+	-	x	/
Annual Aberration	[6.14]	18	17	9	35	4
E-Terms	[6.15]	3	4	0	9	0
Totals		21	21	9	44	4

It is assumed that to compute secant, cosecant, and cotangent functions, one computes the corresponding cosine, sine, or tangent function, then inverts the result by dividing 1 by the result; that l' was computed previously when computing nutation; that seven iterations were used to solve Kepler's equation for E; and that a square root takes as long as a trigonometric function to compute. With these assumptions, annual aberration requires 592 computation units.

Since the value of annual aberration depends on the star's (α,δ) and the position of the Earth in its orbit about the Sun, it will change very little during a single evening of observing, so it can be computed once when the star's coordinates are entered into the control system.

The number of computation units required to compute diurnal aberration using equations [6.16] with r = 1 is 128 units. It is assumed that L_c, the observer's latitude, is entered by hand, or read from a disk file. However, diurnal aberration corrections need not be made, since the coefficients in equations [6.16] are less than $0''.5$ and the other multipliers in these equations are trigonometric functions with values less than or equal to 1.

1d. Apparent Place Correction-Parallax. Using the method of equations [6.17] - [6.18], the number of computation units to compute stellar parallax is 245 units, counting a square root to have the same computational complexity as a trigonometric function, and assuming L_t, e, ν, ε, etc. were computed earlier.

When computing annual parallax by this method, and assuming that C, D, and ε were computed previously when computing annual aberration using equations [6.14], the number of computation units required in using equations [6.19] and [6.21] to compute the additional terms for stellar parallax is 106 units. It is assumed that π is given directly in the star catalog.

If annual aberration is computed using equations [6.14] and annual parallax is computed using the method of equations [6.17] - [6.19], the total number of computation units, allowing for overlapping computations in the two methods, is

834 computation units. When both annual parallax and annual aberration are computed using [6.21], and C, D, and e from [6.14], the total is 698 computation units, saving 136 units.

Equations [6.17] - [6.19] can be used to estimate the amount of time that is necessary for errors to grow to 0".5 . From equations [6.17], the time-dependent quantities are the heliocentric coordinates (X,Y,Z). In equations [6.14], L_t and e depend on T (in units of Julian centuries). During the course of an evening, the T term in e changes by a factor of about $1/(36525 \times 3)$, or about 0.00001 . The change in e, to first order, is roughly 0".0004, and the change in sin e or cos e cannot be greater than that. During the course of an evening, the T term in Lm changes by the same factor, but because this is multiplied by some 36000 degrees, Lm changes by about 0.3 . This is offset by the corresponding term in l', so L_t does not change by more than 0".5 . Therefore, stellar parallax need be computed only once for each different object viewed.

Assuming sec δ, cos δ, and sin δ were computed previously, the number of computation units required to compute the planetary parallax using equations [6.22] is 109 units.

From equations [6.22], the only time-dependent quantity is h, the hour angle. Thus the corrections to (α,δ) will change by at most π (d(sin h)/dt), where π is about 9" per A.U. in distance. During one hour, h changes by 15 degrees, which means that in going from h = 0° to h = 15°, Δα will change by about 2". Thus to keep the error below 0".5, geocentric parallax should be computed every 1/4 hour, or every 15 minutes, per A.U. of distance.

1e. Apparent Place Correction-Refraction. To evaluate the number of calculation units that must be performed to compute refraction, equations [6.23] are used for computing Δα and Δδ, and [6.24] are used for computing R. Assuming that once the trigonometric functions and the denominator in [6.23a] are computed, they need not be recomputed in [6.23b]. The result is a total of 197 computation units.

In Chapter 7, it was argued that refraction should be recomputed every time the pressure or temperature changes by a detectable amount. In addition, refraction should be recomputed whenever the zenith distance changes by an amount capable of changing the corresponding refraction by a tolerable error. Even though the tolerable error for other systematic error sources is only 0".5, considering the accuracy of the refraction calculation, the tolerable error can be 1".0 . This relaxation of the refraction error budget will not seriously affect pointing performance, since most of the other error sources can be held to 0".1 or less, so the total RSS error is still within the total budget. Tracking performance is not affected, since the changes in the errors in computing R are considerably less than 0".5 over long periods of time.

The zenith distance has the greatest effect upon the angle of refraction at the horizon (Z = 90.6° for a ground-based telescope). Rather than evaluating first and second derivatives of R with respect to Z to find the maximum value of dR/dZ, since we know dR/dZ is a maximum at roughly Z = 90°, a simple evaluation of equation [6.24] at Z = 90° and Z = 89° is used instead. Using P = 30 and T = 50, R (Z = 90°) = 2118".2 and R (Z= 89°) = 1481".9, so R changes roughly at the rate of 636 arc seconds per degree change in the zenith angle. Thus to keep the error in R less than one arc second, R must be computed whenever the zenith distance changes by 1/636 of a degree, or every 5.7 arc seconds. The rate of change of zenith distance with time at the horizon can be taken to be no more than 15 arc seconds per second, so refraction must be recomputed roughly three times per second. This is an extremely conservative estimate, since the 15 arc seconds per second rate is experienced only at the Earth's equator, and dR/dZ does not change rapidly until the telescope is pointing very close to the horizon. Photometrists care little about refraction effects, since they do their observing at zenith distances less than 60°, but those who observe occultations occasionally find themselves observing in difficult conditions, including zenith distances in excess of 89°.

1f. Apparent Place Correction–Orbital Motion. The typical method of applying orbital motion corrections is to multiply a constant by the Julian date and add the result. This requires a total of 6 computational units for both RA and Dec. If the stars are close enough to require the calculation to be performed more frequently than once per evening, they probably cannot be resolved by the telescope.

1g. Apparent Place Correction–Proper Motion corrections are usually applied in a manner similar to orbital motion corrections. That is, they require 6 computational units, and need be computed only once per evening per object.

2a. Mechanical Corrections–Conversion of the Encoder Reading. The binary encoder reading (an n-bit integer) must be multiplied by a scale factor to obtain degrees, seconds of arc, or some other physical unit. The scale factor is determined by the gear ratios between the encoder and the telescope axis, and the number of counts per revolution of the encoder. This conversion requires two computation units, and must be performed each time the encoders are read.

2b. Mechanical Corrections–Zero Offset. The addition of a constant to the converted encoder reading represents 1 computation unit. This calculation must be repeated each time the encoders are read.

2c. Mechanical Corrections–Polar Axis Alignment. Evaluation of equation [7.1a] requires the use of three trigonometric functions, one subtraction, three multiplications, and one addition (to apply the correction Δh to h), for a total of 74 computation units. Similarly, evaluation of [7.1b] requires the use of two trigonometric functions (already evaluated in [7.1a]), two multiplications, and two additions (one to apply the correction $\Delta \delta$ to δ), for a total of 6 computation units. Again, this calculation should be repeated each time the encoders are read.

2d. Mechanical Corrections–Azimuth Axis Alignment. The calculation to find the azimuth axis tilt angles **a** and **b** requires 227 computation units. This calculation must be repeated as often as is necessary to keep the error in Z - Z" less than a reasonable amount. From [7.7c], since **a** is small, the greatest dependence of Z is on Z" for values of Z down to about Z = 80°. In this range, dZ"/dh can have a wide range of values, depending on the declination of the star. For example, for Z" = 20°, a = 1°, b = 0, and A = 90°, cos Z = a sin Z" + cos Z" = 0.94566, so Z = $18^\circ.97$ and Z - Z" = $-1^\circ.0252$. For A = 0 and b = 0, cos Z = cos Z", so Z - Z" changes by about one degree in the time it takes to go from A = 90° to A = 0, or six hours. This is a change of one arc second in Z - Z" every six seconds of time. To cover all cases and keep the error below 0".5, the calculation should be repeated once every three seconds.

2e. Mechanical Corrections–Equatorial to Alt-Az Conversion. Equations [7.8] - [7.10] are evaluated in alt-az control systems to convert the apparent position of a star from equatorial to alt-az coordinates. To obtain zenith distance Z, equation [7.8a] is evaluated using previously-computed values of sin φ, cos φ, sin δ, and cos δ. Therefore, two trigonometric functions (cos h and arccos Z), three multiplications, and one addition are performed, for a total of 51 computation units. Similarly, equation [7.8b] is evaluated using one trigonometric function (sin Z), one subtraction, two multiplications, and one division, for a total of 30 units.

These calculations must be performed at the rate used to read the encoders, so that the error (apparent star position - current true telescope position) can be used to generate a motor command.

To compute the azimuth drive rate, equation [7.9b] is used. It contains two trigonometric functions not evaluated previously (arccos Z and sin Z), one division (to obtain cot Z), three multiplies, one addition, and one subtraction, for a total of 55 computation units. Equation [7.9a] is used to compute the altitude (or zenith distance) drive rate. This equation uses two trigonometric functions (arccos A and sin A), and two multiplies, for a total of 48 computation units.

To track a celestial object, equations [7.8] and [7.9] are computed frequently enough so that the drive rate integrated over the time between drive rate computations and updates does not exceed the error budget. Figure 7-1 can be used to find where dZ/dh is changing most rapidly. An example of a high rate of change occurs when $\varphi = 40°$, $\delta = 35°$ and $A = 0°$. At hour angle = 359°, Z = 5°.062 and dZ/dh = - 1.862 "/s, while at hour angle = 0°, Z = 5° and dZ/dh = 0. The time it takes for the hour angle to change by one degree is 240 seconds. If dZ/dh is set to - 1.862 "/s at Z = 5°.062, 240 seconds later, Z = 4°.938 instead of 5°, a difference of 223".68 . This error accumulates in 240 seconds, or at the rate of roughly 1" per second. Thus to keep the error at or below 0".5, [7.9a] should be recomputed twice each second.

Similarly, Figure 7-2 is used to find where dA/dh is changing most rapidly. When $\varphi = 40°$, $\delta = 35°$, and h = 356°, dA/dh is changing very rapidly. At hour angle 356°, A = 326°.359 and dA/dh = 101.890 "/s. If dA/dh is set to 101.890 "/s at A = 326°.359, 240 seconds later, A = 333°.152 instead of 333°.620 at hour angle 357°, a difference of 1684".8. This error accumulates at an average rate of about 7" per second. To keep the error at or below 0".5, [7.9b] should be recomputed 14 times per second. This rough estimate compares favorably with the 20 times per second rate used successfully by a LAGEOS satellite laser ranging system at Goddard Space Flight Center to measure continental drift (Mansfield, 1984). This system works equally well when tracking stars or the LAGEOS satellite.

A given error in azimuth or zenith distance does not translate to the same error in the respective equatorial coordinates. Depending on the latitude where the telescope is located, a larger error in azimuth or zenith distance can be tolerated without affecting the resulting pointing accuracy of the telescope in equatorial coordinates. This would reduce the rate at which the coordinate conversion equations need to be computed. Furthermore, the rates quoted are first order (linear) approximations. If the error in azimuth accumulates at the rate of 7" per second of time worst case, computing the equations 14 times per second would most certainly keep the errors under 0".5, but the errors will not actually grow at this rate. These two factors reduce the control of an alt-az mount to a tractable problem for modern microcomputers.

The final drive rate is for field rotation correction. Although this feature is not currently planned to be a feature of the WMO telescope, it is included for completeness. Equation [7.10] contains one trigonometric function not evaluated previously (sin h), three multiplies, and one addition, for a total of 29 computation units. Equation [7.11] contains two trigonometric functions not evaluated previously (arccos C and sin C), six multiplies, one divide, one addition, and two subtractions, for a total of 62 computation units. Together, [7.10] and [7.11] require 91 units. Figure 7-4 shows that when $\varphi = 40°$, $\delta = 35°$, and h = 356°, dC/dh is changing rapidly. At hour angle 356°, C = 36°.326 and dC/dh = - 111.940 "/s. If dC/dh is set to this value at C = $36^\circ.326$, 240 seconds later, C = $28^\circ.863$ instead of $28^\circ.395$, a difference of 1684".8 . This error accumulates at an average rate of about 7" per second. In Chapter 7, the error budget on field rotation was computed to be 10". To keep the error in C at or below 10", dC/dh should be recomputed roughly once per second.

2e. Mechanical Corrections–Non-Perpendicular Axis Alignment. For the equatorial mount, equation [7.12a] requires only one multiplication and one addition (to apply the correction) or 3 units to compute, since sin δ can be evaluated before pointing and tracking begin. The same is true of [7.12b], since Δh can be computed in advance. This means the correction need be computed only once, since δ does not change during tracking.

For alt-az mounts, equation [7.13a] uses the cos Z found in [7.8a], so only 3 units are required for correcting both A and Z. The correction to A changes at the worst case rate of p sin Z (dZ/dh), which can be as high as 15p "/second at the horizon. To keep the error less than 0".5, the error should be computed 30p times per second. However, this worst case rate occurs where cos Z (in the equation) is approaching zero, which reduces the size of the correction (and its error growth rate) considerably. Therefore, the correction can be computed roughly 2 or 3 times per second. The same is true of the correction to Z.

2f. Mechanical Corrections–Collimation Errors. To compute collimation error corrections for an equatorial mount, C_{ns} is used directly as the correction to declination, and is applied using one addition, or 1 computation unit. The correction to hour angle is computed using equation [7.14]. This requires one multiplication and one division (to compute sec δ from cos δ, which was obtained in evaluating equation [7.8b]) to compute the correction, then one addition to apply the correction, or 6 units.

The alt-az mount correction uses C_{ns} as the correction to Z, and equation [7.15] to compute the correction to A. This requires one division to obtain csc Z (sin Z was obtained previously when equation [7.8b] was evaluated), then a multiplication to complete the evaluation of ΔA, and an addition to apply the correction, for a total of 6 units.

From [7.14] and [7.15], collimation errors are either constants, or depend on trigonometric functions of δ and Z. In an equatorial mount, δ does not change with time when tracking the same object, so collimation errors can be computed once when slewing to a new object, then added in each time the encoder is read. In an alt-az mount, the correction to A depends on dZ/dh. As in the case of non-perpendicular axes, the correction can be computed 2 or 3 times per second to obtain adequate accuracy.

2g. Mechanical Corrections–Tube Flexure. For an alt-az mount, equation [7.16] requires one multiplication and one subtraction, or 3 units, to evaluate, since sin Z was found previously in evaluating [7.9b]. However, for the equatorial mount, the alt-az conversions must be performed, since equation [7.16] requires knowing Z. Equations [7.8] were found earlier to require 81 units. To evaluate [7.10], cos h, sin h, and sin φ were found previously, and cos A was found in [7.8b]. Thus two trigonometric functions are required to find sin A, then 3 multiplies and 1 addition are needed to evaluate [7.10], for a total of 51 units. Equation [7.16] requires two trigonometric functions to find sin Z from cos Z, then one multiply and one subtraction to evaluate and apply the correction, for a total of 47 units. Finally, equations [7.17] require two trigonometric functions to find sin C from cos C, two multiplies to evaluate the equations, and two additions to apply the corrections, for a total of 50 units. The grand total for equations [7.8], [7.10], [7.16], and [7.17] is 81 + 51 + 47 + 50 = 229 units.

From equation [7.16], tube flexure depends on sin Z, hence it changes at a rate of cos Z (dZ/dh). Cos Z is minimized (in absolute value) where dZ/dh is maximized, so again, a calculation rate of 2 or 3 times per second ought to suffice.

2g. Mechanical Corrections-Mount Flexure. This correction is not needed in alt-az mounts. For equatorial mounts, equation [7.18] requires only two multiplications and one subtraction, or 5 units, since sin h was evaluated previously for equation [7.1a] and sec δ can be evaluated at the time the new coordinates are entered. To determine the calculation rate, note that [7.18] varies with sin h. Taking sec δ to be 1, at the meridian $\Delta h = 0$, and at h = +1°, Δh = -18.9 (0.017) = -0".33 . For the example of the AAT (using the -18".9 figure), mount flexure can be recomputed every 364 seconds, or every 6 minutes, to obtain 0".5 accuracy. The calculation rate is different for each telescope, since it depends on the actual tube flexure.

2h. Mechanical Corrections-Servo Lag Error. Using equation [7.19], the servo lag error requires one subtraction to compute I, a multiplication to compute the lag, and one addition to apply the correction, for a total of 4 units. This error must be computed each time the encoders are read and a new motor command is computed.

3. Processor Loading Calculations. The corrections described in Chapters 6 and 7 can be computed in either open loop or closed loop servos to improve pointing and long term tracking accuracy. To be accurate, these calculations must be synchronized with telescope motion to provide accurate current telescope pointing information as inputs to these equations. To point a telescope at an object and track it, the processor must perform the following functions:

1. Compute the desired telescope position
2. Determine the true telescope position
3. Compute the position error
4. Find a set of motor commands which both minimizes the position error and minimizes its rate of growth in the future.

The first function is performed in the reduction from mean to apparent place. The table below summarizes the loading information developed earlier.

Computation	Comp. Units	Comp. Freq.	Comp. Units per Second
1. Annual aberration	698	once	-
2. Stellar parallax (part of Step 1a)			
3. Precession	272	once	-
4. Nutation	524	once	-
5. Orbital motion	6	once	-
6. Proper motion	6	once	-
7. Diurnal aberration	128	once	-
8. Planetary parallax	109	1/(15 minutes)	< 1
9. Refraction	197	3/second	591
		Total	591

The calculations to be performed once per object per evening total 1634 computation units, which can be executed in less than 0.2 seconds on an LSI-11. This time is insignificant, since these calculations are performed immediately after the mean place coordinates of the next object to be observed are entered, and are complete before the astronomer can turn around to check to see if the telescope is moving to the new coordinates.

The following tables summarize the process of converting encoder readings into axis position. In a typical digital servo, the encoders are read, the readings are converted to appropriate coordinates, the result is compared to the desired position, and new commands are sent to the motors, all in sequence. This loop is executed at periodic intervals, so the encoder calibration equations and motor command generation software are executed at the same frequency.

A time of 85 μS was used for each computation unit, to derive the CPU utilization of the LSI-11. Gearing and bearing errors were ignored, as they usually are small enough to ignore until after the major error sources have been modelled. The first two tables give the processor loading data for an equatorial mount. Although this is not relevant to the WMO telescope, most readers will be interested in the results. The computation frequency of 3 iterations per second is seen as the lowest rate consistent with the apparent place and mechanical correction computation frequencies.

Equatorial Mount -- Polar Axis

Computation	Comp. Units	Comp. Freq.	Comp. Units per Second
1. Convert encoder reading to degrees	2	3/s	6
2. Zero offset	1	3/s	3
3. Polar axis misalignment	74	3/s	222
4. Non-perpendicular axes	3	3/s	9
5. Collimation	6	3/s	18
6. Tube flexure	226	3/s	678
7. Mount flexure	5	3/s	15
8. Servo lag	4	3/s	12
		Total	963

Equatorial Mount -- Declination Axis

Computation	Comp. Units	Comp. Freq.	Comp. Units per Second
1. Convert encoder reading to degrees	2	3/s	6
2. Zero offset	1	3/s	3
3. Polar axis misalignment	6	3/s	18
4. Non-perpendicular axes	3	3/s	9
5. Collimation	1	3/s	3
6. Tube flexure	3	3/s	9
7. Servo lag	4	3/s	12
		Total	60

The total processor utilization for error modelling with an equatorial mount is (963 + 60) x 85 = 87 milliseconds per second, or 8.7%. This can be reduced by two-thirds if tube flexure is small enough to be ignored.

The comparable figures for an alt-az mount are given below.

Altitude-Azimuth Mount -- Azimuth Axis

Computation	Comp. Units	Comp. Freq.	Comp. Units per Second
1. Convert encoder reading to degrees	2	14/s	28
2. Zero offset	1	14/s	14
3. Azimuth tilt	227	1/3s	76
4. Convert to alt-az	81	14/s	1134
5. Drive rate	55	14/s	770
6. Non-perpendicular axes	3	14/s	42
7. Collimation	6	14/s	84
8. Servo Lag	4	14/s	56
		Total	2204

Altitude-Azimuth Mount -- Altitude Axis

Computation	Comp. Units	Comp. Freq.	Comp. Units per Second
1. Convert encoder reading to degrees	2	2/s	4
2. Zero offset	1	2/s	2
3. Convert to alt-az (already done above)			
4. Drive rate	48	2/s	96
5. Non-perpendicular axes	3	2/s	6
6. Collimation	1	2/s	2
7. Tube flexure	3	2/s	6
8. Servo Lag	4	2/s	8
		Total	124

Altitude-Azimuth Mount -- Field Rotation Axis

Computation	Comp. Units	Comp. Freq.	Comp. Units per Second
1. Convert encoder reading to degrees	2	1/s	2
2. Zero offset	1	1/s	1
3. Field rotation	29	1/s	29
4. Rotation rate	62	1/s	62
5. Servo Lag	4	1/s	4
		Total	98

The total processor utilization for error modelling with an alt-az mount is (2204 + 124 + 98) x 85 = 207 milliseconds per second, or 20.7%.

After the control software has found the desired and true telescope positions, the position error E is found by subtracting the desired position from the true position. Next, the current drive rate is corrected by the error as follows:

$$ S = K_m \left(R + \frac{E}{P} \right) $$

where S = motor speed command (steps per second, or volts)
K_m = steps per arc second (stepper motor), or
volts per arc second (servo motor)
R = the basic axis drive rate (arc seconds/second)
E = position error (arc seconds)
P = time between motor speed updates (seconds)

If the computed motor speed is significantly different from the current motor speed resulting from the previous command update, the motor must be ramped up or down. In addition, if the computed speed is greater than the maximum motor speed, slewing will have to be performed, or, in the case of an alt-az mount near the zenith, tracking of the object must cease until it appears on the other side of the cone defining the area of the sky through which the mount cannot track.

The total computational load to compute the position error and resulting motor command should require about 50 computation units. When this is repeated 3 times per second on each of the two axes on an equatorial mount, it requires 25.5 ms per second, or 2.6% of the LSI-11 processor. When computed 14 times per second on the azimuth axis, 2 times per second on the altitude axis, and once per second on the field rotation axis, it consumes 72.25 ms per second, or 7.3% of the LSI-11 processor.

One approach to synchronizing telescope motion and error calculations is to use timer interrupts to repeat motor commanding at well defined intervals. In such a system, the interrupts of concern to the programmer are clock interrupts indicating that it is time to perform a given set of calculations. Encoders are read and motors are commanded without interrupts, since the data registers for these devices are updated continuously.

In an alt-az mount, encoders are read and motor commands are sent at a rate of 17 times per second (14 for azimuth, 2 for altitude, and 1 for field rotation). In addition, the system clock interrupts arrive at a rate of 60 per second. This does not include interrupts that result from updating a command and status page on the command console terminal. If a telescope position display showing UTC, sidereal time, raw encoder readings, zenith distance, azimuth, right ascension, declination, hour angle, and drive rates for azimuth, zenith distance, and field rotation is updated once per second, about 300 characters must be transferred per second, with one interrupt per character. The processor must, therefore, be capable of handling about 400 interrupts per second.

In the LSI-11, the interrupt vectoring to the software which handles interrupts is embedded in the processor and bus architecture. This keeps the interrupt latency small, which permits a large number of interrupts to be serviced each second. The typical interrupt service routine is about 25 instructions. Using an average instruction time of 10 microseconds, the servicing of 400 interrupts per second consumes 10% of the processor. In addition, there is an interrupt latency of about 50 μS per interrupt while the Q-Bus arbitration is performed and the jump to the address in the interrupt vector is performed. At 400 interrupts per second, the latency totals 20 Ms, or 2% of the CPU, for a total of 12% of the CPU used for interrupt servicing. Although the rate at which encoders are read for the equatorial mount is slower (6 times per second instead of 17), the CPU loading for interrupt servicing on the equatorial mount is about the same as for the alt-az mount, since most interrupts are for servicing screen updates and the system clock.

Interrupt handling is not the only function to be performed when generating a display page. For the items listed above to be displayed, about 15 ASCII conversions from floating point numbers must be made for each update. About 400 instructions must be executed for each conversion, and assuming the average LSI-11 non-arithmetic instruction takes about 10 μS, 15 conversions per second requires 60 mS. Additional overhead to put in cursor positioning commands is about 10 mS, for a total of 70 mS per second, or 7% of the CPU.

Although most systems would not require that all the error correction computations described earlier be performed at the prescribed rates, for those systems requiring very high accuracy, the CPU loading can be high. The table below summarizes the CPU loading of the benchmark LSI-11, based on the assumptions made previously.

Item	Equatorial % CPU	Alt-Az % CPU
1. Apparent place	5.1	5.1
2. Error modelling	8.7	20.7
3. Motor command	2.6	7.3
4. Interrupt servicing	12.0	12.0
5. Display page generation	7.0	7.0
Subtotals	35.4	52.1
6. Operating system overhead (+20%)	7.1	10.4
7. High level language overhead (+30%)	10.6	15.6
Totals	53.1	78.1

The operating system overhead will vary considerably, and has been estimated somewhat conservatively. Real-time operating systems will have about 5 to 10 % overhead, whereas a non-real-time system, like Unix, will have considerably more. The overhead for the high level language may also be too high, but the value used is realistic and typical of compiled languages. Interpretive versions of languages, such as BASIC, run much slower (5 to 15 times slower), so unless the processor loading for the system is quite low, a compiled version of the applications software should be used.

One does not want to select a processor that can just barely keep up with the required calculations, that is, one in which the required calculations consume 100% of the processor resources. A good rule of thumb when sizing a system is to have 50% reserve, but a reserve of only 20% can usually be tolerated. Therefore, a single LSI-11 is probably capable of performing all the alt-az mount calculations at the required rates. If one wants to use a slower processor, the following options are available:

1. Drop some of the calculations
2. Reduce the rates at which the calculations are performed
3. Code the applications software in assembler language
4. Offload the processor of some of its assigned functions

The first two options are the ones taken by Boyd and Genet with the APT control system, which only computes precession in an environment in which only one pulse train is being generated at a time, and the number of interrupts per second is very low. In the WMO system, with its tighter requirements on pointing and tracking accuracy, a faster processor is used which is capable of handling all the computations and interrupts at the required rates. Coding the software in assembler language is not recommended, since the value of the additional time wasted in debugging assembler language code is many times more the value of smart controller cards (the fourth option) or a faster processor, and the resulting assembler language program is many times more difficult (and costly) to maintain and enhance over the life of the system.

Another computer resource to be considered, besides that of the processor, is the bus. Since all buses have a maximum rate at which data can be moved across them, there is a maximum bus bandwidth, in bytes per second, for each bus. Since the amount of data (encoder readings, motor commands, etc.) moving over the computer bus is small compared to the number of instructions moving over the bus each second, and the processor manufacturer designs the computer bus to accomodate the instruction rate of the processor plus a reasonable amount of data, the bus loading should be equal to or less than the processor loading. This may not be the case when large amounts of instrument data move across the bus, for example, when doing multi-channel high speed photometry. Therefore, bus utilization should be considered when high data rate instruments are added to the computer bus.

H. COMPUTER HARDWARE SELECTION CRITERIA

1. Processor Speed. Based on the estimates of processor utilization made above, any processor that is to be considered for the WMO control system should have at least the speed of an LSI-11/2. If estimates show that roughly 100% of a particular processor is required to do a job, since such estimates are often a bit low, and since "collisions" among real-time tasks that need the processor make the actual CPU loading vary considerably, one should choose a processor that will be loaded on average only about 50% or less.

Another consideration is that the system will be evolving, with new functions being added over time. Therefore, a computer should be purchased which permits the processor to be upgraded easily and inexpensively. An example of this is the LSI-11. The basic processor is the LSI-11/2, which was the benchmark for processor loading used above. A processor which is roughly 2.5 times as fast is the LSI-11/23, which simply plugs into the same slot normally occupied by the LSI-11/2 processor. The fastest processor in this family is the LSI-11/73, which is nine times as fast as the LSI-11/2, and which is roughly comparable to the MC68000 processor used in many small computers today. DEC also recently announced the MicroVAX, which uses the LSI-11 Q-Bus, and has roughly 30% of the speed of a VAX-11/780, or about the same speed as the 11/73. One feature to look for, then, is not only the speed of the processor in the computer purchased initially, but also the range of processors available for future upgrades.

To compare a processor under consideration with the LSI-11, the times for floating point operations should be compared, not the time for a MOVE or integer add. One could do all arithmetic in fixed point (integer) format using a small 8-bit processor, but the resulting software would be more difficult to develop and test. Considering the low cost of fast 16-bit processors and the high cost of programmer time, it is usually less costly to select a faster processor that is more than capable of performing the required task.

2. Arithmetic Hardware. Often, hardware for performing integer or floating point arithmetic is an option on microcomputers. For example, the KEV-11 EIS/FIS option for the LSI-11/2 is about $165. The 8087 arithmetic co-processor for the 8088 (the processor used in the IBM personal computer) is roughly the same price. These hardware arithmetic options are extremely cost-effective, considering the high density of arithmetic instructions in the instruction mix of telescope control software. The penalty for performing floating point arithmetic in software is typically a factor of five to ten in processor utilization. Given the processor utilizations listed above, which assumed the presence of floating point hardware, the advantage of obtaining the optional hardware is obvious.

3. Interrupt Hardware. In the previous section, it was estimated that there would be a few hundred interrupts per second, most of which occur in updating a real-time display of telescope pointing status once per second, which requires one interrupt per character displayed. Many early-generation microprocessors sample all devices in a round-robbin fashion to see which device needs I/O servicing when an interrupt flag is raised. This polling of devices consumes too much time when several hundred interrupts per second are generated in a real-time system.

A better method is to use vectored interrupts. The LSI-11 implements vectored interrupts within its bus structure. When a device requires interrupt servicing (typically when a data transfer between memory and a peripheral device is completed) it uses an interrupt signal line shared with other devices. When the processor is ready to service an interrupt, it manipulates control lines to transfer a vector address in low memory, which is stored on the peripheral interface card, from the device to the processor. The contents of the vector address are then loaded into the program counter, which causes the processor to begin executing the interrupt service routine. The LSI-11 uses a stack architecture to keep track of interrupted program status. Similar vectored interrupt hardware is available for the

8086, 68000, and other 16-bit processors, and a few 8-bit processors as well.

4. Standard Bus Structure. A telescope control system requires that bus interface cards be built or purchased for non-standard devices. Almost all computers have interface cards available for printers or disk drives, or these features reside on the processor board, but only a few have cards available for shaft encoders and stepper motors. By choosing a computer that uses a bus for which standard off-the-shelf interface cards are available to interface non-standard devices, time can be spent building a telescope control system and doing astronomy, instead of spending the same time building and debugging custom interface cards. Using a standard bus also usually permits selection from a wide range of processors, which permits the kind of processor growth mentioned above.

Standard buses for which there are many cards commercially available are the STD bus, LSI-11 Q-Bus, S-100 bus, Multibus, Versabus, and the VME bus. The latter two buses tend to be used primarily with the Motorola 68000 processor, but 68000 processor cards are also available for the Multibus, STD bus, and the Q-Bus. The Multibus was invented by Intel for its firmware development systems, but several other manufacturers have jumped on a good bandwagon, with the result that the Multibus is as strong a contender for consideration as any other bus.

I. SOFTWARE DEVELOPMENT FACILITIES

As was discussed earlier, software development can be a major cost component of any telescope control system budget. Procurement of good hardware and software tools to aid software development can yield significant software development cost savings. Selection of a processor should entail evaluation of the software development environment that is available with each different manufacturer's processor, as well as the ability of a particular processor to handle the computational load. The greatest system development cost savings are realized if components of the software development environment can also serve as necessary components of the operational system. The individual elements of the software development environment are discussed below.

1. Development Hardware. To develop software for an automated telescope system, if one has access to a large computer facility with a cross-assembler for the telescope's processor, one need not procure special software development hardware or software. Large computers with cross assemblers often are not available, or are located far from the observatory that houses the telescope control computer. In many cases, therefore, the software development environment must be purchased as part of, or in addition to, the operational environment. Software is often the most tedious, time-consuming, and expensive part of the whole project. A little money spent on a good software development environment repays itself very quickly, and many times over.

The minimum hardware configuration for software development should support the following requirements:

1. Rapid entry and correction of source code
2. Rapid production of legible hard copy of source code, link maps, and data file listings
3. Rapid compiling, assembling, and linking of programs
4. Roughly 2-10 million bytes of mass storage capability, with at least one-quarter million bytes on-line at any given time
5. Ability to copy source code in case of medium or peripheral device failure
6. Adequate memory for executing editing, copying, compiling, and other utility programs, as well as for holding the operational programs
7. Integration of the operational hardware configuration with the software development system, to allow software testing and integration with the final hardware configuration.

Requirement 1 is most easily met with a monochrome video terminal with at least 24 lines of 80 characters each. The terminal should be capable of operating with a standard RS-232C serial interface at 9600 baud (bits per second per data line). Although the temptation is to purchase a hard copy keyboard terminal which also satisfies Requirement 2, this is false economy. Most commercially available operating systems support full screen text editors or word processors, which allow a programmer to be much more productive in entering and correcting his source code than he could be using a hard copy terminal. Considering that most video terminals are about $1000-$1500, with many models well below $1000, the expense is not great. The video terminal is easily incorporated into the operational system, and can be used to replace several expensive dials and meters for displaying telescope position, time, and other parameters. For an additional $1000-$1500, a high resolution graphics capability can be added to several manufacturers' terminals to show star fields, or to replace more expensive strip chart recorders for displaying data (with the computer's disk used for recording the data permanently).

There are several small, fast, high quality dot matrix printers available for about $300-$800 capable of speeds of 100 characters per second or more with line widths of 80 or 132 columns, and with a graphics capability. Such a device would satisfy Requirement 2.

Requirements 3-5 are met with a disk drive. Although one could use digital or audio cassettes as the mass storage medium, experience with both digital cassettes and floppy disks on systems requiring 1-10 man-months of software development effort has shown that the savings of software development time far outweigh the extra cost of the disk. The time to make a one-line change in a program on cassette, re-compile it, and re-link it can often be as much as 45-60 minutes, versus about 30 seconds to five minutes to make the same change on a disk-based system. When integrating and testing custom hardware devices and interfaces, and the custom software device drivers, many small code changes need to be made before the system works. Schedules and budgets can benefit tremendously from the high programmer productivity resulting from the fast turnaround offered by a disk-based system.

A typical low-cost system might include two high density 8-inch or 5.25-inch floppy drives. Each drive holds about 1/2 MB. Typically, one drive is used to hold the operating system and associated utilities, such as the system boot blocks, editors, assemblers, compilers, program linkers, system libraries, and file copy utilities. The other drive is used to hold the programmer's source code and related object and executable modules. When the software has been developed and tested, and the system becomes operational, an operational system disk can be prepared holding the boot blocks, operating system, device drivers, and the telescope control programs, since language compilers and other software development tools are not needed during observing operations. Freed from storing the source code and the intermediate object files of the control programs, the other drive is available for storing data gathered by the instruments, catalogs of objects to be observed that evening, or any other files that may be useful.

Hard disks using the "Winchester" disk technology are available at relatively low cost for systems requiring more than the minimum amount of storage. These disks hold about 5 MB - 50 MB, and are the most reliable direct access high speed mass storage available today. Their reliability stems from the fact that they are sealed against contaminants which shorten the life of disks made using other disk technologies.

Requirement 6 is usually met by purchasing the maximum amount of memory that the processor can address, from 32 kilobytes (KB) up to about 256 KB. Newer processors can address 8 MB of memory or more, but many of the processors in general use today have a limit of 64 KB. Memory prices have been reduced dramatically in the last few years, so that today, it is more economical to purchase a large amount of memory than it is to lower programmer productivity by causing the programmer to run out of available memory at crucial moments.

Requirement 7 is normally met by developing the software on the same machine that will be used to control the telescope. A machine with adequate cabinet space and power supply reserve for peripherals is needed, so that both software development peripherals and telescope control peripherals can be installed, powered up, and functioning simultaneously. If software is developed on a different machine (which is not recommended), a relatively high speed data link between the two machines is necessary, to minimize the time required to modify and test software. It should take no more than five minutes, and preferably even less time, to make a simple change to the source code and load a new program version.

2. Development Languages. In most systems built today, software is the most expensive component. Booch (1983) describes the current situation as "the software crisis". He cites a study which indicates that for systems which include both hardware and software, the proportion of costs was divided in 1965 in the proportions 85% : 15% between hardware and software respectively, but in 1970 the proportions were 35% : 65%, and by 1985 they will be 10% : 90%. The Department of Defense spent over $3 billion on software alone in the early 1970's, and Booch quotes a study which predicts it will grow to $32 billion by 1990.

Booch presents the results of another study which showed that of all software dollars spent by the Department of Defense in 1973, data processing (written primarily in COBOL) used 19% of the total, scientific software (written primarily in FORTRAN) used 5%, with the largest segment being embedded computer systems, which used 56%. Other indirect software costs used the remaining 20%. Embedded computer systems are those systems which use computers to do something other than compute and display (or print) numbers. That is, the **primary** purpose of the system is not data processing. A telescope control system is an example of an embedded computer system, since the real purpose of the system is to actuate motors to point a telescope, even though some numbers are computed and displayed on the operator's screen to achieve this.

The symptoms of the software crisis, as enumerated by Booch (1983, p.6), are as follows:

"Responsiveness: Computer-based systems often do not meet user needs.
Reliability: Software often fails.
Cost: Software costs are seldom predictable and are often perceived as excessive.
Modifiability: Software maintenance is complex, costly, and error prone.
Timeliness: Software is often late and frequently delivered with less-than-promised capability.
Transportability: Software from one system is seldom used in another, even when similar functions are required.
Efficiency: Software development efforts do not make optimal use of the resources involved (processing time and memory space)."

To solve this problem, in recent years a collection of methods has evolved into a system development methodology called structured programming (see, for example, Dijkstra (1976) and Tausworthe (1977)). Also, a new programming environment, centered around the Ada programming language, has been adopted by the Department of Defense and a few other organizations as the only acceptable environment in which to build embedded systems. The emphasis of published papers in the field has been on software, because, as mentioned above, advances in hardware technology and the current software job market have made software the most expensive component of a system, especially of large systems (see Appendices A and C).

The main goals in selecting a software development language are as follows:

1. Generate efficient code that executes quickly in a real-time control environment
2. Minimize the initial software development costs
3. Minimize the total system life cycle costs

Although assembler language can be used to generate more efficient code than high level languages, programmer productivity is much higher using high level languages, so initial software development costs are much lower when a high level language is used. Since it is easier to understand the purpose of any given section of code written in a high level language, it is easier to find errors in the code, and to make enhancements to the code after the system is operational. Also, after the original programmer is no longer available for software maintenance, there is a better chance of finding programmers conversant in a well-known high level language than in the assembler language of any given processor. It takes a new programmer less time to acquaint himself with the telescope control software if it is written in a high level language than if it is written in assembler language. All this means code written in a high level language has a lower life cycle cost than code written in assembler language.

Device drivers and some routines which are executed frequently and need to be efficient are best written in assembler language. All other code should be written in a high level language.

The relative advantages and disadvantages of some of the more popular high level languages are summarized in the paragraphs below. There is no one language which is clearly the best for implementing a telescope control system, and experts disagree over the usefulness of any particular language for a given application. Each system implementer must select the language that is best suited to his software development and operational environments, and to the programmers developing the software.

The Ada programming language was invented specifically for developing embedded computer systems, and to address the problem of the software crisis mentioned earlier. Ada employs many modern software engineering language constructs which are designed to force a programmer to develop correct, readable, and maintainable code. It was designed for real-time applications, and uses features of the language itself to control inter-task synchronization and communication, rather than relying on a programmer's knowledge of a particular operating system to perform these operations. This allows the code to be transported to different machines without modification. To enforce this transportability, the Department of Defense has trademarked the name "Ada", and only the DoD certifies particular compilers to be Ada compilers. Therefore, there is only one standard version of the language.

To date, only a few Ada compilers have been certified. One highly publicized early attempt to build an Ada compiler in a very short period of time resulted in a compiler that took forever to compile code that executed very slowly. In contrast, modern commercially-available Ada compilers are available which compare favorably with FORTRAN in both compile time and execution speed (Jaworski, 1983), yet address the problems of real-time control and life cycle cost in a way that FORTRAN and many other languages cannot hope to do. The availability of an Ada compiler has been given very heavy weight in the WMO computer procurement deliberations.

The BASIC language is probably the most widely-known high level computer language. Its most popular form is an interpreter, which translates each line of code into machine instructions as the program executes. Although interpreters are useful for developing and testing software quickly (since there are no compile and link phases when a program is changed), the execution speed is 2-50 times slower than a compiled version of the same language. Early versions of BASIC limited variable names to two characters, which hindered any attempt to make the code readable and maintainable. Modern disk-based versions usually do not suffer from this drawback. In addition, most BASIC versions are not structured. This means that certain fundamental program structures which make programs easier to understand, test, and maintain, such as IF-THEN-ELSE, DO-WHILE, and DO-CASE, are not available in these versions, so the programmer tends to fill a routine with GO TO statements which deter others from understanding the program. Many early versions of BASIC also lacked the ability to divide a program into subroutines, and many do not use double precision (8-byte) or even single precision (4-byte) real variables. Although BASIC is available in compilers as well as interpreters, the compilers often do not optimize the machine code for execution speed.

However, there are some versions of BASIC compilers, such as the Basic09 mentioned in Section IV, which are quite sophisticated, and which provide all the tools needed to program a real-time control system. Since so many different versions of BASIC are available, if BASIC is considered as the programming language of choice, the version to be purchased should be investigated thoroughly before the decision to purchase it is made.

FORTRAN is one of the oldest programming languages in use, so there is a large base of programmers available to do the software development and maintenance. It is available as a compiler, often with execution speed optimization.

Variable names are up to six characters long, with some newer versions allowing 31-character names. Many newer versions also permit structured IF, DO, and CASE blocks, which lowers both development and life cycle software costs. Most real-time operating systems support both FORTRAN and a large library of FORTRAN-callable routines to perform real-time software functions. FORTRAN encourages the use of subroutines, and many operating systems that support FORTRAN allow overlaying of subroutines to minimize the use of memory.

Pascal is a relatively new language which was invented as a structured language, that is, it contains many of the features of Ada that tend to enforce good programming practices. In fact, Pascal was one of the models used to develop Ada. It is available as a compiler, and is in wide use on microcomputers. Although it was not intended for real-time control applications, many real-time operating systems support Pascal. One advantage that Pascal has over other languages is that both the programming language and the machine code bit patterns that the compiler generates have been standardized. As a result, there are several "p-code microengines" available which are designed to execute this standard machine code very rapidly. This reduces the execution overhead time that is usually incurred when using a high level language. Pascal lacks many of the real-time and embedded computer system support features of Ada, which often makes developing a control system in Pascal very difficult.

The last language to be considered is FORTH. It was invented as a real-time control language at the National Radio Astronomy Observatory, and one of its first applications was telescope control. It is increasing in popularity among microcomputer owners because of its efficient use of memory, and the fact that it is self-extensible, that is, the programmer can define new commands in the language. Its main disadvantage is that FORTH uses reverse Polish notation, so that it is not very easy to read and understand a FORTH program. Also, it is not as popular as any of the preceeding three languages, so there are fewer programmers familiar with the language. The language structures do not enforce good programming practices, which tends to exacerbate, rather than ameliorate, the software crisis.

Whatever language compiler is chosen, it should be fully supported by the operating system, and should be capable of using the operating system features for real-time programming.

3. Development Utilities. To develop software, one needs to enter the source code into a computer, and construct an executable program from it. Once entered, the source code must be capable of being modified easily. To perform these functions, one usually uses a source code editor to enter and modify source code, and store it in files located in mass storage. A high level language compiler or assembly language assembler translates the source code into a machine code object module. The resulting module is then processed by a linker program that combines it with other user-written object modules and system library object modules to produce an executable program. In many systems, these functions are handled by individual operating system utilities. However, all of these functions are incorporated into high level language interpreters, such as BASIC.

The source code editor should be easy to use, and should allow changes to be made rapidly. The best source code editors are full-screen editors intended to be used with CRT terminals, since the programmer sees the results of changes very quickly. Such editors tend to be similar to word processors, in that they can move large blocks of lines at a time, and can "cut and paste" characters, words, or paragraphs quickly and easily. If word processing software is available inexpensively, the computer can be used not only for software development and for controlling the telescope, but also for preparing a journal article describing the results of one's research. Since programmers are not trained typists, and secretaries often do not possess the specialized knowledge required to enter source code correctly and efficiently, good source code editors can enhance programmer productivity enormously.

The language compiler and linker must be procured along with the operating system to ensure that the programs generated by these utilities can be executed by the operating system.

Other useful utilities that are usually supplied with operating systems include programs to copy a file from one device to another device, build a library of object modules, display a list of all the files stored on a particular disk, and dump all or selected portions of memory, or of a file.

All of these utilities are needed for managing code and data files during software development.

J. OPERATING SYSTEMS

The operating system is a program which manages the computer resources, including access to peripheral devices and the processor. The operating system usually includes the device drivers, which handle the flow of data between main memory and a peripheral device, and some form of task scheduler, which determines when programs are allowed to execute.

The operating system can be designed to provide the programmer with support for running programs in a particular environment. For example, a time-sharing operating system provides equal periods of time to each of several users in a round-robin fashion. An interpreter, such as BASIC, often has the device drivers and other rudimentary functions of operating systems built into it, so that no other operating system is needed. Compiled languages, such as FORTRAN, generally require the services of a separate operating system.

A real-time operating system typically allows more than one program to reside in memory simultaneously and to appear to the user to be executing in parallel. The operating system provides a means for tasks to synchronize themselves to each other, to external events, and to one or more timers. In this way, a calculation can be performed periodically, or as the result of the receipt of an interrupt from a particular peripheral device.

Various operating systems are discussed below with regard to their suitability to a real-time control environment. Since there is no good objective means for comparing operating systems, these evaluations are, of necessity, subjective.

One of the most popular operating systems for microcomputers is CP/M (control program for microcomputers). This is a small, simple operating system capable of executing only one program at a time. It provides an adequate set of utilities for software development, but it is not designed for the real-time control environment. Some operating systems have appeared recently which allow several simutaneous users to use CP/M, but these systems do not permit a single user to run multiple programs in parallel. CP/M has begun to fade in popularity since IBM selected MS-DOS as the operating system for its PC.

Another popular microcomputer operating system is Unix. This operating system was designed for the software development environment, but older versions were not well-suited to real-time applications because they were slow in executing. The latest version of Unix, System V, is designed for real-time applications, but currently it is available for only a limited number of processors. This situation should not last long.

A relatively new microcomputer operating system is MS-DOS, which was originally designed to run on the IBM personal computer, but which will run on several others as well. Because it can run only one program at a time, MS-DOS is not suited to real-time applications.

A new operating system designed for microcomputers that appears to have good real-time features is S1 from Multi Solutions, Inc. This operating system is available for the 68000, Z-80, 8080, and 8085, and as of this writing, plans are to make it available for the 8086/88, 80186, 80286, 16032, and 32032 processors as well. The operating system is available for $250 to $1100, depending on the processor used, and has a full array of utilities, such as an editor, linker, loader, and networking software, and supports a relatively large number of languages, including Fortran 77, Pascal, C, BASIC, COBOL, PL/1, Ada, LISP, and SNOBOL (the latter four languages are scheduled for late 1984). Real-time features include multitasking, communications between tasks, timers, semaphores, and locks, as well as other features, such as windowing, terminal support, printer and plotter support, and an extensive file system that can read CP/M, MP/M II, CP/NET, Concurrent CP/M, MS-DOS, UNIX, XENIX, FLEX, IBM 3741, and DEC Files-11 file formats. Language compilers are optional at extra cost, which ranges from $90 up to $850, depending on the language and the target processor.

There are two real-time operating systems available from Digital Equipment Corporation for use on its PDP-11 and LSI-11 computers. RT-11 supports two programs, or tasks, one of which is given a higher execution priority than the other. Individual FORTRAN subroutines may be connected directly to interrupts, or they may be triggered by a timer, rather than by being explicitly called by another routine. A full array of operating system features for real-time control is available in FORTRAN-callable routines, including the ability to set and wait for timers, pass data to the other task, and execute a task periodically. RT-11 is designed to execute on all PDP-11 and LSI-11 processors, including the LSI-11/2. It is designed to consume 5 - 8 KB of main memory.

RSX-11M is a complex multi-tasking real-time operating system which permits up to 256 tasks to execute in parallel. It supports local and global event flags for signalling input/output (I/O) completion or some other significant event, queuing of I/O requests to device drivers without waiting for I/O completion, and global common areas for inter-task communication. Data messages may be sent directly between tasks, or data may be sent by reference (only the address of a data block is sent). A wide variety of timers is available for scheduling tasks to run periodically, or tasks may be connected directly to a particular interrupt. RSX-11M consumes about 50 KB of memory and costs about $2700 for the Micro/RSX version, so it is intended only for sophisticated systems. It is ideal for the real-time control environment, and executes on all PDP-11 and LSI-11 processors, except the LSI-11/2.

MicroVMS is a downsized version of the VAX/VMS operating system intended for execution on the MicroVAX. It has all of the features of the large, sophisticated, and mature VMS operating system, except for PDP-11 compatability mode and batch queue processing. VMS features which are supported include demand memory paging, which maps 1 billion bytes of virtual space into up to 4 million bytes of physical space, and a complete and extensive set of real-time multitasking features. Anyone who has used VMS on a VAX realizes the large gap in performance and features that exists between VMS and all the other operating systems mentioned thus far.

When the wide array of available processors is evaluated for suitability to the task of real-time telescope control, each processor should be evaluated not only on its ability to perform all required calculations in the required time, but also on the availability of software and hardware to work in harmony with the processor to develop the telescope control software and support its execution in a real-time control environment.

K. SELECTION OF THE LSI-11.

By now, you have no doubt guessed that I have selected the LSI-11 to serve as the control computer. The main reasons for this selection are as follows:

1. I have 11 years' experience as a PDP-11 programmer. The PDP-11 and LSI-11 have the same instruction set, so programs written for one machine will execute on the other. I am very familiar with DEC operating systems, compilers, utilities, hardware, assembly language, and philosophy, so it was decided to use a product that held promise for success.

2. A surplus LSI-11 system from Newman Computer Exchange in Ann Arbor, Michigan, was purchased very inexpensively. The 12-slot card cage, dual power supply, processor, memory, and quad serial port together cost only about $2000. This system is shown in Figure 14-8.

3. The open frame design of the Q-Bus card cage and cabinet that were purchased permit the routing of a large number of cables into Interface cards with ease. This is not always as convenient in chassis of computers which are marketed as "personal" computers.

4. A good operating system and language compiler were purchased inexpensively with which I had previous experience. RT-11 was chosen as the operating system (RSX-11M was preferred, but it was too expensive) and FORTRAN as the development language. Although an excellent Ada compiler is now available for the VMS and MicroVMS operating systems, as of this writing, it is not available for either RT-11 or RSX-11M. RT-11 comes with a very broad collection of utilities which save countless hours of software development time. It also comes with a wide array of device handlers, which helped to get the system up and running quickly.

5. As mentioned earlier, as processing needs grow, the processor can be upgraded quickly, easily, and inexpensively.

6. One of the three text editors available with RT-11 is the same DEC standard full-screen editor that I use at work on VAX's and PDP-11's. It works with DEC VT-100 and VT-52 CRT terminals, so a VT-100 was

purchased for the WMO control system. This combination produces a very powerful text editing system. Several manufacturers make graphics upgrade boards for the VT-100, and other kinds of boards are also available for the VT-100, including a voice recognition system.

7. There are thousands of programs written for the PDP-11. For example, a copy of RUNOFF, a text formatting program similar to a word processor, was purchased for $20 from DECUS (the DEC users' society). The DEC standard editor and RUNOFF were used to write my Master's thesis and this book.

8. Appendix H contains a list of manufacturers who make Q-Bus compatible interface cards for a wide variety of applications. From the length and content of this list, it appears there will be no trouble connecting strange devices (shaft encoders, stepper motor controllers, astronomical instruments, etc.) to the computer. Those interested in image processing should be aware that Matrox makes a wide range of graphics and image boards for the LSI-11. For under $10,000 , one can assemble a high resolution (512 x 512 x 12-bits) image processing system that rivals those costing five times as much. Furthermore, Sky Computers makes an array processor for the LSI-11 for under $5,000 which offers the performance to do two dimensional Fourier transforms and other image processing calculations rapidly.

14-8 LSI-11 Telescope Control Computer

The LSI-11 fulfilled all of the computer selection criteria at an affordable price, and my extensive experience with DEC machines made it the obvious choice. It is now more than two years since the computer was purchased, and with each

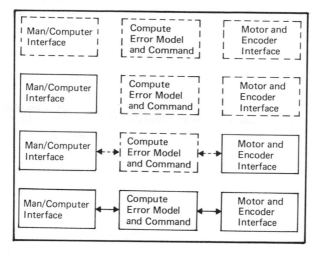

14-9 System Development Strategy

passing week (usually during which there is an announcement from DEC or someone else of yet another Q-Bus compatible product) the decision looks better. However, for those not so wedded to DEC, an excellent choice for a project of the magnitude of the WMO telescope would be a 68000 system based on the Multibus using the S1 operating system and the Ada programming language.

The key to computer selection is to think out every feature of both hardware and software that is needed to make the system work, and then buy a computer that does not lack any of these features. Many have made the mistake of waiting until the computer was brought home to discover that the hardware or software lacked some feature that was assumed to be included. For example, any operating system billed as being "real-time" does not necessarily have all the features that are needed to make a particular system work. Most computer vendors will allow a prospective buyer to peruse the manuals before making a large investment in hardware and software.

L. SYSTEM DEVELOPMENT AND EVOLUTION

It should be noted that the telescope described in these two chapters has not yet been built, although the project is in an advanced stage of design, and most of the anticipated problems have been solved on paper. To help ensure the success of this project, the system will be built in a methodical fashion. Figure 14-9 shows the basic functions of the system, and how the system development will proceed. The system consists of (1) the observer/computer interface (hand paddle, CRT keyboard, and CRT screen), (2) the error and command processing and servo implementation, and (3) the computer/telescope interface.

In this system, the observer/computer interface will be developed first. It is probably the simplest part of the system to design and code, and the easiest to debug. Doing an easy part first helps the system implementer get accustomed to the software development environment. By developing the CRT display pages first, a method will then exist of displaying various intermediate results to help test the rest of the software in the system as it is developed. This is very important. When developing software for a real-time system, often one must see what is going on inside a program. If a WRITE statement is placed in the program to display or print intermediate results, the relatively long time it takes to perform this extra I/O may throw off the timing of the real-time software. By first writing the "background" or lower-priority task that is used in the operational system to display data, data from a higher priority task under development can be sent to the lower priority task for display without disturbing the critical timing of the high priority task.

The motor and encoder interface drivers will be developed next. The encoder interface hardware and software can be tested first by slewing the telescope by hand without the motor connected, and watching the raw bits change on the CRT screen. Next, a simple command generator will be developed which commands the motors to run at a constant speed. Motor control can be tested using the encoders. This will allow verification that the raw bit pattern changes by the right amount and direction as the telescope moves. The next step is to get the motor to respond to keyboard inputs from the terminal. The final step of this phase is to get the motors to respond to hand paddle inputs. This will include the stepper motor ramping for slewing, and the alt-az drive rates. Once all this is finished, about one-third of the total lines of code, two-thirds of the subsystems shown in Figure 14-5, and three-quarters of the total software development effort (measured in hours) will have been completed.

The last phase is to put in the reduction to apparent place, encoder calibration, and error calculations. At this stage, a simple power series calibration will be used. After some field experience is gained, the error models discussed in Chapter 7 will be implemented, and only if field experience dictates will the extended Kalman filter be used. There will still be several months of tweaking and tuning to do, even though the software development for telescope control will be essentially complete. Once the telescope is going, attention will be turned to the development of the high speed photometer. A more detailed evolution plan is presented in the next chapter.

This approach is an example of the "build a little, then test a little" philosophy, since each new capability is tested before the next one is even coded. If all the code were developed at once, there would be thousands of lines of code to peruse to find the bug(s) that prevents proper operation. This way, one builds on software that works, which tends to isolate the location of the next irksome bug.

In the following chapter, specific details of the hardware and software design of the WMO telescope are discussed.

Chapter 15. **PORTABLE TELESCOPE
HARDWARE AND SOFTWARE DESIGN**

In the previous chapter, the system requirements were defined, the top level design was developed, and the basic system evolution plan was described. Now, the detailed hardware design can begin. This involves identifying each functional building block of the system, finding the available options to perform that function, and then performing a trade-off study to select the best option. The resulting system should then be compared to the original functional requirements to ensure that it will give the desired level of performance.

In the following sections, each major function of the Winer Mobile Observatory (WMO) telescope is listed. Where it would be instructive, options are identified, and the reasons for choosing each option are explained. The reader is reminded that what follows is a first attempt at a design that has not yet been built, although the computer system is in hand.

A. DETAILED SERVO DESIGN

The easiest method of implementing pointing and tracking functions on an equatorial mount is to "stop the sky" in hardware. This simplifies the software by allowing it to maneuver in equatorial coordinates. This is the approach taken by Boyd and Tomer.

A similar approach is taken in the WMO telescope, but the alt-az drive rates (shown in Figures 7-1 and 7-2) do not permit the sky to be stopped in hardware very easily. Instead, the approach is to use software to generate the alt-az drive rates needed to stop the sky in one routine, then add corrections to these rates generated by the error correction routines. The resulting drive rate is then sent to the motor. The frequency with which this is done is the effective bandwidth of the servo, with the control software serving as the servo filter.

There are two major control software elements in the system--a closed loop (for tracking), and an open loop (for slewing). A closed loop is used for tracking since the alt-az drive rates for both axes depend on the current pointing angles, and small errors in position quickly become large errors when the drive rates are high. Because this servo must operate over a wide range of drive speeds, it will be capable of "locking in" and generating correct drive rates despite relatively large errors in position. Therefore, a simple open loop can be used for slews. At the end of a slew, the tracking software will be activated to bring the telescope onto the target. These control elements consist of separate routines which reside in memory at all times, and are called when needed.

For the tracking servo, there are two factors which determine how often the motor speed must be updated: (1) the natural frequency of the telescope, and (2) the rate at which both modelled and alt-az drive rate errors accumulate an error in position. The goal is to make the motor speed update rate very much different from the natural frequency of the telescope. This is to avoid subjecting the mechanical structure to speed changes (forces) at a rate that causes it to oscillate continuously. Since telescope natural frequencies tend to be in the range of 1-5 Hz, speed changes should occur either slowly (once every 10 seconds or more), or rapidly (10 or more times per second).

In the previous chapter, it was determined that the alt-az drive rate errors accumulate faster than errors due to either astronomical or mechanical corrections. These error accumulation rates require that the azimuth drive rate be computed 14 times per second, and the altitude drive rate be computed twice per second. To correct for drive rate errors at other than the telescope's natural frequency, speed changes must be made at a rate which is higher than this frequency. Since the altitude drive natural frequency is likely to be near 2 Hz, the altitude motor speed update rate may need to be increased in the future. This will be part of the tuning process that is a necessary component of any telescope control project.

When new RA and Dec coordinates are entered, the apparent place is computed. These coordinates are next converted to alt-az coordinates, then the slew software is activated. A motor speed profile which includes ramp up and down slopes is computed to get the telescope to the new position in the least amount of time, subject to ramping constraints which are determined empirically. This profile is computed to first order using the azimuth and altitude that correspond to the new position's equatorial coordinates at the moment the motor speed function is computed. A second order correction is then added to account for the changes in alt-az coordinates that occur during the time required to execute the slew. The motor speed function is stored in a table containing one motor speed entry for each 1/60 second of time the slew maneuver is performed.

The slew is then performed using a routine that executes 60 times per second. Each time it is executed, the routine fetches the next entries in the motor speed tables, and sends them to the motors. This is done completely open loop, without reading the encoders. After the slew is complete, the tracking servo software is re-entered, which handles the final lockup on the target using the encoders.

After the slew is complete, the tracking servo software is activated. At the appropriate intervals, the planetary parallax (if needed) and refraction corrections are recomputed, and the apparent RA and Dec are updated. The encoders are also read at the appropriate intervals, and corrections are applied to compute the current telescope pointing angles. Each time this is done, the apparent equatorial coordinates of the target are converted to alt-az coordinates. The difference between the actual and desired positions is then found, and converted to a step rate change designed to reduce the error to zero by the next execution loop of the servo software. Finally, the drive rate for that axis is computed, and the error correction rate is added to it. The resulting rate is sent to the motor controller card. An enhancement to keep in mind for the system tuning stage is to add a term that depends on the rate of error growth (as measured by two successive encoder readings). This added complexity would make the tracking a bit smoother.

Plans at this stage are to use fixed rates for slew and set commands entered on the hand paddle. The software will inject appropriate ramps when executing these commands. The rates used will be adjusted in accordance with field experience.

This approach to the pointing and tracking functions represents a reasonable balance between hardware and software complexity, while it promises to meet or exceed the performance requirements.

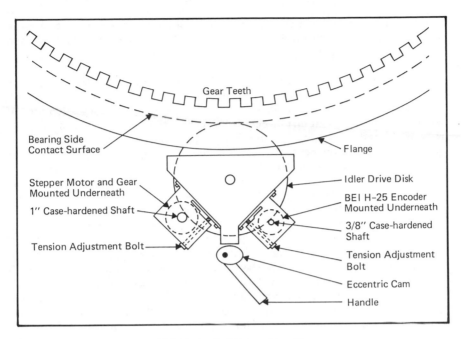

15-1 Mechanical Drive Assembly

B. DRIVE TRAIN DESIGN

1. Drive Train Mechanical Design. The top level design discussed in the previous chapter employs a stepper and friction disk drive. Figure 15-1 shows the azimuth bearing and its drive. The size and location of the outer flanges of the bearing (which is already in hand), and the need to have part of the telescope mount overhung the top flange of the bearing preclude driving the bearing at the flange. Instead, an intermediate roller disk will drive the outer wall between the top and bottom flanges.

The drive disk will be made of 1-inch thick hardened steel so that a force of several hundred pounds can be used to bring the disk into contact with the bearing. It will be supported on a 1-inch diameter case hardened steel shaft riding on roller bearings. The roller bearings are press-fitted into steel plates that form the drive assembly housing to which the motor and encoder are bolted. The assembly is brought into contact with the bearing using an eccentric cam pushing on the assembly from behind.

Options for driving the disk are: (1) direct drive using a shaft passing through the center of the disk and attached to it, or (2) driving the disk with a roller in friction contact. A disadvantage of the first option is that the disk must be relatively large for clearance between the motor and the moving edge of the azimuth bearing flange. This means that if the motor drives a shaft attached to a 12-inch diameter disk, the effective gear reduction for a 45-inch diameter azimuth bearing is only 45/12 or 3.75:1. Although this would work, it does not take full advantage of the characteristics of the friction drive, which is virtual freedom from backlash, tooth-to-tooth gear errors, periodic gear errors, and other typical gear errors. Hence, this last stage in the drive will reduce the errors occuring in earlier stages by the effective reduction ratio of the final stage without introducing significant new errors. This means that a final reduction ratio higher than 3.75 is more desirable.

The second option permits more latitude in selecting the reduction ratio, since the disk serves only as an idler. The reduction ratio is the bearing diameter (45 inches) divided by the drive shaft diameter (1-inch). The shaft is precision ground and case hardened to prevent dirt or dust from scoring the surface and reducing the accuracy of the reduction. The 1-inch diameter is a good compromise between the strength needed to bear the contact force, the accuracy required to give a constant gear reduction without introducing errors, and a good reduction ratio from drive components earlier in the drive train. Note that the shaft-to-disk contact force is applied by slotting the motor box, bolting it to the main drive box, then adjusting the contact force with the two additional hardened bolts working against the slot.

A shaft angle encoder in a separate box is mounted on the other side of the main drive box using the same slot and bolt arrangement. It uses a 3/8-inch hardened and precision ground steel shaft. Here the contact force need not be large to drive the encoder, and the smaller shaft produces a higher gear reduction which results in greater encoder resolution. To avoid encoder "gearing" errors, the smaller shaft must have very little runout, so it is ground to high precision after hardening.

The drive for the altitude axis is similar to that for the azimuth axis. In place of the azimuth bearing, a 30-inch diameter steel flywheel ring which was found in a junk yard will be accurately machined and ground, and attached to the main telescope tube assembly. A drive assembly using only the rollers attached to the motor and encoder will be pressed directly against the disk.

The drives will not be loaded when the trailer is being transported, since movement of the roller against the driven disk or bearing when the trailer hits a bad bump will tend to score the roller or disk. Scoring introduces periodic errors which are difficult to model, especially if the scoring occurs on the roller.

2. Motor and Motor Gearing Selection. To simplify the drive design, the computer interface for the motors, and the control software, only one motor per axis will be used.

Three motor configuration options were considered:

> 1. A 200-step per revolution Superior Electric Slo-Syn motor (or equivalent), coupled to the one-inch drive shaft through a 359-tooth high accuracy (Thomas Mathis or Byers) worm and worm gear.
> 2. A 50,000-step per revolution Compumotor directly coupled to the one-inch drive shaft.
> 3. A several thousand step (2,000 to 12,000) per revolution Mesur-Matic motor, with a nutating gear reducer coupling the motor to the shaft.

Option 1, the stepper driven at full (or preferably half) steps through a worm, has a major disadvantage. Such motors have large drops in torque starting around 1000 steps per second when driven with conventional stepper drivers, and require a very sophisticated driver to operate above 5000 steps per second. The 359-tooth worm gear is used to make each motor half step represent 0".1 rotation of the azimuth axis. This is done to obtain smooth tracking. At 5000 half steps per second, the maximum speed is 500" per second, or about 0.14 degree per second. This would be a slow slew speed, so separate slew motors would be needed, complicating the drive considerably. Even if full steps are used (which would make the motion jerky) the slew rate would only be 0.28 degree per second.

Option 2, the Compumotor directly driving the shaft, has the wide range of speeds needed to drive an altazimuth mounted telescope. The shaft-to-bearing reduction of 45 gives 50,000 x 45 = 2,250,000 steps per revolution, or 0".576 per step. This step size is a little larger than desired, because it might result in a jerky motion. However, this is a viable option.

To insure enough torque to move a 30-inch telescope, the motor would have to have a torque rating of about 500 oz-inches, which costs about $1700 per axis for the motor and its microstepping controller. The motor is capable of step rates up to 200,000 per second with little loss of torque, and 500,000 per second at greatly reduced torque. The latter step rate would permit a theoretical maximum slew rate of 80 degrees per second with ramping--truly impressive performance.

The main disadvantage of the motor, aside from its high price, is the motor shaft deflection error. If the change in load torque is a significant fraction of the pull-in torque, a potentially large shaft position error can result. The other problem of microstepped motors, step size variation due to torque ripple, is also a problem here, because the 45:1 final friction drive reduction divides the 5' worst case sinusoidally varying error down to about 7". In the WMO system, the error budget is already tight without such a large contribution. If the final friction drive reduction were 200:1 or higher, a more desirable step size would result, and shaft position and step size errors would be acceptable. However, this is not the case, so the Compumotor driving the friction roller directly is not the best approach for this application.

Option 3 is the Mesur-Matic motor and microstepper drive that can be programmed (by a PROM) to have as many as 12,800 steps per revolution. The first step is to choose the effect of one motor step on the azimuth axis of the telescope. Step sizes used successfully by observatories in high accuracy drives vary from 0".125 to 0".05. To try to keep the step rates higher than most of the anticipated mechanical resonances in the drive and mount, each step will be 0".05.

The next step is to choose the reduction gear ratio between the motor and the shaft. With 1,296,000" per circle, if each step is 0".05, there are 25,920,000 steps per circle. With a reduction of 45:1 in the final friction drive, this means the motor/gear combination produces 576,000 steps per revolution of the one-inch drive shaft. Since the Mesur-Matic motor/microstepper drive combination produces a maximum of 12,800 steps per revolution, further gear reduction is necessary.

Probably the best gear reduction for this application will be a product which Mesur-Matic has announced, but which is not yet available at the time of this writing (normally I am skeptical of products that don't yet exist, but this appears to be in an advanced stage of development). This is their Series 900 Gear Reducer, which works on a nutating gear principle. In this design, an inner gear attached to the input shaft with n teeth "wobbles" (nutates) around inside a case with n+1 teeth that engage most of the n teeth of the inner gear. The effective reduction is n:1. This type of gear would work best with a reduction of about 100:1, although ratios from 10:1 to over 250:1 are feasible. It remains to be seen what ratios will be available. The announced price is $495. Advantages of such a gear are automatic wear compensation, effectively zero backlash, and good accuracy. Although I have not seen specifications for the final product, an accuracy of 0.067 degrees and a repeatability of 0.01 degrees have been achieved in prototypes. When this is divided by the factor of 45 in the final drive stage, the maximum error contribution of the nutating gear reducer is 5".36, while typical errors are expected to be around 2". This represents a large fraction of the total error budget for pointing and tracking, but it is still acceptable. (Note that Byers and Thomas Mathis usually quote around 10" total accuracy for their stock worm gear drives.)

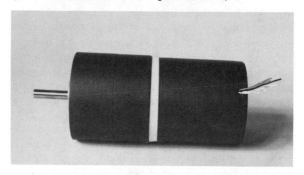

15-2 Telescope Motor and Gear Assembly

15-3 Mesur-Matic Model 3640A Microstepping Motor Controller

Assuming a reducer gear ratio of 100:1, the motor can have relatively low output torque. It should have roughly 5760 steps per revolution to achieve the 576,000 steps per revolution of the one-inch shaft. If a Model 3640A Microstepping Driver/Translator is used with a division factor of 29, then there will be 29 x 200 = 5800 steps per revolution. With the 100:1 reducer, this will give 580,000 steps per revolution of the one-inch drive shaft, making each step 0".05 . This driver is capable of handling pulse rates up to 1 MHz, with actual slew rates being as high as 500,000 steps per second. If used with a Mesur-Matic permanent magnet stepper, the result is an adequate speed range for tracking, setting, and slewing. The Series 900 Gear Reducers are designed to connect directly to the Mesur-Matic Series 340 steppers. Figure 15-2 shows the Mesur-Matic Series 340 stepper motor and Series 900 gear combination, and Figure 15-3 shows the Model 3640A microstepping motor controller.

Although some gear reduction could be used with the Compumotor to reduce the errors due to dynamic load torque and torque jitter to tolerable levels, the resulting step size would be extremely small, since Compumotor only offers motors with 25,000 and 50,000 steps per revolution. This places severe demands on the pulse rate generator interface, and reduces the slew speed considerably. Although the Compumotor is an excellent choice for many telescope drives, in this application, the flexibility of the Mesur-Matic choices in microsteps per revolution tips the scales in favor of the latter.

3. Motor Controller Computer Interface Selection. Stepper drivers typically require two input lines. One carries a series of pulses that determines the motor's speed (the motor steps once for each pulse received). The other controls the motor's direction. The purpose of the motor interface is to provide the pulse rates needed to drive the motor over the range of rates needed for accurate tracking.

The long term tracking accuracy requirement is 5" per hour, which represents 1 part in 10,800. The short term tracking accuracy requirement is a 2" circle of confusion, which means the pointing angle in equatorial coordinates should not change more than 2" between two successive motor speed commands. For the azimuth axis, there are 14 motor speed commands generated each second, and for the altitude axis, two each second.

Any motor interface card can offer only a finite number of motor speeds. If that number is too small, the difference between two "adjacent" speeds may be so coarse that in the half second between commands to the altitude motor, the speed selected in the last command cycle produces an error of more than 2" before the next command is sent. To prevent this from happening, motor speed resolution (the number of different motor speeds available) should be greater than or equal to the long term tracking resolution requirement (10,800).

From the commercial boards for the LSI-11, the options considered for the pulse rate generator were:

    1. The ADAC 1604/POC pulse output controller, supplemented by a separate high pulse rate output board

    2. The ADAC 1412DA D/A converter and a separate voltage-to-frequency (V/F) converter (speed output), plus four lines from a separate TTL I/O board (direction output)

The ADAC 1604/POC board was designed specifically for controlling steppers. It has four output pulse channels, with each channel containing two outputs (one for "up" pulses and one for "down" pulses), a separate open collector driver for both up and down DC levels, and a user-defined software controlled output bit. In addition, there are two single-bit inputs per channel. In the WMO drive system, the up pulses output line and user-defined output bit line would be used for motor speed and direction, respectively, and the two input bits per channel would be used for limit switches. The output frequency can be varied from 10 Hz to 50 kHz by varying the pulse widths (10 µS to 50 mS) and pulse off times (10 µS to 50 mS) independently, using on-board 11-bit counters.

At first, this board seemed ideal for this application, but upon further investigation, several drawbacks were discovered. One is the maximum pulse rate. If 50,000 steps of size 0".05 are made per second, the maximum slew rate is only 2500" per second, or 5/6 degree per second. That is not a very fast slew rate, though track and set rates should not exceed this speed. (If the Compumotor option in the previous section had been selected, with each step being 0".576, 50,000 steps per second is 28,800" per second, or 8 degrees per second, which is a very rapid slew rate.) The low end of 10 Hz (0".5 per second) is a bit high, but tolerable. Some alt-az tracking rates are as low as 0".005 per second, which really demands that the computer generate each pulse in software. To make a smooth transition between software-generated pulses and interface card generated pulses, a pulse rate less than 10 Hz is needed, so that the computer is not burdened with making pulses, for example, at a rate of 9.586 per second, or whatever. Another drawback of the board is that one cannot access the entire range of pulse rates from software. Instead, one of four speed ranges is selected by a hard-wired jumper. The main drawback of the board, however, is that the 11-bit counters offer a motor speed resolution of only 1 part in 2048. This is inadequate to meet the long term tracking accuracy requirement of one part in 10,800.

A way around the slew rate and jumpered speed range problems of the 1604/POC board is to use a custom switching board with the 1604/POC board. This custom board would switch in a frequency from a different source for slewing. One possibility is to use a 1 MHz oscillator and a simple 4-bit counter to divide the 1 MHz by any number from 1 to 16. The output of this circuit (e.g., 7492 TTL) would be switched in place of the 1604/ POC output. The additional circuit would be built on an MDB MLSI-1710 bus foundation module card that provides all the address decoding and interrupt vector circuitry for the Q-Bus interface, as well as wire-wrap spaces for your own circuits. The speed range jumper switching would also be handled on this board. There is no cure, however, for the motor speed resolution problem.

The second option is to use a 12-bit D/A converter to feed a V/F converter. The Burr-Brown VFC32 has a full-scale output of 0.5 MHz when a full-scale input of 10 volts is applied. The ADAC 1412 D/A board can be ordered with 0-5 VDC or 0-10 VDC outputs, among others. If the 0-5 VDC option is ordered, the D/A and V/F combination covers the pulse range of 0-250,000 steps per second in 4096 equal steps of 61.035 pulses per second per count. At 0".05 per step, this represents speeds of roughly 3" per second up to 12500" per second (3.5 degrees per second), which is a typical slew rate. If the ADAC board is ordered with 0 - 10 VDC output, then there are 4096 speeds, ranging from about 6" per second up to about 7 degrees per second (500,000 steps per second). The problem with this approach is that it, too, lacks adequate motor speed resolution, since the 12-bit D/A offers a resolution of only 1 part in 4096.

Since there is no off-the-shelf Q-Bus pulse generator with the required resolution, the interface card is custom made. The main disadvantage to the custom approach is that it consumes a great deal of time (better spent on integrating the system) to design the board, build it, document the design, and debug it (especially the latter). However, after considering all these factors, the decision was made to build custom boards for the elevation and azimuth axes, and to use the ADAC 1604/POC for the focus. The latter does not require pulse rates higher than 50,000 per second, or a wide range of pulse rates, so the ADAC board is ideal. The use of the ADAC board will allow the altitude and azimuth drives to be tested while the custom boards are being debugged. Later, the other ADAC board ports can be used to control the filter and diaphragm wheels on the photometer. Use of the ADAC board will also provide limit switch inputs, which would otherwise have to be designed, built, debugged, and documented.

To avoid using exotic circuitry, such as ECL chips (which require transmission line techniques for impedance matching of chip interconnections), the oscillator will run at 20 MHz. Using a 32-bit divide-by-n counter, pulse rates from 20 MHz down to 0.005 pulses per second can be generated. Provisions will be made for controlling when the pulses are sent to the motor controller, for sending individual pulses under direct software control, and for a direction bit to tell the controller to turn the motor clockwise or counter-clockwise.

Assuming step sizes of 0".05, the pulse rates will range from one million arc seconds per second (278 degrees per second) down to 0".00025 per second. A microstepped motor has a practical limit of about 500,000 steps per second, which translates to 25,000" per second (about 7 degrees per second). A maximum slew of 180 degrees will take about 26 seconds, plus ramp up and ramp down time. This is very good slewing performance.

At a drive rate of 15" per second, the pulse rate is 300 steps per second, requiring a divisor of 66667, which gives 14".999925 per second. The next divisor, 66666, gives 15".00015 per second. The difference is 0".000225 per second, or one part in 66,666 , which more than meets the one part in 10,800 criterion. At slew speeds, a divisor of 40 gives 500,000 steps per second. This gives speed change increments of roughly 10,000 steps per second for each successive integer divisor, which should be adequate resolution to ramp the motor by incrementing or decrementing the 32-bit divisor by one several times over a period of a few seconds.

Therefore, from the standpoint of step rate range and resolution, the custom board is the only approach which meets the performance requirements. By using a 4-channel commercial card for the focus motor, all the drives can be used for testing until the custom cards are debugged.

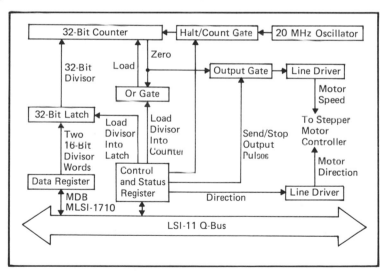

15-4 Stepper Motor Rate Generator

Figure 15-4 is the block diagram of the custom board. The circuit is built on an MDB MLSI-1710 bus foundation module, which contains all of the Q-Bus address decoding and bus interface logic, plus a large board area for the custom circuit. A

20 MHz crystal oscillator sends TTL-level pulses to the "count down" input of a 32-bit counter through a gate which is controlled by a bit in the control and status register (CSR) of the MDB bus interface card. The 32-bit frequency divisor is loaded through two registers in the interface into a 32-bit latch, with a CSR bit determining when the new divisor is loaded into the latch output. This allows the divisor to be loaded from memory with two 16-bit Q-bus transfers without having an incorrect value loaded into the counter between the times the high and low order 16-bit groups are transferred. When the correct new divisor is on the input side of the latch, a control bit can be toggled to place the correct 32-bit divisor on the counter input lines.

When the counter reaches zero, it sends out a pulse which reloads the divisor in the latch into the counter. This pulse is also sent to the motor controller through an output gate controlled by a CSR bit. For example, if the number 20,000 is loaded into the latch, the counter is reloaded, then the halt/count gate is set to count, every 20,000th pulse of the 20 MHz oscillator would reload the counter and be sent out to the motor controller. This means the motor would receive 20,000,000 divided by 20,000 , or 1,000 pulses per second.

Additional bits in the CSR are used to reload the counter through an OR gate, stop or allow pulses to reach the motor controller through the output gate, and set the motor direction. Line drivers on the pulse and direction outputs send the pulses to the motor controllers. High speed Schottky TTL chips are used for the counters and comparators. Simple 8-bit ripple counter chips (74LS193) form the 32-bit counter, since phase delays and a few extra pulses are not really critical, and synchronous counters capable of 20 MHz operation are harder to obtain.

Once the circuit is debugged, several other boards will be made so that spares will be on hand in the field if a failure occurs.

4. Position Encoder Selection. Of all the encoder types, the optical encoder appears to give the best performance for the least cost. My own ability to obtain several high resolution (15 and 16 bits) optical absolute encoders inexpensively on the surplus market demonstrates the viability of this option. However, the electronics in these encoders were often bad, or at least suspect, which is probably what caused them to be declared surplus. Most telescope builders would find themselves wasting too much time diagnosing and repairing surplus encoders.

My inclination over the past ten years or more has been to favor absolute encoders over incremental. This is because every time the power is turned off, incremental encoders catch "instant amnesia". When the electronics are powered up every night, at least one star must be sighted to recover the zero offset calibration constant for the incremental encoders.

To avoid this problem, one can buy an incremental encoder with a zero track that gives a pulse at a predefined zero point of the encoder disk. If the telescope is stored in a "home" position at the end of an observing session, then when the telescope is moved from the "home" position at the start of the next observing session, the zero pulse can be used by the computer to zero out the raw position count.

If an error occurs in an incremental encoder, it cannot be detected or corrected. For example, if a positive noise pulse occurs on the line from the encoder, the raw position count is too high. If a genuine signal pulse is mixed on the line with a negative noise pulse, the counter may not respond to the good pulse. Either way, a random error is introduced into the estimated position, and no way exists to sense the fact the error occurred, much less to correct it. A succession of errors of the same type tend to accumulate until the error is large enough to affect the pointing accuracy of the telescope.

An absolute encoder, on the other hand, never allows errors to accumulate. If a bad reading is received by the computer for some reason, the next time the absolute encoder is read, the error is corrected. However, absolute encoders do have some disadvantages:

1. They are considerably more expensive (up to 10-20 times) when purchased new than incremental encoders.
2. They require a more elaborate (and typically more expensive) computer interface.
3. There are many more signal lines to be run from the encoder on the telescope to the interface card in the computer.
4. Although surplus absolute encoders are relatively easy to find, they often have one or more of their bits burned out, or some other electrical malfunction, which must be found and repaired.

Recently, Genet brought to my attention the BEI H-25 incremental encoder, which has up to 50,800 counts per revolution (almost 16 bits) and costs about $450. Although this is more expensive than the typical surplus encoder, the price is low enough to justify purchasing new encoders that will almost certainly work the first time, that have only a few signal lines to the computer interface card, and for which LSI-11 interface cards are available. The encoders will be ordered with the zero track pulse, and a "home" position will be defined that will be used to reset the counter on the interface card. The H-25 can be ordered with open collector outputs and pullup resistors to aid noise immunity. This option, when used with heavy guage (16-18 awg) wires, should afford enough noise immunity to bring noise-induced errors down to an acceptable level with proper grounding and shielding.

5. Position Encoder Computer Interface Selection. Of all the output options offered for the BEI H-25 encoder, there are two that would be useful to the WMO telescope. The first is interpolation logic, using the standard quadrature outputs on two lines. This produces 12,700 pulses per revolution on each quadrature output line, with the two outputs offset in phase by 90 degrees. The Buckminster Corporation makes an LSI-11 interface card with four channels of two-line quadrature input, and on-board pulse steering logic that multiplies the counts by four to 50,800. This board sells for about $1300. With the 3/8-inch friction roller coupled to the encoder, there are 6,096,000 encoder counts per revolution of the azimuth axis, so each encoder count represents roughly 0".2 .

The second option is to order the H-25 with both interpolation logic and pulse steering logic, which puts out a string of 50,800 pulses per revolution on two lines. One line has pulses on it if the encoder shaft is rotated CW, and the other line has pulses on it if the encoder is rotated CCW. This is a $60 option on each encoder, but permits the use of a standard counter interface (as long as the counter can count both up and down). A commercial LSI-11 counter card suitable to this application does not appear to exist. Most commercial counter cards, such as the ADAC 1604/OPI, only count up. One could, in theory, use two counter ports on the 1604/OPI board and take the difference between them, but the software to sort out the two counts would be more complex, and, given the number of times per second the encoders would be read (about 17), the additional loading on the CPU would not be insignificant. An alternative would be to build a custom counter card, but there are already enough custom boards in this system, so the Buckminster board will be used, configured as two 32-bit counters.

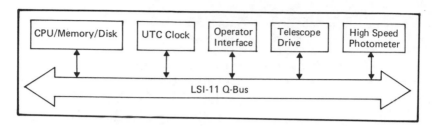

15-5 Telescope Control System Block Diagram

C. COMPUTER SYSTEM HARDWARE

Figures 15-5 through 15-11 show the major hardware subsystems down to the board or major component level. Figure 15-5 shows the major subsystems of the WMO computer system. The LSI-11 Q-Bus and card cage accept two sizes of circuit boards. The larger size is about 8 inches by 10 inches, and is known as the "quad-height" board, because there are four sets of card edge fingers that mate with Q-Bus connectors. The half size board, which measures roughly 8 inches by 5 inches, is known as the "dual-height" board, because there are only two sets of edge connector fingers. All components of the computer are mounted on separate Q-Bus plug-in boards, including the processor, memory, disk interface, and serial interface ports. There are no active computer components hard-wired to the Q-Bus, except the power supply, and even that is easy to disconnect or upgrade. Each subsystem in Figure 15-5 consists of one or more interface cards in the Q-Bus card cage located in the van connected by one or more cables to hardware located at the telescope or in the van. Each subsystem is described below in detail.

1. **Control Computer.** The control computer components are shown in Figure 15-6. The processor currently in the card cage is equivalent to the LSI-11/2. The current plan is to upgrade to the 11/73 in a few years, once the basic tracking software is functioning well. The LSI-11/2 supports 64 KB of memory, whereas the 11/73 supports 4 million bytes (MB). In a few years, DEC will start using the new 256 Kbit dynamic RAMs in its memories. It will then be feasible to put 1 MB of memory on a single dual-height Q-Bus card. The processor and memory will be purchased and the upgrade will be done all at once. If a good Micro/VAX processor is available as a separate unit, and the Micro/VMS operating system and Ada or Fortran 77 are reasonably priced, the upgrade will use the 32-bit processor and the more sophisticated operating system.

The DSD-880 floppy/Winchester disk unit and controller are already part of the current system. This unit provides 30 MB of storage on the Winchester unit, and 1 MB of storage on the double-sided double-density 8-inch floppy drive. Optical disks, which recently became available, may be inexpensive enough in 5-10 years to provide a very large amount of storage (1 billion bytes or more) at a reasonable price. This will make it both practical and economical to store on-line the entire SAO catalog, several other large catalogs, and many years of data.

2. **Operator Interface.** The various devices that form the operator interface are shown in Figure 15-7. The standard multi-port Q-Bus serial interface is the DLV-11J quad serial port. This interface supports only the data and ground lines of the RS-232C protocol--no modem control is provided. Instead, the device driver

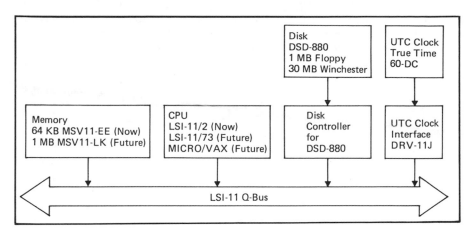

15-6 Control Computer

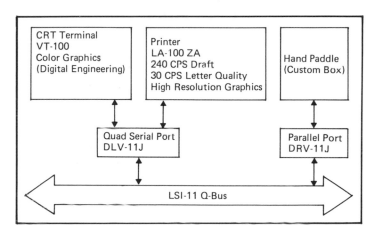

15-7 Operator Interface

regulates data flow using the XON and XOFF characters in the data stream. There are two devices connected to this interface card: a CRT terminal and a printer.

The CRT terminal is a DEC VT-100. This allows use of the DEC standard full-screen text editor that comes with the RT-11 operating system. Selanar, DEC, and Digital Engineering all make plug-in graphics upgrade boards for the VT-100. Digital Engineering also offers an option to replace the monochrome monitor with a color monitor for high resolution color graphics. The VT-100 will be used at first for software development. Later, when field operations commence, it will display control system status, and serve as the means of entering pressure, temperature, latitude, and longitude (for computing sidereal time and various corrections). If field experience dictates, the graphics board can be added after the basic system is operational.

The printer is a DEC LA100-ZA, which is a dot matrix printer that prints at 240 characters per second in "draft" mode, and 30 characters per second in "near letter quality" mode. It can store up to five different type fonts simultaneously and switch between them under either software or front panel control. One of the available fonts is a mathematical and Greek symbols font. Two of the five font slots are set up to accept removable font cartridges, so that any number of fonts can actually be used on the same page. The LA100 also has a high resolution graphics capability. The printer is absolutely essential for software development and for preparing journal articles, and the graphics capability can be used in the van during observing operations to produce a hard copy plot of the data.

The hand paddle consists of a box with pushbuttons for direction (N, S, E, and W) and speed (slew, set, and guide). One option that was considered is to place circuitry into the box that generates ASCII characters when individual buttons are pressed, and converts these characters from parallel to serial format for transmission to a serial input on the DLV-11J quad serial port. However, this circuitry would be subjected to temperatures outside of the normal commercial range, so the components would need to meet military temperature specifications. The result would be needless complexity and expense. Instead, a simple multiconductor cable connects the hand paddle to a DRV-11J 64-bit parallel port.

3. Telescope Drive. Figure 15-8 shows the telescope drive components. The pulse output interface card (ADAC 1604/POC) will be connected to all motor controllers directly until the custom motor interfaces work, at which point the ADAC card will be used as the motor interface only for the focus stepper. This application does not require the fine steps and high step rates of the elevation and azimuth axes, so a conventional stepper interface card, motor, and motor controller can be used.

The BEI H-25 optical incremental shaft encoder will be used on the azimuth and elevation axes. The interpolation and pullup resistor options will be ordered, but not the pulse steering logic. This will make the encoders compatable with the Buckminster encoder interface board. The focus will be run open loop, since temperature variations night to night move the focus position. Limit switches will prevent overtravel, and define the zero point of focus.

The two BEI encoder zero track pulses will come in on the 1604/POC inputs, which are also used for limit switch inputs from the elevation and focus assemblies. Note that the only motion limit in azimuth is cable wrap-up. A limit switch input for azimuth will be added only if this becomes a problem.

4. Navigation and Timing. Highly accurate location and time information are needed to compute local sidereal time, which is used to compute local hour angle from RA. Although both position and time information can be obtained by sighting stars, it is easier and faster to know the time and position independently, so that there are fewer unknowns to be found, which means fewer stars need be sighted before commencing operations. To obtain hour angle to an accuracy of 0".5, both longitude and Coordinated Universal Time (UTC) must be known to within 0".25. Since there are 15 arc seconds per time second, this means that UTC must be accurate to within 0s.017.

Time information will be provided by the UTC clock shown in Figure 15-6. The required level of time accuracy is not obtainable using short-wave WWV broadcasts, because the path of the signal from the transmitter to the observing site is unknown and changing. Therefore, very low frequency (60 kHz) broadcasts on WWVB will be used to ensure that a ground wave of known path length is received. The True Time Model 60-DC receiver has circuitry which computes the UTC to the second using

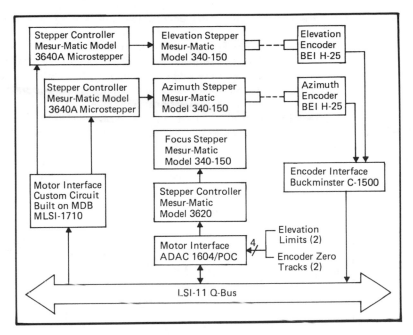

15-8 Telescope Drive

the distance from Fort Collins as an input. The accuracy of the True Time receiver is 0s.0005 when the receiver has a good signal. If the signal is lost (which often happens just before sunrise or just after sunset), a quartz crystal clock inside the receiver is used to "flywheel" through signal dropouts. This signal can be received almost anywhere in the conterminous 48 states, which is a real advantage to a portable telescope. The 1 kHz frequency standard output will be fed to a counter which is reset whenever the seconds count changes. This counter value will be read on the same DRV-11J as the date and time (with seconds resolution).

Obtaining accurate position information is more difficult than obtaining accurate time. A minute of longitude is roughly equivalent to a statute mile at U.S. latitudes, so 0".25 is roughly equivalent to 20 feet. An accuracy of 3 feet can be achieved by observing both channels of several Transit satellites over a period of several days, and reducing the data on a large mainframe computer two weeks later when the exact satellite ephemerides are made known for the periods of observation. That is not much help to a portable telescope that needs an accurate position fix within minutes of arrival at the observing site. By observing one channel of a Transit satellite over the roughly five-minute period the satellite is in view, an accuracy of about 200 meters (8" of longitude) can be obtained. However, one may have to wait up to 45 minutes for a Transit satellite to come into view.

The Global Positioning System, which uses satellites in much higher orbits than the Transit system, usually provides three or more satellites in view all the time from most places on the Earth, so a position fix accurate to 50 feet (on the military encrypted channel) or 150 feet (on the public channel) can be obtained in seconds. GPS receivers, by their nature, also furnish time information accurate to better than one millisecond. GPS receivers made by Trimble Engineering and Stanford

Telecommunications are capable of these accuracies, but they are expensive ($13,000 and higher).

Owing to the expense of satellite navigation receivers, USGS maps will be used to obtain position, which will then be entered by hand on the VT-100 terminal. Roughly 95% of all features shown on these maps are accurate to within 40 feet of the North American datum, though experience has shown that buildings are depicted less accurately than road intersections and rivers (Dunham, 1984). Although use of the maps is more time consuming than a navigation receiver, the results are probably slightly more accurate. The problem with using maps is that last minute changes in plans may move the observing site as much as several hundred miles to an area not covered by maps in the observer's possession. It would be prohibitively expensive, in terms of both cost and space in the van, to keep a complete set of USGS maps.

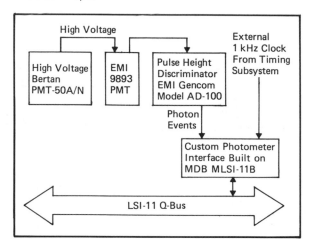

15-9 High Speed Photometer

5. High Speed Photometer. Figure 15-9 shows the high speed photometer block diagram. Since the pointing and tracking accuracies are expected to be high enough to keep the target centered in the diaphragm, a flip mirror for centering may not be necessary. The photometer head contains the high voltage power supply, the PMT, Fabry lens, aperture, and pulse height discriminator electronics. Provision will be made for inserting filters manually, but no motorized filter wheel is currently planned. The PMT will be an EMI 9893 bi-alkali end-on tube, which has very low dark current at room temperatures. This tube is suitable for photon counting without cooling. The Bertan high voltage power supply provides the required voltage regulation.

If the research program were UBV (slow) photometry, the ADAC 1604/OPI counter board could be used to count the photon events. In this case, time synchronization would not be critical. The integration periods would be one to several seconds, so at the end of each integration, an interrupt could be generated, the data moved to memory, and then to the disk, all under direct control of software in an interrupt processing routine. The clock could be read once each reading with only one second resolution.

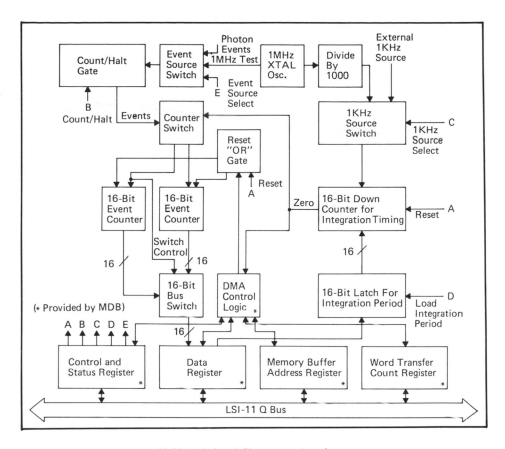

15-10 High Speed Photometer Interface

The instrument requirements for data handling and time synchronization for observing minor planet occultations are much more stringent than for UBV photometry. One system designed to fulfill modern research requirements in this field is that of Schnurr and A'Hearn (1983). This system, which is based on an Apple computer and custom hardware, has 0.001 second photometer integration periods, and 0.001 second time resolution and synchronization to WWV signals. With 1,000 integration periods (data points) per second, if each data point were to generate an interrupt, as it would under ordinary programmed I/O from an ADAC 1604/OPI card, the computer would quickly become overburdened simply handling interrupts.

The solution is to use a "smart" peripheral card that controls event counter resetting, time synchronization, and data transfers without requiring software execution for every data point. The functional diagram of this card is shown in Figure 15-10, and is built on the MDB Systems MLSI-11B direct memory access (DMA) bus foundation module. The card comes from MDB with the Q-Bus DMA control logic and room for custom circuitry. The external inputs to the card are the photon event pulses from the photometer's pulse height discriminator, and a 1 kHz pulse source. At first, this timing source will be a simple crystal oscillator, and the system clock will be used for time tagging the data. Eventually, both the 1 kHz timing pulses and accurate time will come from the True Time 60-DC time code receiver.

The card is initialized by device driver software on request from a Fortran control program. The device driver loads the integration period in milliseconds (up to 32767), the number of data points to transfer, and the address in memory to which to transfer them. When the operator signals the control program to start taking data, it commands the device driver to start the data collection process, which, in turn, sets a bit in the control and status register to initiate data transfers.

Each time a pulse from the 1 kHz clock is received, the integration timing counter is incremented. When this counter equals the integration period (in milliseconds) that was sent to the card during initialization, it directs the counter switch to route all further photon events to the second photon event counter. The contents of the stopped photon event counter are transferred through the bus switch and data register on the Q-Bus to the address in memory previously specified. This is done under the control of the MDB-supplied DMA control logic. If the integration period is specified as 1 ms, then this happens on every pulse from the 1 kHz clock. The memory buffer address is then automatically incremented by 2 bytes by the DMA logic, the Q-Bus register holding the number of data points to transfer is decremented by 1, and the second event counter counts photon events until the next integration period has ended. When this occurs, the first counter is reset and all further photon events are routed to it, and the contents of the second counter are transferred to the (previously incremented) memory address.

This process continues until the Q-Bus register holding the number of data points to transfer is zero. At this point, the card generates an interrupt. Up to this point, the entire process has been handled by hardware, without one instruction in software being executed for this activity. This has allowed the CPU to perform other functions in between the DMA data transfers. If the number of data points between interrupts is set to 250, and the sampling rate is once per millisecond, the computer only has to deal with an interrupt four times per second. When this interrupt is received, the interrupt handling software reads the clock, writes the clock and photometer data to disk, and reinitializes the buffer address and transfer count on the photometer card. Two memory buffers are used, so that new data can be transferred to memory by the photometer DMA card while the previous set of data is being written to the Winchester disk. The DMA card is set up to release the bus after every data transfer, so it does not monopolize the bus between transfers. The Q-Bus is organized so that bus cycles are stolen from the CPU without its being aware of the transfer, so no special software is needed to perform the data transfers, only to handle the interrupt when the entire data transfer of 250 16-bit words is complete.

This is yet another example of the benefits to be obtained from using a smart peripheral card and a standardized bus. A smart card allows a microcomputer to handle very high data rates, and the use of the Q-Bus means that a card containing all of the Q-Bus DMA logic is available off the shelf. Only the relatively simple logic shown in the upper half of Figure 15-10 needs to be designed, built, and tested by the user to implement all these rather complex functions. The hard part, the manipulation of the bus control lines to steal bus cycles from the CPU, has already been designed, built, and tested by MDB.

D. SOFTWARE DESIGN

1. Operational Considerations. A telescope control system is a type of command and control system. Such systems typically receive and execute operator commands typed on a keyboard. They also display the state of the system and the results of commands on a video screen.

There are two basic ways to command a system from a terminal. The first is to display a menu of choices and prompt the operator to choose one. He enters his choice, then if there are more commands to be entered, a second menu is displayed with commands related to the first choice the operator made. This is repeated until the entire command is entered, then the commanded operation begins. This entire sequence can be depicted on a state diagram, in which each menu or the final selected operation is a state of the system, and the menu entries represent the set of paths to other states of the system. This method of commanding has the advantage that a relatively inexperienced operator can be guided through all the possible choices to give the desired command to the system. A major disadvantage is that it may require a relatively long time to step through all of the menus to enter the command.

The other method of commanding a system is to define a command language. This is a set of keywords or position-dependent words or values which completely defines each command. The advantages of a command language are (1) entering a command takes less time than for a menu-driven system, since one need not step through a whole sequence of menus to enter a command, and (2) one is not confined to a particular set of pathways between any two system states, one merely enters the command to get to a particular state, usually regardless of the current state. The major disadvantage of a command language is that inexperienced or infrequent operators may not know or may not remember the command they need to enter to make the system perform the desired function.

Several systems have been built recently that use a command language with an additional HELP command. An example is the popular text editor WORDSTAR. When one types HELP, a list of the valid commands in the language appears on the screen. When one types HELP <command>, where <command> is one of the valid commands in the list, a description of the functions performed by that command and the syntax for entering the command are displayed. This approach gives the user the same help they would obtain from a menu-driven system, with the advantages of a command language.

Assuming this latter approach is used, the method of presenting information to the telescope operator must be considered. A non-scrolling page with information in fixed locations has been used successfully in several satellite control systems at Goddard Space Flight Center. Each WMO screen page has 24 lines of 80 characters each, divided into five areas: (1) a header line displaying calendar date, Julian date, UTC, and local sidereal time, (2) a 20-line data area, (3) a command input line, (4) a command response line, and (5) an alarms line. A page for monitoring telescope pointing angles for an alt-az mount might display the following information in the 20-line data area:

1. Current telescope azimuth (from calibrated encoder)
2. Current telescope altitude (from calibrated encoder)
3. Object's apparent and mean right ascension
4. Object's apparent and mean declination

5. Object's mean position epoch
6. Object's apparent hour angle
7. Drive rates for azimuth, altitude, and image rotation
8. Atmospheric pressure and temperature

Additional pages will display raw encoder readings, error constants, values for each apparent place correction, a list of stars to sight for encoder calibration, etc. When the system is first put into operation, only a few pages will be defined in the data base. As experience with the system is gained, more pages will be defined. A data base definition language entered using the standard text editor will be employed. At the beginning of each evening, the data base will be read and the display tables set up. By regenerating the display tables each time from scratch, new pages can be added very easily.

This approach provides a system that is easy to use by those whose orientation is the use of telescopes, not the programming of computers. In addition, a video terminal displaying different pages allows the system to grow and evolve without incurring the high costs of replacing and enhancing dedicated hardware, such as dials, meters, 7-segment displays, and special pushbuttons, to provide the command and control capabilities. The only pushbuttons in the entire system will be those on the hand paddle and a large "kill" button that physically disconnects power from the telescope.

The servo loop portion of the control program is a list of calculations performed in sequence, culminating in a command sent to a motor. The calculations are repeated periodically, so that they are synchronized with each other. The complete control system, however, is a set of several asynchronous processes occurring with no temporal synchronization with each other. Often, these independent processes require some sort of communication among themselves. It is the job of the operating system to provide this communication. The processes occurring in a telescope control system are as follows:

1. Reception and interpretation of operator inputs, and generation and display of responses
2. Reading the UTC clock and computing sidereal time
3. Fourteen times per second, reading the azimuth encoder, computing the actual azimuth position, computing the desired azimuth position, computing the motor command, and sending the command to the azimuth motor
4. Twice per second, reading the altitude encoder, computing the actual altitude position, computing the desired altitude position, computing the motor command, and sending the command to the altitude motor
5. Updating the page on the video screen
6. Optionally, running an extended Kalman filter to update the error constant estimates
7. Reading the condition of limit switches, detecting alarm conditions, and displaying alarm messages on the video screen
8. Receiving guiding inputs from the hand paddle or the video terminal keyboard, correcting the actual position calculation, and moving the telescope in response to the operator's commands
9. Receiving focussing commands from the hand paddle or the video terminal keyboard, and sending commands to the secondary mirror motor

When designing the software, one should remember that Items 2 - 4 should be given highest priority; if execution of this software is delayed while awaiting completion of a lower priority item, tracking accuracy will suffer.

The items listed above occur while tracking a celestial object. Additional functions which are performed when one is not tracking are slewing, calibrating the encoders, performing a trend analysis of the error constants, analyzing instrument data, searching a star catalog data base for objects to observe, and preparing the manuscript of a journal article. The software should be designed in a manner to accomodate all of these functions separately, yet allow one to switch from one to another quickly and easily.

2. Top-Level Software Design. The selection of the LSI-11/73 processor permits the choice of either the RSX-11M or RT-11 operating systems. The former operating system permits each of the nine asynchronous tracking processes to be a separate program, or task, and assigned its own execution priority. This flexibility is quite beneficial, since it allows the system performance to be tuned quite easily. However, the purchase price of RSX-11M (over $3500) rules it out for the WMO telescope.

RT-11 permits two tasks to execute in parallel. The higher priority, time-critical calculations are performed in the "foreground" task, while the other items are handled in the "background" task when the processor is not being used by the foreground task.

The foreground task consists of three routines and a library of special functions. The main routine is very short. It simply assigns input/output (I/O) channels, and sets a timer to control the execution of the other routines. Unlike most FORTRAN programs, in which the main routine calls its subordinate routines in a well-defined sequence, RT-11 allows a FORTRAN subroutine to be declared to be a completion routine, which is executed when previously requested I/O completes or some other event occurs. The remaining two routines perform the control of each axis, and are activated by a timer. These routines use a set of library routines to compute sidereal time, diurnal aberration, refraction, flexure, collimation error, the motor command, etc. When a particular value is used by both routines, such as the sidereal time, it is computed by the more-frequently executed routine, and placed in a common area for the other routine to use. When each routine is done with its calculations, the last actions to be performed before exiting the routine are sending the background task the results of calculations used in the display pages.

The background task consists of a main routine and seven subordinate routines, one for each of the asynchronous processes not performed by the foreground task. The main routine defines the I/O channels and sets up the I/O for the video screen keyboard, hand paddle, and limit switches. It also reads in encoder calibration constants from the disk and sets a timer for updating the video screen and checking for and reporting alarm conditions. When a timer times out, or input is received from the keyboard or hand paddle, the operating system activates the appropriate subroutine as a completion routine. One routine is dedicated to receiving data from the foreground task and placing it in common storage for the video update and alarm routines to use. The limit switches are double-pole switches, so that power to the motor is physically disconnected, as well as reporting to the computer that a limit has been reached. Logic circuits will be installed on the limit switches to allow commands to reach the motor that tell it to back off the limit.

The design approach described above solves the problem of assigning priorities to different asynchronous processes, and permits the use of a high level language to develop the software, since the RT-11 operating system offers FORTRAN support of its real-time services. A more detailed software design is not presented here, because it requires extensive knowledge of the RT-11 operating system.

Separate diagnostic programs for each type of interface card will be written, if they are not available from the manufacturers of the cards. RT-11 FORTRAN supports PEEK and POKE commands that permit reading from, or writing to, respectively, a particular memory or device address. RT-11 FORTRAN also permits one to connect a subroutine to an interrupt and thereby write an interrupt service routine in FORTRAN. These facilities can be used to diagnose problems in the custom hardware, and to learn how to communicate with a custom device. After this experience is gained, a standard RT-11 device driver will be written in MACRO-11 assembler language for each device to be handled.

E. ASSESSMENT OF SYSTEM PERFORMANCE REQUIREMENTS

The major system performance requirements presented in Chapter 14 are reviewed below to assess the ability of the system just described to meet these requirements.

1. Portability: The telescope will be mounted on a trailer for portability. Preparing the telescope for transport consists of hitching the trailer to the tow bar of the van, securing a cover over the telescope, and driving away. The mount and primary and secondary mirror supports will not be torn down for transport, and the optics will remain in place on the trailer at all times, even during transport. By using an alt-az mount, the center of gravity and total mass of the trailer are both kept low enough to make a 30-inch aperture telescope towable by a car or small van. A generator located inside the van provides power in remote locations.

2. Setup Time: This requirement is probably the most difficult to meet. It will require several months of experience with the system in the field before it will be determined how realistic this requirement is.

3. Optics: The first phase of the telescope system will use a 12-inch aperture Cassegrain telescope to test the basic system concept. Once the system is working well, the investment will be made in a set of 30-inch folded Cassegrain (Nasmyth mount) optics. Good optics of this size will meet the optical requirement.

4. Telescope Pointing Accuracy: The main impediment to pointing accuracy is the accuracy of atmospheric pressure readings (0.1 inches of Hg, or about 6" of refraction at the horizon). The total error should be less than ±15 arc seconds if enough stars were sighted during the encoder calibration process. The atmospheric pressure measurement only becomes a factor at very low altitudes. The following table lists the error sources and their expected absolute accuracies:

Apparent Place Corrections	Error
1. Annual aberration	0".1
2. Stellar parallax	0".1
3. Precession	0".1
4. Nutation	0".1
5. Orbital motion	1"
6. Proper motion	1"
7. Diurnal aberration	0".1

8. Planetary parallax	0".5
9. Refraction	2" typical
	6" worst case

Mechanical Corrections	Error
1. Zero offset	3"
2. Azimuth tilt	3"
3. Drive rate	0".5
4. Non-perpendicular axes	2"
5. Collimation	3"
6. Tube flexure	3"
7. Servo lag	2"
8. Hour angle	0".5

(based on time and location accuracy;
the result is used in other calculations)

The RSS error value obtained when all these errors are combined is 6".7 when the refraction error is held to 1" or better, and 6".99 when the refraction error is 6". It was assumed that mechanical errors that do not vary with location, such as tube flexure, could be well understood after a few month's observing, while errors which change night to night with changes in the trailer location, such as azimuth tilt, will be larger. The accuracy of computing the apparent place corrections is predictable, but the accuracy of measuring and estimating the encoder calibration constants is not yet known. Many of these errors can be as large as 10" and still meet the overall pointing accuracy requirement.

5. Telescope Pointing Time: The slew rate with the proposed drive system is about 7 degrees per second. At that rate, it will require 26 seconds to slew 180 degrees at the maximum slew rate. Ramping should add about 20% when ramping from tracking rate (300 steps per second) to slew rate (500,000 steps per second) and back down. This should take about 5.2 seconds. Counting the ramping, it should take about 31 seconds to complete a slew. Note that as higher slew rates are used, the added ramp time grows while the total slew time falls, so the percentage of total time spent ramping increases.

6. Long Term Tracking Accuracy: The error modelling proposed above is expected to be accurate enough to meet the long term tracking accuracy requirement.

7. Short Term Tracking Accuracy. The motor speed resolution, encoder resolution, servo loop repetition rate, and the expected natural frequency of the telescope all combine to lead to the expectation that the short term tracking accuracy requirement will be met. The following table lists the error sources and their expected short term accuracies:

Apparent Place Corrections	Error
1. Annual aberration	0".1
2. Stellar parallax	0".1
3. Precession	0".1
4. Nutation	0".1
5. Orbital motion	0".5
6. Proper motion	0".5
7. Diurnal aberration	0".1

8. Planetary parallax	0".5
9. Refraction	1" typical
	6" worst case

Mechanical Corrections	Error
1. Zero offset	1"
2. Azimuth tilt	1"
3. Drive rate	0".5
4. Non-perpendicular axes	0".5
5. Collimation	1"
6. Tube flexure	0".5
7. Servo lag	0".5
8. Hour angle	0".5
(used in other calculations)	

The RSS error value obtained when all these errors are combined is 2".46 when the refraction error is held to 1" or better, and 3".32 when the refraction error is 6". Again, it was assumed that mechanical errors that do not vary with location, will be more stable than those that do. To obtain smooth tracking, these error sources need not be known absolutely to these values. Instead, the errors in computing the corrections must not change over the course of a few seconds by these values.

8. Data Input and Control Device: The hand paddle will be built with simple pushbuttons which meet the temperature and humidity requirements, and will be housed in a sealed enclosure to keep out dust and dirt. It will be positioned in a convenient place, and will have a cable long enough to enable it to be moved about easily.

9. Computer Environment: The van computer will be kept within its environmental operating limits by using heating and air conditioning inside the van.

10. Commands: A command language will be defined using English words familiar to most observing astronomers. A HELP facility will be provided to guide new or inexperienced users through the system.

11. Extraneous Light Control: The optics will be placed inside a well baffled tube, rather than a Serrurier truss, to keep out extraneous light. If foreign particles carried by the wind, such as sand, prove to be a problem, a thin mylar optical window will be installed to seal the telescope tube.

All of the requirements listed in Chapter 14 will be met, or are likely to be met, by the proposed system. The next step is to proceed with building the system. As progress is made, the requirements and design will be modified as problems are encountered and experience is gained with the system.

EPILOGUE

It is our hope that after reading this book, you will be inspired to start or continue your own telescope control project, and that you will now have the technical knowledge and the commitment to a design and implementation approach that will lead to success.

This entire book can be summarized in a few key points:

1. Any control system works best with well designed and well aligned mechanical components. Do what you can to minimize the moment of inertia and maximize the stiffness of the telescope optical assembly and mount. A high natural frequency of the mechanical structures will make your final system easier for the software to control and a pleasure to use.

2. Think about the intended use of your telescope. The many differences in optical design, mount type, mechanical layout, controlling processor, implementation language, and overall expense of the various telescope control systems described in this book are due almost exclusively to differences in the observing programs in each example. If you don't think first about what you want to do with your telescope, you will probably spend more money, time, and energy than required.

3. Consider the tradeoff between hardware and software carefully. Why waste your evenings developing complex software (or hardware) when you could be out observing?   Your own special talents in hardware and software should be factored into this tradeoff. If you are better at building mechanical components and are a bit inexperienced in software, consider Tomer's system using two motors and gear systems per axis. The software for this system is relatively simple. If you are experienced in both hardware and software, an approach like that in Chapter 13 may be more suitable.

4. Keep your design simple. Even the top professionals have problems getting a simple design to work. There is no use burdening yourself with additional work that isn't needed. Which would you rather do--spend time out under the stars using the system you built, or spending it staring at a video screen debugging a complex design?

5. Build a little, then test a little. Implement your system in many small, well defined stages. Limit the number of places you have to look for a bug--in both hardware and software.

6. Hang in there!   Perseverence is usually rewarded with success. Don't be afraid to ask for help. This is a new field, so those of us in it often enjoy trading techniques and helping others get started.

Good luck in your project!

Appendix A—ESTIMATES OF TELESCOPE CONTROL SYSTEM COSTS

Using modern microcomputers, telescope control systems can be developed very inexpensively. An example of a modern project to retrofit a new system to an existing telescope by a commercial concern is the system installed in 1983 by DFM Engineering, Inc. on the CTIO 1.5-m telescope. Similar systems cost about $50,000 (Melsheimer, 1983), including the dual-processor computer (about $10,000), new motors and gearboxes (about $5000), terminal and displays (about $1000), software (about $25,000), and installation (about $5000).

An example of a retrofit done by an amateur is described by Tomer and Bernstein (1983). Their computer upgrade of an existing trailer-mounted 12-inch Cassegrain telescope cost a total of about $1000 plus the cost of an Apple II+ computer with disk. The simplest system one can build, using a single board computer and two stepper motors, can be assembled for under $500.

In contrast, professional observatories a decade ago were faced with far higher costs. The following projects were successfully completed using late 1960's and early 1970's minicomputer technology.

(a) $50,000 plus software (Meeks, 1975, p.23)

(b) $120,000 plus software (Linnell and Hill, 1975, p.60)
The price represents 10% of the cost of the whole observatory.

(c) $150,000 to $200,000 plus software (McCord, Paavola, and Snellen, 1975, p.160)

(d) "well above half a million (dollars)" (Beaumont and Wolfe, 1975, p.275)

Note that these estimates are for hardware only. Since purchase orders were not issued for applications software, software costs tended to be ignored. In the last two decades, the cost of computer hardware capable of telescope control has dropped by an order of magnitude or more, if modern estimates of computer system hardware and software costs are used (Basili, 1980). Accordingly, at professional observatories which must pay a programming staff, software is now the most expensive component of most systems that include both hardware and software.

COMPUTER HARDWARE USED IN EXISTING CONTROL SYSTEMS

Computer Hardware Used in Existing Control Systems

Examples of modern hardware configurations include the following:

(a) Two 6502 microprocessors in a custom circuit (Melsheimer, 1983)

(b) An Apple II+ (Tomer and Bernstein, 1983)

(c) One 6809 on a Peripheral Technology PT-69 single board computer, driving custom circuitry (Fairborn Observatory APT)

(d) LSI-11 with 64 KB memory, driving Compumotor microsteppers at the Lowell Observatory (White, 1984)

By way of comparison, the hardware in the following examples represents the late 1960's and early 1970's technology that was available at the time these systems were built.

(e) An 8-bit Multi-8 computer made by Intertechnique (France) with 12K bytes of memory, ASR 33 teletype, 32-bit I/O board, and 8-line interrupt board (Bourlon and Vin, 1975, p.63). Similar configurations can be assembled for under $2500 in hardware costs at current prices.

(f) A 16-bit HP 2100 with 8K of memory (Van der Lans and Lorenson, 1975, p.253). A complete 16-bit computer with roughly comparable speed and 64K bytes of memory with a dual floppy disk and hard copy operator's console is available for about $6000 at current prices.

(g) A system employing four simple microprocessors to handle the dome, hour angle, declination, and RA and DEC tracking rates (Fridenberg, Westphal, and Kristian, 1975, pp.218-242). Note that many of today's inexpensive microcomputers have the same processing capability and far more memory than these early systems.

Appendix C—
ESTIMATES OF COMPUTERIZED CONTROL SYSTEM DEVELOPMENT TIME

Published estimates of programmer time for designing, coding, de-bugging, and documenting telescope control software written at professional observatories a decade ago are given below. These span several years and many different hardware and software approaches. Note that these projects were concerned with high accuracy control of large telescopes, and the software was developed without many of today's software development tools, such as good editors, high level language compilers, and even disk drives in some cases (paper tape was used on many of the smaller minicomputers).

(a) "more than a man-year": (Ingalls, 1975, p.24)

(b) "approximately 2 man-years": (Taylor, 1975, p.28)

(c) "worked on it part-time for 3 years ... written about 50,000 machine language instructions": (Hill, 1975, p.60)

The standard conversion rate in the software industry is 10 designed, coded, de-bugged, and documented lines of code per man-day, which translates to 5000 man-days or roughly 2 1/2 man-years.

(d) "We spent a good amount of time programming ...": (Bourlon, 1975, p.68)

(e) "100,000 FORTRAN statements ... of which about a quarter are currently active.": (Moore, Merillat, Colgate, and Carlson, 1975, p.95).

Using the same conversion rate, 25,000 lines of code is 1 1/4 man-years.

(f) "something less than one man year": (Paavola, 1975, p.160)

(g) "Everyone seems to be taking between one and two man years to program his system, almost regardless of the size of the system ...": (Rather, 1975, p.390)

(h) 1.5 man-years (Melsheimer, 1983)

Even with today's advanced development tools, if a system is designed from scratch, it is complex, and almost all functions are handled in software instead of allocating some of them to hardware, it could take 1-2 man-years to program a telescope control system. To demonstrate how expensive this is at a professional observatory which must pay for software, the following cases are presented.

1. Software consultant - expensive, but fast @$50/hour x 1.0 man-years (2080 hours), Cost = $104,000

2. Full-time professional programmer - also expensive @ $30/hour ($15/hour salary + 100% overhead) x 1.5 man-years, Cost = $93,600

3. Full-time astronomy graduate research assistant - slower, not as experienced with the computer or with programming techniques @ $7000/year x 2.0 man-years   Cost = $14,000

Since the hardware costs for a retrofit control system can be as low as $5,000-$15,000 today, software is a significant fraction of the total system costs, even when the least expensive labor is used. This is confirmed by the proportion of system costs attributed to software in the DFM Engineering, Inc. commercial system (see Appendix A).

The cost of labor is unimportant to those of us who program for the fun of it. However, the time required to get these systems to work suggests that simplicity is the best approach, and the less software that is required, the better.

THE BASIC SERVO PARAMETER IN EXISTING CONTROL SYSTEMS

Those who are designing closed loop control systems might be interested in the history of the basic approach used in systems built a decade ago. Successful computerized telescope control systems have used a variety of servo techniques. Often the same system will be called different names at different observatories, so that what is a velocity servo to one engineer is a position servo to another. The following types of servos have been implemented successfully:

1. Position only

    Kitt Peak NRAO 36-foot radio telescope (Moore, 1975)

2. Digital position and analog velocity

    University of Hawaii 88-inch telescope (Harwood, 1975)

    ESO 3.6-meter telescope (van der Lans, 1975)

    CFH telescope (Beaumont, 1975)

    Anglo-Australian 3.9-meter telescope (Wallace, 1975 and Bothwell, 1975)

    MMT 4.5-meter synthetic aperture telescope (Stephenson, 1975)

    Kitt Peak National Obervatory -- various telescopes (Paffrath, 1975)

3. Digital position (pointing) or digital velocity (tracking)

    INT 98-inch telescope (Beale, 1975)

4. Open loop, no feedback

    Haystack aperture synthesis interferometer (Burke, 1975)

WARNING! Some of the company names, addresses, or telephone numbers in this list are based on data sheets which are dated.

1. Aerotech, Inc.
   101 Zeta Drive
   Pittsburgh, PA   15238
   (412) 963-7470

   Steppers and drivers
   Servomotors and controllers

2. Astrosyn America, Inc.
   14349 Victory Boulevard
   Van Nuys, CA   91401
   (213) 785-2197

   Steppers and drivers

3. B&B Motor and Control Corp.
   96 Spring Street
   New York, NY   10012
   (212) 966-5777

   Variable speed motors;
   torque motors, synchronous and
   stepping motors and drivers;
   drive system clutches and brakes

4. Buckminster Corporation
   99 Highland Avenue
   Somerville, MA   02143
   (617) 864-2456

   Servo motors, encoders, power
   amplifiers, complete systems;
   motor interface cards

5. Compumotor Corporation
   1310 Ross Street
   Petaluma, CA   94952
   (707) 778-1244

   Very small angle steppers
   (50,000 steps per revolution)
   and drivers

6. Cybernetic Micro Systems
   445-203 South San Antonio Road
   Los Altos, CA   94022
   (415) 949-0666

   IC intelligent stepper
   controllers which interface
   to 8-bit micros

7. Eastern Air Devices
   Dover, NH   03820
   (603) 742-3330

   Steppers, synchronous and
   induction motors, gearmotors
   and reducers

8. Hurst Manufacturing Corp.
   P.O. Box 326
   Princeton, NJ   47670
   (812) 385-2564

   Steppers

9. Inland Motor
   501 First Street
   Radford, VA   24141
   (703) 639-9047

   Torque and servo motors; linear
   amplifiers; linear force motor;
   alternators/generators; brushless
   motors

10. Measur-Matic Electronics Corp.
    50 Grove Street
    Salem, MA   01970
    (617) 745-7000

    Steppers and controllers;
    microstepper drivers to 12,800
    steps; Multi-bus smart
    controllers

11. Motion Science, Inc.
    1485 Kerley Drive
    San Jose, CA 95112

    Excellent servo motor system
    with STD Bus or RS-232
    interface, microstepper drives

12. Oriental Motor U.S.A. Corp.   Inexpensive stepper motors
    2701 Toledo St., Suite 702
    Torrance, CA 90503
    (213) 515-2264
13. Rogers Labs   Apple stepper interface cards
    2710 S. Croddy Way   and complete motor/card kits
    Santa Ana, CA 92704   at reasonable prices
    (714) 751-0442
14. Sigma Instruments, Inc.   Steppers
    170 Pearl Place
    Braintree, MA 02184
    (617) 843-5000
15. Silicon General, Inc.   DC servomotor controller ICs
    11651 Monarch Street   (Application Note SG1731)
    Garden Grove, CA 92641
    (714) 892-5531
16. SIL-WAKER (America)   IC stepper controller which
    653 Las Casas Ave.   interfaces to 8-bit micros
    Pacific Palisades, CA 90272
    (213) 454-4772
17. The Singer Company   Steppers, drivers, and
    Kearfott Division   synchronous motors
    1150 McBride Avenue
    Little Falls, NJ 07424
18. The Superior Electric Company   Steppers, controllers, vibration
    Bristol, CT 06010   dampers, synchronous motors
    (203) 582-9561
19. Torque Systems   Servomotors and controllers
    36 Arlington St.
    Watertown, MA 02172
    (617) 924-6000
20. Warner Electric Brake   Steppers, driver cards,
    and Clutch Co.   vibration dampers
    Beloit, WI 53511
    (815) 389-3771
21. Winfred M. Berg, Inc.   Gears, belts, and many useful
    499 Ocean Avenue   drive components
    East Rockaway, LI, NY 11518
    (516) 599-5010

WARNING! Some of the company names, addresses, or telephone numbers in this list are based on data sheets which are dated.

1. Analog Devices
   P.O. Box 280
   Norwood, MA 02062
   (617) 329-4700

   Resolver, synchro, and Inductosyn electronic modules

2. Astrosystems, Inc.
   6 Nevada Drive
   Lake Success, NY 11040
   (516) 328-1600

   Synchro/digital converters, multispeed converters for synchros geared in groups

3. Baldwin Electronics, Inc.
   1101 McAlmont Street
   Little Rock, AK 72203
   (501) 372-7351

   Full line of incremental and absolute optical encoders, up to sub-arcsecond resolution

4. Buckminster Corporation
   99 Highland Avenue
   Somerville, MA 02143
   (617) 864-2456

   Encoder to computer interfaces for the LSI-11 and STD Bus

5. Disc Instruments, Inc.
   2701 S. Halladay Street
   Santa Ana, CA 92705
   (714) 549-0343

   Inexpensive incremental optical encoders up to 1000 counts per turn

6. Elm Systems Encoders Division
   1101 Brown Street
   Wauconda, IL 60084
   (312) 526-5003

   Inexpensive incremental optical encoders up to 12700 counts per turn

7. Farrand Controls
   99 Wall Street
   Valhalla, NY 10593
   (914) 761-2600

   Inductosyns (both rotary and linear)

8. Interface Engineering, Inc.
   386 Lindelof Avenue
   Stoughton, MA 02072
   (617) 344-7383

   Synchro/digital converters; binary angle/sine, cosine converters; coordinate conversion modules

9. Litton Systems, Inc.
   Encoder Division
   20745 Nordhoff Street
   Chatsworth, CA 91311

   Full line of incremental and absolute optical encoders

10. Northern Precision Labs., Inc.
    202 Fairfield Road
    Fairfield, NJ 07006
    (201) 227-4800

    High accuracy resolvers

11. Trans-Tek, Inc.
    Route 83
    Ellington, CT 06029
    (203) 872-8351

    Inexpensive high-accuracy variable capacitor; linear displacement and velocity transducers

Instruction times were derived from the figures given in the 1980 Microcomputer Processor Handbook from Digital Equipment Corporation, Maynard, MA for the KEV-11 EIS/FIS integer and floating point arithmetic hardware. Because of the addressing modes generated by the LSI-11 FORTRAN compiler running under the RT-11 operating system, worst case times are used. All times quoted are in microseconds (μS).

Instruction	Time
Floating point add	85.15
Floating point subtract	85.45
Floating point multiply	164.15
Floating point divide	275.05

Timing Ratios

Floating point   ADD/SUB /MUL/DIV
                 1 / 1 /1.9 /3.2

Conventions:
Each computation unit = 85 μS
Each add = 1 unit
Each subtract = 1 unit
Each multiply = 2 units
Each divide = 3 units
Each trigonometric function = 8 adds + 7 multiplies
                           = 22 units
                           = 1.87 milliseconds

Interrupt Latency (with the KEV-11 arithmetic hardware)
    During memory refresh   135 μS
    Not during memory refresh   44 μS
Arbitrarily chosen latency to reflect the probability
of an interrupt occuring during memory refresh   50 μS
Average non-numerical instruction generated in FORTRAN   6 μS

WARNING! Some of the company names, addresses, or telephone numbers in this list are based on data sheets which are dated. Much of the information on this list was obtained from The DEC Professional Magazine, P.O. Box 362, Ambler, PA 19002. This magazine is a goldmine of information on DEC-compatable equipment and software products (data bases, text processors, spreadsheet calculators, systems, independent repair specialists, etc.)

1.	Able Computer 1732 Reynolds Avenue Irvine, CA (714) 979-7030	Communications interfaces and other peripherals
2.	ADAC Corporation 70 Tower Office Park Woburn, MA 01801 (617) 935-6668	A/D, D/A, data acquisition boards; analog I/O and power control boards; counter and pulse output boards; complete LSI-11 systems
3.	Advanced Computer Systems, Inc. 250 Prospect Street Waltham, MA 02154 (617) 894-3278	Complete LSI-11 systems
4.	Andromeda Systems, Inc. 9000 Eton Avenue Canoga Park, CA 91304 (213) 709-7600	Controllers and interfaces, mass storage subsystems, complete computer systems
5.	Bubbl-Tec 6800 Sierra Court Dublin, CA 94566 (415) 829-8700	Bubble memory RX01 disk emulator
6.	Buckminster Corporation 99 Highland Avenue Somerville, MA 02143 (617) 864-2456	Motor and encoder interfaces for DC servomotors and incremental encoders
7.	Central Valley Management Company, Inc. 585 Manzanita Avenue, Suite 7 Chico, CA 95926 (916) 895-8322	Large disk subsystems
8.	Charles River Data Systems Four Tech Circle Natick, MA 01760 (617) 655-1800	Standard DEC boards, disk and tape systems, complete LSI-11 systems
9.	Chrislin Industries, Inc. 31352 Via Colinas, No. 101 Westlake Village, CA 91362 (213) 991-2254	RAM memory, complete systems

10. Computer Extension Systems, Inc. Extension and memory cards
    17511 El Camino Real   Pulse height analyzer
    Houston, TX 77058   Complete line of interface
    (713) 488-8830   cards
11. Computer Technology   Q-Bus DMA card
    6043 Lawton Avenue
    Oakland, CA 94618
    (415) 451-7145
12. Computer Technology   Floppy disk controller
    3014 Lakeshore Ave.
    Oakland, CA 94610
    (415) 465-9000
13. Cyberchron Corporation   Expansion memory
    5768 Mosholu Avenue
    Riverdale, NY 10471
    (212) 548-0503
14. Datacube, Inc.   Graphics and image processing
    670 Main Street   board
    Reading, MA 01867
    (617) 944-4600
15. DATARAM Corporation   1/2" tape drive controllers,
    Princeton Road   many other cards
    Cranbury, NJ 08512
    (609) 799-0071
16. dataware, incorporated   Used DEC equipment
    1500 Northwest 62nd Street
    Suite 512
    Fort Lauderdale, FL 33309
    (305) 771-7600
17. Datel Systems, Inc.   A/D, D/A boards, complete
    11 Cabot Boulevard   data acquisition systems
    Mansfield, MA 02048
    (617) 828-8000
18. decComp   Individual system modules, disk
    14752 Sinclair Circle   and tape subsystems, hardware
    Tustin, CA 92680   and software support
    (714) 730-5116
19. DECMATION   CP/M for LSI-11 plug-in card
    3375 Scott Blvd., Suite 422
    Santa Clara, CA 95051
    (408) 980-1678
20. Digi-Data Corporation   1/2" tape drives and
    8580 Dorsey Run Road   controllers
    Jessup, MD 20794
    (301) 498-0200
21. Digital Engineering   Graphics boards to upgrade
    630 Bercut Drive, Dept. 217   DEC VT-100 or ADM 3A
    Sacramento, CA 95814   terminals to high resolution
    (916) 447-7600   graphics
22. Digital Equipment Corporation   Inventor of the LSI-11 and
    Microcomputer Products Group   Q-Bus, only makers of LSI-11
    77 Reed Road   CPUs; complete line of boards
    Hudson, MA 01749   and systems

23. Distributed Logic Corp.   1/2" tape controllers
    12800-G Garden Grove Boulevard
    Garden Grove, CA   92643
    (714) 534-8950

24. Dynatech Corp.   Q-Bus backplanes
    1225 East Wakeham Avenue
    P.O. Box 1019
    Santa Ana, CA   92702
    (714) 558-8755

25. Dynus, Inc.   Disk controller
    3190 K Airport Loop Drive
    Costa Mesa, CA   92626
    (714) 979-6811

26. EECO   Paper tape reader/punch
    1441 East Chestnut Avenue   interface board
    Santa Ana, CA   92701
    (714) 835-6000

27. EMULEX   Peripheral interfaces and
    3545 Harbor Blvd.   communications products
    P.O. Box 6725
    Costa Mesa, CA   92626
    (800) 854-7112

28. Emulogic Microprocessor   Cross assemblers, Pascal/C
    Development Systems   cross compilers,
    Three Technology Way   in-circuit emulators
    Norwood, MA   02062
    (617) 329-1031

29. First Computer Corporation   Disk and tape subsystems,
    645 Blackhawk Drive   complete LSI-11 systems
    Westmont, IL   60559
    (800) 292-9000

30. Gen/Comp   Disk, printer, and paper tape
    6 Algonquin Road   controllers
    Canton, MA   02021
    (617) 828-2008

31. Interactive Microware, Inc.   RAM disk emulator
    P.O. Box 771, Dept. 72
    State College, PA   16801
    (814) 238-8294

32. Interfaces Limited   Systems and supplies; term-
    (412) 941-1800   inals and printers
33. Marway Products, Inc.   Power controllers
    2421 South Birch Street
    Santa Ana, CA   92707
    (714) 549-0623

34. Matrox Electronic Systems, Ltd.   Monochrome and color
    5800 Andover Ave.   graphics and image
    T.M.R., Quebec H4T 1H4   processing boards
    Canada
    (514) 735-1182

35. MDB Systems, Inc.   Wide range of controller and
    1995 N. Batavia Street   interface cards, card cages,
    Orange, CA   92665   and boxes
    (714) 998-6900

36. National Instruments
    12109 Technology Blvd.
    Austin, TX 78727
    (800) 531-5066

    IEEE-488 (HPIB) bus interface
    cards

37. Netcom Products, Inc.
    430 Toyama Drive
    Sunnyvale, CA 94086
    (408) 734-8732

    Enclosures

38. Newman Computer Exchange, Inc.
    P.O. Box 8610
    Ann Arbor, MI 48107
    (313) 994-3200

    Used DEC equipment; CPUs,
    peripherals, hardware,
    complete systems

39. PERITEK
    5550 Redwood Road
    Oakland, CA 94619
    (415) 531-6500

    Graphics display generators

40. Pick Computing Machinery, Inc.
    57 North Main Street
    Hartford, WI 53027
    (414) 673-6800

    Complete LSI-11 systems

41. Plessey Peripheral Systems
    Computer Products Division
    1674 McGaw Avenue
    Irvine, CA 92714

    Memories, voice recognition
    system for the VT-100

42. Qualogy
    (formerly Data Systems Design)
    2241 Lundy Avenue
    San Jose, CA 95131
    (408) 727-3163

    Very nice floppy or tape plus
    Winchester drives with
    excellent in-unit diagnostics;
    very good business reputation

43. Scanoptik, Inc.
    P.O. Box 1745
    Rockville, MD 20850
    (301) 762-0612

    Diagnostic memory card
    (traps and stores failures)

44. Scientific Computer Systems, Inc.
    2438 30th Street
    Boulder, CO 80301
    (303) 447-0353

    Bus grant cards

45. Selanar Corporation
    2403 De La Cruz Boulevard
    Santa Clara, CA 95050
    (408) 727-2811

    Graphics plug-in for
    the VT-100

46. Sigma Information Systems
    6505C Serrano Avenue
    Anaheim, CA 92807
    (714) 974-0166

    Terminal interfaces, full line
    of products

47. Sky Computers, Inc.
    Foot of John Street
    Lowell, MA 01852
    (617) 454-6200

    Array processors

48. Spectra Logic
    1227 Innsbruck Drive
    Sunnyvale, CA 94086
    (408) 744-0930

    Disk and tape controllers

49. Standard Engineering Corporation   CAMAC modules
    44800 Industrial Drive
    Fremont, CA   94538
    (415) 657-7555

50. thomas business systems, inc.   Used DEC equipment
    (305) 392-2007

51. Unitronix Corporation   Tape drives and complete
    197 Meister Avenue   systems
    Somerville, NJ   08876
    (201) 231-9400

52. U.S. Design Corporation   Very nice hard disk/tape combo
    5100 Philadelphia Way
    Lanham, MD   20706
    (301) 577-2880

53. Virtual Microsystems   CP/M emulator board
    2150 Shattuck Ave.
    Berkeley, CA   94704

54. Western Peripherals   1/2" tape controller; data
    1100 Claudina Place   cartridge tape controller
    Anaheim, CA   92805
    (714) 991-8700

55. Xacom Technology, Inc.   Removable Winchester disk
    560 Forbes Boulevard   drives
    South San Francisco, CA   94080
    (415) 952-1512

56. Xebec Systems, Inc.   Peripherals and controllers
    2985 Kifer Road
    Santa Clara, CA   95051
    (408) 988-2550

57. Xylogics, Inc.   Disk controller
    42 Third Avenue
    Burlington, MA   01803
    (617) 272-8140

1. Atec
   P.O. Box 19426
   Houston, TX   77224
   (713) 468-7971

   25 cards plus accessories
   12, 14, and 16-bit
   synchro/digital converters,
   incremental shaft encoder

2. Amtec
   12740 28th Avenue NE
   P.O. Box 27128
   Seattle, WA   98125
   (206) 363-0217

   6 cards; closed-loop servo
   control (two different types),
   encoder interfaces; the servo
   controllers and encoder
   interfaces are smart and
   easily used

3. Analog Devices
   Route 1 Industrial Park
   P.O. Box 280
   Norwood, MA   02062
   (617) 329-4700

   About 6 cards; a number of A/D
   and D/A converters; one A/D
   converter is very smart;
   tradition of high quality A/D

4. Antona Corp.
   13600 Ventura Blvd.
   Suite A
   Sherman Oaks, CA   91423
   (213) 986-6835

   12+ cards and complete systems

5. Applied Micro Technology, Inc.
   P.O. Box 3042
   Tucson, AZ   85702
   (602) 622-8605

   About 50 cards plus complete
   systems; programmable clock,
   modem card, speech synthesizer,
   special video cards

6. Augat
   Interconnection Components Division
   33 Perry Avenue
   P.O. Box 779
   Attleboro, MA   02703

   Bare boards

7. Baradine Products Ltd.
   P.O. Box 86757
   North Vancouver, British Columbia,
   Canada   V7L 4L3

   Six CMOS cards and mainframes;
   multi-tasking software

8. BDS Microsystem Designs Ltd.
   28 Pinewood Close
   St. Albans, Herts.   AL4 0DS
   England

   Three cards

9. BICC-Vero Electronics
   171 Bridge Road
   Hauppauge, NY   11788
   (516) 234-0400

   Blank cards, connectors, and
   hardware

10. Blue Chip Computers, Inc.
    7648 Heather St.
    Vancouver, B.C.   V6P 3RI
    Canada

    CMOS cards, nice variety

11. Bubbl-Tec
    6800 Sierra Court
    Dublin, CA 94566
    (415) 829-8700

    Memory cards; can store
    complex programs without disk;
    bubble memory cards

12. Buckminister Corp.
    99 Highland Avenue
    Somerville, MA 02143
    (617) 864-2456

    Encoder interface, and smart
    servo control cards

13. CAL-TEX Computers, Inc.
    7080 East Trimble Road   504
    San Jose, CA 95131
    (408) 942-1424

    Single (large; not STD size)
    board computer with STD bus
    output, hard disk controller,
    serial port, many extras

14. Campbell Scientific, Inc.
    P.O. Box 551
    Logan, UT 84321
    (801) 753-2342

    About six cards; cassette tape
    interface, clock, Z-80 CPU,
    memory

15. Circuits and Systems, Inc.
    Two Main Street
    Hollis, NH 03049

    Several very nice cards. The
    timer/counter and I/O card is
    very useful for many tasks

16. CompuPro Division
    3506 Breakwater Ct.
    Hayward, CA 94545

    Basic CPU, memory, and other
    cards; same high quality as
    other Bill Godbout products.

17. Computer Dynamics Corp.
    105 S. Main Street
    Greer, SC 29651

    Daisy wheel printer interfaces.

18. Contemporary Control Systems, Inc.
    4949 Forest Ave.
    Downers Grove, IL   60515
    (312) 963-7070

    Over 24 cards and accessories
    and complete systems. Interval
    timer. Very helpful foundation
    module card that has address
    decoding and buffer chips in
    place, plus wire-wrap area for
    your own special chips. Saves
    time when prototyping special
    cards.

19. Cytec Corp.
    107 North Washington Street
    East Rochester, NY   14445
    (716) 381-4740

    Large scale switching.

20. Data Translation
    100 Locke Drive
    Marlboro, MA 01752

    About six cards. A/D and D/A.

21. Digital Dynamics, Inc.
    16795 Lark Ave.
    Los Gatos, CA 95030

    About 10 cards, general
    purpose.

22. Douglas Electronics
    718 Marina Blvd.
    San Leandro, CA 94577
    (415) 483-8770

    Very nice selection of five
    different types of bare
    boards, handy extender board.

23. DY4 Systems, Inc.
    888 Lady Ellen Place
    Ottawa, Ontario K1Z 5M1
    Canada

    18 cards plus systems and
    accessories. True multi-pro-
    cessing capabilities. Wide
    selection of high-quality
    cards.

24. E&L Instruments, Inc.
    61 First Street
    Derby, CT 06418

    Breadboarding system.

25. Environmental Systems Corp.
200 Center Drive
Knoxville, TN 37912

BASIC enhanced for real-time control. Can be purchased in ROM. Single board system with STD extension.

26. Electrologic, Inc.
1359 28th Street
Signal Hill, CA 90806
(213) 595-0551

Industrial module control board to control Opto-22 and similar devices, including high power AC. Very nicely done.

27. Ellar Engineering
495 South Arroyo Parkway
Pasadena, CA 91105

Mainframes.

28. Enlode, Inc.
1728 Kingsley Ave.
Orange Park, FL 32073
(904) 264-4405

Over 10 highly useful cards. Real-time clock, keypad entry, cassette drive, display systems. Top quality, yet reasonably priced. Unusually good documentation.

29. Euteknic Associates
1488 Ramon Dr.
Santa Clara, CA 95051

Fiber optics serial interface.

30. GW3 Incorporated
7623 Fullerton Road
Second Floor
Springfield, VA 22153
(703) 451-2043

TMS-9995 single board computer, Eyring Research Institute's PDOS real-time disk OS and BASIC

31. Infinity, Inc.
29429 Southfield Road, Suite 5
Southfield, MI 48076

Complete, low-cost STD systems. US distributor of Pulsar single board computer. East coast representative for many STD companies. Large packet of STD information for the asking.

32. Integrated Technologies, Inc.
444 W. Maple, Building F
Troy, MI 48084
(313) 362-4466

Highly intelligent servo control cards.

33. Intersil Systems
1275 Hammerwood Ave.
Sunnyvale, CA 94089
(408) 743-4442

Over 50 cards, complete systems, accessories. Large selection.

34. Intra Computer
101 West 31 Street
New York, NY 10001
(212) 947-5533

RS-232 interfaces.

35. I/O Control, Inc.
517 East Lincoln Highway
Langhorne, PA 19047

Over a dozen cards. Nice dual-axis stepper controller.

36. Ironics, Inc.
117 Eastern Heights Drive
P.O. Box 356
Ithaca, NY 14850
(607) 277-4060

Single board computer with STD interface. Speech synthesizer with built-in words used in control. Color video board.

37. JF Micro
Star Route 1
Box 1174-D
Pasco, WA 99301
(509) 547-3397

Over 20 general purpose boards.

38. Jonos Ltd.
    920-c E. Orangethorpe
    Anaheim, CA   92801

Portable STD computer with dual
3-inch disks, built-in monitor
and keyboard.

39. Kadak Products Ltd.
    206-1847 West Broadway Avenue
    Vancouver, B.C.   V6J 1Y5
    Canada

Multi-tasking control language
for CP/M systems.

40. Lang Systems, Inc.
    1392 Borregas Avenue
    Sunnyvale, CA   94086

Battery-backed memory.

41. Lincoln Instruments, Inc.
    456 W. Montana Street
    Pasadena, CA   91103

A/D card.

42. Matrix Corp.
    1639 Green Street
    Raleigh, NC   27603
    (919) 833-2837

Over a dozen cards. Complete
systems. Smart single and dual
stepper controllers. Stepper
translators. Multi-tasking
software. FORTH.

43. Matrox Electronics Systems Ltd.
    5800 Andover Ave.
    T.M.R., Quebec   H4T 1H4
    Canada
    (514) 735-1182

Wide selection of video cards
for both STD bus and the LSI-11
Q-bus.

44. Micro-Aide, Inc.
    San Dimas Commerce Center
    482-F W. Arrow Hwy.
    San Dimas, CA   91773
    (714) 592-3804

Over a dozen cards.
Battery-backed clock.

45. Microcomputer Systems, Inc.
    1814 Ryder Drive
    Baton Rouge, LA   70808
    (504) 769-2154

Over a dozen cards. CMOS and
TTL types. Speech synthesizer.

46. Micro-Link Corp.
    14602 North U.S. Highway 31
    P.O. Box 517
    Carme, IN   46032
    (317) 846-1721

Over two dozen cards. Complete
systems.

47. MicroStandard Technologies, Inc.
    P.O. Box 319
    New Lebanon, OH   45345
    (513) 687-1395

Very nice portable STD computer
system with dual 5-inch disks
(8-inch external disk), 9-inch
monitor, and detachable VT-100
style keyboard. Card cage has
10 slots, and can be expanded.
Comes complete with large
library of software.

48. Micro/SYS
    1367 Foothill Blvd.
    La Canada, CA   91011
    (213) 790-7267

Over two dozen cards plus
complete systems.

49. Miller Technology, Inc.
    647 N. Santa Cruz Ave.
    Los Gatos, CA   95030
    (408) 395-2032

Low cost CPU cards with simple
BASIC in ROM.

50. Mostek, Inc.
    1215 West Crosby Road
    Carrollton, TX   75006
    (214) 466-6000

Over two dozen cards. Complete systems. Co-founder of the STD bus with Pro-Log.

51. Mullen Computer Products, Inc.
    Box 6214
    Hayward, CA   94545

Very nice line of high quality STD boards. Opto-isolated I/O. Tiny BASIC single-board computer. Card cages and motherboards.

52. Octagon Systems Corp.
    5150 W. 80th Ave.
    Westminister, CO   80030
    (303) 426-8540

Complete single-card control systems with control BASIC in ROM and EPROM storage of programs.

53. Opto-22
    15461 Springdale Street
    Huntington Beach, CA   92649
    (714) 891-5861

The inventor and main provider of the industry standard power control modules. High and low level AC and DC input and output modules.

54. Pro-Log Corp.
    2411 Garden Road
    Monterey, CA   93940
    (408) 372-4593

Over 40 cards and accessories. Pro-Log and Mostek started the STD bus, and they have a complete line of high quality cards.

55. Pulsar Electronics, Ltd.
    Australia
    (see Infinity, Inc. above)

Their single-board computer has 4 Mhz Z-80, 64K RAM, 2K ROM monitor, two RS-232 ports, and battery-backed clock/ calendar. One of the best buys in the STD business.

56. Quasitronics, Inc.
    211 Vandale Dr.
    Houston, PA   15342
    (412) 745-2663

About a dozen general purpose cards and systems.

57. Robotrol Corp.
    1250 Oakmead Parkway
    Suite 210
    Sunnyvale, CA   94086
    (408) 778-0400

Over a dozen analog I/O cards, including one very intelligent one.

58. Samco
    4417 Longworthe Square
    Alexandria, VA   22309

One of the few STD 6502-based (Apple CPU) cards.

59. Scanbe Division of Zero Corp.
    3445 Fletcher Ave.
    El Monte, CA   91731

Top quality motherboard/card cages at very reasonable prices. Includes cages with as many as 24 slots for the more complex systems.

60. Spurrier Peripherals Corp.
    10513 Le Marie Drive
    Cincinnati, OH   45241
    (513) 563-2625

Over a dozen cards and accessories, including a 6809-based CPU card.

61. Startronix, Inc.
    104 S. Rogers
    Irving, TX   75060

Over a dozen cards, including multi-processor CPU arrangements, and a Motorola 68000-based CPU card.

62. STD Systems, Inc.   Keypads, I/O, and accessories.
    16 E. Franklin Street
    Danbury, CT 06810
    (203) 744-5713

63. Systek   Over a dozen cards, including
    1023 N. Kellogg Street   8088 CPU (IBM PC CPU) and
    Kennewick, WA 99336   32- and 64-bit floating point
    (509) 586-9114   math cards. Runs CP/M-86.

64. Technitron, Inc.   STD cards and systems.
    Doman Road
    Camberley, Surrey
    England

65. Tetronics   Cassette control board.
    322 East Deepdale Road
    Phoenix, AZ 85022
    (602) 866-1926

66. TL Industries, Inc.   About a dozen cards and
    2573 Tracy Road   systems. 6809-based CPU.
    Northwood, OH 43619
    (419) 666-8144

67. Transwave   A truly outstanding single-
    Cedar Valley   Board control computer with
    Box 489   TINY BASIC, real-time clock,
    Vanderbilt, PA 15486   ROM utilities, and EPROM
    (412) 628-6370   programmer all built
                                               in (and many other features).
                                               One of the best-written
                                               system manuals. Also other
                                               I/O cards and accessories.

68. Versatile Logic Systems Corp.   Video memory.
    87070 Dukhobar Road
    Eugene, OR 97402
    (503) 485-8575

69. Ward Systems, Inc.   A/D card.
    2608 Valley Ave.
    P.O. Box K
    Sumner, WA 98390
    (206) 863-6387

70. Whedco, Inc.   A number of very useful STD
    6107 Jackson Road   control cards. Their stepper
    Ann Arbor, MI 48103   controller card is the smartest
    (313) 665-5473   one in the business. Also
                                               stepper translators. New
                                               high-performance (60,000
                                               steps/sec) translator under
                                               development.

71. Wintech Systems, Inc.   CMOS cards including Opto-22
    Box 121361   interface board.
    Arlington, TX 76012
    (817) 274-7553

72. Xitex Corp.   Cards and complete systems.
    9861 Chartwell Drive   TRS-80-to-STD interface.
    Dallas, TX 75243
    (214) 349-2491

MANUFACTURERS OF HARDWARE RELATED TO TELESCOPE CONTROL

WARNING!   Some of the company names, addresses, or telephone numbers in this list are based on data sheets which are dated.

Computers and Software

1. Peripheral Technology
   3760 Lower Roswell Road
   Marietta, GA   30067
   (404) 973-0042

   Single board 6809 microcomputer

2. Microware
   5835 Grand Avenue
   Des Moines, IA   50312
   (515) 279-8844

   OS9 real-time operating system; BASIC09 control system language

3. Multi Solutions, Inc.
   660 Whitehead Road
   Lawrenceville, NJ   08648
   (609) 695-1337

   Multitasking real-time operating system for most 8- and 16-bit microprocessors; complete set of utilities, including editors, high level languages, etc.

Environmental Sensors

4. Electronic Instrumentation
   and Technology, Inc.
   1439 Shepard Drive
   Sterling, VA   22170
   (703) 450-6660

   Temperature probe

5. National Semiconductor Corporation
   2900 Semiconductor Drive
   Santa Clara, CA   95051
   (408) 732-5000

   Pressure transducers

6. Omega Engineering, Inc.
   One Omega Drive
   Box 4047
   Stamford, CT   06907
   (203) 359-1660

   Temperature probes and systems

7. SenSym
   1255 Reamwood Avenue
   Sunnyvale, CA   94086
   (408) 744-1500

   Pressure transducers

Time Receivers

8. Spectracom Corporation
   320 N. Washington Street
   Rochester, NY   14625
   (716) 381-4827

   Complete clock system using WWVB

9. True Time Instrument Company
   429 Olive Street
   Santa Rosa, CA   95401
   (707) 528-1230

Boards, receivers, and complete
clocks using WWVB; receivers
for WWV

Position Receivers

10. ITT Decca Marine
    P.O. Box G
    U.S. 1 and St. Joe Road
    Palm Coast, FL   32037
    (904) 445-2400

Model 801 receives both Transit
satellite and Loran-C

11. Magnavox Advanced Products
    and Systems Company
    2829 Maricopa St.
    Torrance, CA   90503
    (213) 328-0770

Transit satellite receiver

12. Navigation Communication
    Systems, Inc.
    20100 Plummer St.
    Chatsworth, CA   91311
    (213) 349-1250

Wherefinder II satnav receiver

13. Stanford Telecommunications, Inc.
    1195-T Bordeaux Drive
    Sunnyvale, CA   94086
    (408) 734-5300

Global Positioning System (GPS)
receiver, Model 5026 Time
Transfer Receiver/Processor

14. Trimble Navigation
    1077 Independence Avenue
    Mountain View, CA   94043

Global Positioning System
receiver (about $13,000)

Appendix K—APT CONTROL ALGORITHMS

A. INTRODUCTION

For those who wish to develop their own APT control software, this appendix provides a number of useful algorithms or routines. Our special thanks to Louis Boyd for providing these. He first presented them at the 1984 IAPPP Big Bear Symposium.

B. COMPUTER MATHEMATICS

Some of the routines which follow require the use of integer division to force an integer output. In the algorithms in this article an integer constant is shown with no decimal point (e.g., 9) while a floating point number, even though it represents an integer value, will be shown with a decimal point (e.g., 9.). In addition, subtraction, and multiplication there is no difference in the result, but the division of two integers yields an integer which carries no fractional part. For example, 9./2.=4.5, but 9/2=4.

If either number is floating point, the result will be floating point so that 9/2.=4.5. This should be kept in mind when converting these algorithms to a particular computer language.

Radians are used to measure angles to enhance the speed of computing trigonometric functions, of which there are many. Decimal hours are used instead of radians in some algorithms involving time calculations. It should be remembered that sidereal hours and solar hours are not the same.

C. MODIFIED JULIAN DATE

Modified Julian Date (MJD) is used in these calculations rather than Julian date because of the limited precision of BASIC09's real numbers. The BASIC09 compiler has nine significant digits of precision, and the use of the full Julian date would cause unnecessary loss of resolution. The Julian date can easily be obtained by adding 2400000.5 to the MJD. The following algorithm is based on the Almanac for Computers. It is valid for dates between 1901 and 2099 as the year 2000 is a leap year.

Parameters	Type	Values		(1)
YEAR	integer	1901 - 2099	Universal Date	
MONTH	integer	1 - 12	January = 1	
DAY	integer	1 - 31	Day of month	
HOUR	integer	0 - 23	Universal time	
MINUTE	integer	0 - 59		
SECOND	integer	0 - 59		
MJUL	real	Modified Julian Date with fraction		

MJUL = -678987.+367.*YEAR+INT(INT(275*MONTH/9)+DAY-INT(7*(YEAR+INT((MONTH+9)/12))/4))+(HOUR*3600.+MINUTE*60+SECOND)/86400.

D. LOCAL MEAN SIDEREAL TIME

Local Sidereal Time is equivalent to the right ascension of a stellar object along the local meridian. Local Mean Sidereal Time (LMST) differs from Local Apparent Sidereal Time (LAST) in that it is not corrected for nutation of the Earth. The difference is at most about 20". For the APT, LMST is sufficiently accurate and the calculation of nutation is complex. This algorithm is from the Almanac for Computers. Note that only the integer part of the modified Julian date is actually used in the calculation. While the LMST could be calculated from the decimal Julian date directly, accuracy would suffer because of the roundoff error in the computer. MOD stands for modulo, and MOD(A,B) is equivalent to (A / B - INT(A / B))*B.

Parameters	Type	Values	(2)
HOUR	integer	0 - 23 Universal Time	
MINUTE	integer	0 - 59	
SECOND	integer	0 - 59	
MJUL	real	Modified decimal Julian Date	
LMST	real	Local Mean Sidereal Time in decimal sidereal hours	

Constants	Type	Values
LONG	real	Local longitude in decimal sidereal hours

LMST = MOD(6.67170278+.065709823*(INT(MJUL)-33282.)+
1.00273791*HOUR+MINUTE/60.+SECOND/3600.)-LONG,24)

E. POSITION OF THE SUN

This is a relatively low precision calculation of the right ascension and declination of the Sun. It accounts for the tilt of the Earth's axis and uses an approximation to account for the ellipticity of the Earth's orbit. It will generally be correct to about one degree. Far better accuracy may be obtained from the trigonometric series calculations in the Almanac for Computers, but at the expense of greatly increased computation time. This calculation is good enough to calculate where the Sun is below the horizon to determine whether it is dark enough to operate the photometer. In the future, it will be used for avoiding the Sun when doing infrared photometry in the daytime. This would also be a good routine for Sun trackers on solar power devices.

Parameters	Type	Notes	(3)
MJUL	real	MJD	
RTA	real	RA in radians	
DEC	real	Dec in radians	
T	real	Julian centuries from 0 Jan 1900	
MAOS	real	Mean anomaly of the Sun in radians	
TLOS	real	True longitude of the Sun in radians	

HALFPI=1.57079633
TWOPI =6.28318531
T=(MJUL-15019.5)*2.73785079E-5
MAOS=6.25658358+628.301946*T-2.61799388E-6*T*T
TLOS=MOD(MAOS+.03344051*SIN(MAOS)+.000349066*SIN(MAOS+MAOS)+
4.93169,TWOPI)

```
RTA=ATN(.91746*TAN(TLOS))
(* QUADRANT CORRECTION *)
IF TLOS>HALFPI THEN RTA=RTA+PI
    IF TLOS>HALFPI+PI THEN RTA=RTA+PI
    ENDIF
END IF
DEC=ASN(.39782*SIN(TLOS))
```

From the RA and Dec, the zenith angle may be calculated (see Zenith Angle) and when it exceeds a certain angle (100 degrees used in the Phoenix APT) it is dark enough to operate the telescope.

F. POSITION OF THE MOON

The calculation for the position of the Moon is simpler than that for the Sun, but has even less accuracy. It operates reliably for keeping the telescope 10° or more from the Moon when taking readings. This algorithm was extracted from "Celestial Basic" (Burgess, 1982) and compacted.

Parameters	Type	Notes	(4)
MJUL	real	MJD + fraction	
RTA	real	RA in radians	
DEC	real	Dec in radians	
BD	real	Base date	
LM	real	Longitude of the Moon in radians	

```
P2=6.28318531
BD=MJUL-36934.5
LM=5.43090264+.229971506*BD
RTA=MOD(LM+.109756775*SIN(LM-1.9443666E-3*BD-4.46356263),P2)
DEC=.403946*SIN(RTA)+.0398*SIN(RTA-3.11888592+9.24221652E-4*BD)
```

Avoidance of the Moon is accomplished as follows: If SQ(RTASTAR - RTAMOON)+SQ(DECSTAR - DECMOON)<.03 then the star is within 10° of the Moon. Note that .03 is approximately 10° converted to radians squared. In calculating the RA difference of the Moon and the star, the shortest distance must be taken.

G. ZENITH ANGLE

The zenith angle calculation is used in determining airmass for each observation and for determining whether twilight has occurred. This is a simple spherical trigonometry problem, and is limited in accuracy by the use of mean instead of apparent sidereal time. It ignores refraction, since the APT never operates near the horizon.

Parameter	Type	Notes	(5)
RTA	real	RA of the object in radians	
DEC	real	Dec of the object in radians	
LMST	real	Local Mean Sidereal Time, 0-24 hours	
COSZ	real	Cosine of the zenith angle	

Constants	Type	Notes
SLAT	real	Sine of the local latitude
CLAT	real	Cosine of the local latitude

```
COSZ=SLAT*SIN(DEC)+CLAT*COS(DEC)*COS(.2618*LMST-RTA)
```

The cosine is chosen as the output for this routine as it saves a step in calculation, depending on whether conversion to airmass (which requires the secant of Z) or determining the zenith angle (which requires the arccosine of COSZ) is to be performed. Constants were chosen for the sine and cosine of the latitude to speed calculation by eliminating two trigonometric operations. These could be calculated if the program were to be made versatile for a portable instrument.

H. HELIOCENTRIC CORRECTION

Several sources give algorithms for the calculation of heliocentric correction, but this one has been optimized for speed and should yield accuracy to a few seconds.

Parameters	Type	Notes	(6)
RTA	real	RA of the star in radians	
DEC	real	Dec of the star in radians	
MJUL	real	MJD with fraction	
CORRECTION	real	Correction in decimal days	
COSDC	real	Cosine of the star Dec	
T	real	Base date	
LOTS	real	Longitude of the Sun in radians	
RADIUS	real	Radius vector of the Sun in mean Earth radii	

COSDC=COS(DEC)
T=(MJUL-15019.5)*2.73785079E-5
LOTS=4.88162794+628.331951*T+5.279621E-6*T*T
RADIUS=.999720458/(1.+.016719*COS(LOTS+4.956105))
CORRECTION=-.0057755*(RADIUS*(COS(LOTS)*COS(RTA)*COSDC+
        SIN(LOTS)*(.39794248*SIN(DC)+.91741037*SIN(RTA)*COSDC)))

The correction should be added to the calculated Julian date to obtain the heliocentric corrected Julian date. Beware of limited computer precision when working with fractional Julian dates.

I. PRECESSION TO CURRENT COORDINATES

Precession is a large enough value (several minutes of arc) that it must be taken into account when hunting for navigation stars. It is easiest to work with star coordinates at an epoch for which there are good catalogs. As the Phoenix APT works stars of magnitude 8.0 or brighter, epoch J2000.0 coordinates were chosen because of the availability of the Sky Catalog 2000.0 which contains all stars which will be observed. This algorithm precesses from J2000.0 coordinates to current coordinates.

Parameter	Type	Notes	(7)
MJUL	real	MJD with fraction	
RTA	real	Input and output RA in radians	
DEC	real	Input and output Dec in radians	
DAYS	real	Days from epoch 2000	
RDIF	real	Change in RA	
DDIF	real	Change in Dec	

DAYS=51544.-MJUL
RDIF=DAYS*(6.112E-7+2.668E-7*SIN(RTA)*TAN(DEC))
DDIF=DAYS*(2.655E-7*COS(RTA)
RTA=RTA-RDIF
DEC=DEC-DDIF

These calculations do not take into account the change in direction of precession for extended periods of time, but they will suffice for APT use without correction well into the next century. Nutation, stellar aberration, and other apparent place corrections are small enough that they can be ignored.

J. DETERMINATION OF OBSERVABILITY OF A STAR

The APT must make a decision as to whether a star is observable within a predetermined area above the telescope. A cone with a 45° zenith angle was selected to be the observable window. There may be other limitations, such as the limit in Dec placed by the selection of stars, and any other boundaries could be set. This algorithm calculates the rising and setting time for an object within a cone of a fixed zenith angle.

Parameters	Type	Notes	(8)
RTA	real	RA of the star in radians	
DEC	real	Dec of the star in radians	
RISE	real	Rising sidereal time	
SET	real	Setting sidereal time	
ANGLE	real	Absolute value of hour angle of rising or setting object	

Constants	Type	Notes
SINLAT	real	Sine of the local latitude
COSLAT	real	Cosine of the local latitude
COSZMAX	real	Cosine of zenith angle of obs. cone
RAD_HRS	real	12/π converts radians to hours

ANGLE=ABS(ACS((COSZMAX-SINLAT*SIN(DEC))/(COSLAT*COS(DEC))
RISE=MOD(RTA-ANGLE*RAD_HRS+24,24)
SET=MOD((RTA+ANGLE)*RAD_HRS,24)

This algorithm has a number of uses in that it can calculate the time of sunrise, sunset, moonrise, and moonset if COSZMAX is set to zero. Note that the output times are in local apparent sidereal time which differs slightly (insignificantly) from local mean sidereal time.

These are all of the algorithms related to positional astronomy which are required to run an APT.

This appendix contains all the software needed to operate an automatic photoelectric telescope of the type described in Section V. All the procedures are written in Microware Basic09 except RAMP, which is in editor-assembler for the 6809 processor. Most of this software was originally written by Louis J. Boyd of the Fairborn Observatory (West). Portions of the software peculiar to the hardware used at Fairborn Observatory (East) were written by Russell M. Genet and Lloyd W. Slonaker, Jr. The authors express their thanks to Louis Boyd for permission to use his programs in this book.

The intent in including this software is to provide complete working examples of control software at three levels of complexity. The simplest, the procedures QJOY2, DIGITAL, MOVE, and RAMP provide a complete set of software to move the telescope with a control paddle. Only motions in three speeds in RA and Dec are supported. See Chapter 10 for detailed hardware descriptions.

Intermediate level procedures allow automatic initialization of the telescope and automatic slewing to stored or entered coordinates. These instructions should be of interest to the casual observer of double stars, deep sky objects, etc. To control the telescope to this extent is a fairly complex task, and a dozen procedures are involved as shown in the figure below. Although all of the various procedures (except JOY4) have been previously described, a brief description of how each works is given here to augment the figure.

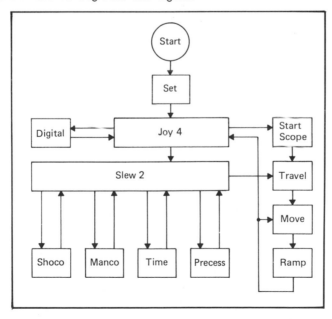

In the system for which this particular software was written the processor board has a built in clock calendar. While quite accurate, on occasion the time needs to be reset using SET (one second of error in time translates to a 15 arc second position error). JOY4 is the main control program for the slew to coordinates control program. To initialize the position of the telescope, STARTSCOPE moves the telescope to the southeast limit switches, an accurately known and calibrated position. SLEW2 is then used to provide choices of preprogramed objects or for manual entry via MANCO and SHOCO of new objects (Epoch 2000.0 coordinates). Coordinates are automatically brought to the current epoch with PRECESS, and TRAVEL, MOVE, and RAMP are used to actually determine the motion to be made and then make it. JOY4 also provides an option for direct manual control via DIGITAL and the control paddle.

The slew to coordinates ensemble has been found useful for two things: First, it can be used to align and check out the system. Second, one needs only to enter the coordinates of an object and the system slews to it and a precise scientific visual evaluation can be made by sorting the objects into such taxinomic classifications as: 1. Wow!, 2. Outstanding!, 3. You've seen one fuzzball you've seen them all, and 4. that faint smudge is supposed to be M xx?

Finally, there is the automatic system that finds and photometrically measures objects on its own. Here MAIN has overall control. This particular version of MAIN is written for a single photometric band ("V" of the Johnson UBV system) and was used by the Fairborn Observatory during the Fall of 1984 (concurrent with the writing of this book). This section of the overall program works nicely, but its creation has been a learning process, and it will be modified as more is learned and our needs evolve. Since it is unlikely that anybody will duplicate our system exactly these listings make an excellent guide but should not be considered our (or your) final product.

The listings which follow are actual printouts of running, debugged systems. Experienced computer programmers will appreciate the fact that we have not attempted to "enhance" the quality of the listings by re-typing. We have also had enough experience with computers and software to feel it our duty to recognize that there are subtle differences from one situation to another that only become apparent when you make the new system try to run. Therefore, study these listings so that you know not only **what** is happening but **how** it is being done.

For ease in location, the listings are alphabetically ordered by procedure name and a summary list is:

1. BUILDFILE	12. MEAS	23. STARFILE
2. COEFFICENTS	13. MOVE	24. STARTSCOPE
3. DATREAD	14. PRECESS	25. STOPSCOPE
4. DIGITAL	15. PTCLK	26. SUNANGLE
5. HELIO	16. QJOY2	27. THRESH
6. HUNT	17. RAMP	28. TILDARK
7. JOY4	18. SET	29. TIME
8. LOCK	19. SHOCO	30. TRANSFORM
9. LUNAR	20. SHOWTIM	31. TRAVEL
10. MAIN	21. SLEW2	32. ZENITH
11. MANCO	22. SOLAR	

Many of the above procedures can be used as listed in a simple manual, slew to coordinates, or automatic control system. In fact, about two-thirds of the procedures were used "as is" on a 6809 processor using Basic09 but with vastly different support hardware than the original system for which they were written.

1. BUILDFILE

```
(* THIS PROGRAM ALLOW MANUAL ENTRY OF NEW STARS INTO
(* THE FILE 'STARFILE' LOCATED ON THE 'CHD' DRIVE
(* THE STARS REMAIN IN THE ORDER ENTERED AS THERE IS NO
(* REQUIREMENT IN THE MAIN PROGRAM FOR A PARTICULAR ORDER
BASE 0
TYPE STARIN=INAM:STRING[10]; IRTA,IDEC,IMAG:REAL
TYPE GROUPIN=NAMEIN:STRING[10]; DIAPHRAGM:INTEGER; ISTAR(4)
 :STARIN
DIM FILE,OUTPATH:INTEGER
FILE=3
DIM STARNAM(5):STRING[5]
STARNAM(0)="CHECK"
STARNAM(1)="  SKY"
STARNAM(2)=" COMP"
STARNAM(3)=" VARI"
DIM STARG:GROUPIN
DIM NUMS:STRING[10]
NUMS="0123456789"
OPEN #FILE,"STARFILE"
LOOP
  PRINT "CURRENT GROUPS ON FILE:"
  MAXGROUP=0
  SEEK #FILE,0
  WHILE NOT(EOF(#FILE)) DO
    GET #FILE,STARG
    IF MOD(MAXGROUP,4)=0 THEN PRINT
    ENDIF
    PRINT USING "I6>,': ',S10<",MAXGROUP,STARG.NAMEIN;
    MAXGROUP=MAXGROUP+1
  ENDWHILE
  PRINT
  PRINT
  GROUP=MAXGROUP+1
  INPUT " QUIT ",A$
EXITIF A$="Y" THEN
ENDEXIT
  INPUT "PRINTOUT OF CURRENT FILES ",X$
  IF X$="Y" THEN
    OUTPATH=1
    SEEK #FILE,0
    CLOSE #1
    OPEN #OUTPATH,"/P1"
    GROUP=0
    WHILE NOT(EOF(#FILE)) DO
      GET #FILE,STARG
      GOSUB 10
      PRINT
      GROUP=GROUP+1
    ENDWHILE
    CLOSE #1
    OPEN #OUTPATH,"/TERM"
  ENDIF
  INPUT "ADD TO LIST (Y/N) ",X$
  IF X$<>"Y" THEN
```

```
REPEAT
  INPUT "NUMBER OF GROUP TO MODIFY ",GROUP
UNTIL GROUP>=0 AND GROUP<=MAXGROUP
SEEK #FILE,GROUP*SIZE(STARG)
GET #FILE,STARG
SEEK #FILE,GROUP*SIZE(STARG)
GOSUB 10 \(* DISPLAY CURRENT RECORD
ENDIF
REPEAT
  PRINT "ENTER COMMON NAME OF STAR GROUP"
  PRINT "  "; NUMS
  IF X$<>"Y" THEN
    PRINT "  "; STARG.NAMEIN
  ENDIF
  INPUT K$
  IF K$<>"" THEN
    STARG.NAMEIN=K$
  ENDIF
  PRINT "DIAPHRAGM (0,1,2) ";
  IF X$<>"Y" THEN
    PRINT "("; STARG.DIAPHRAGM; ")";
  ENDIF
  INPUT K$
  IF K$<>"" THEN
    STARG.DIAPHRAGM=VAL(K$)
  ENDIF
  FOR I=0 TO 3
    PRINT STARNAM(I); " :"
    IF I<>1 THEN PRINT "  STAR NAME [HDNNNNNNNN] OR [SAONNNNNNN]"
      PRINT "  "; NUMS
      IF X$<>"Y" THEN
        PRINT "  "; STARG.ISTAR(I).INAM
      ENDIF
      INPUT K$
      IF K$<>"" THEN
        STARG.ISTAR(I).INAM=K$
      ENDIF
    ELSE
      STARG.ISTAR(I).INAM=""
    ENDIF
    IF I=1 THEN PRINT "(ENTER ZERO'S FOR DEFAULT COMP-VAR CENTERING

    ENDIF
    IF X$<>"Y" THEN
      RTA=STARG.ISTAR(I).IRTA
      DEC=STARG.ISTAR(I).IDEC
    ENDIF
    RUN MANCO(RTA,DEC)
    STARG.ISTAR(I).IRTA=RTA
    STARG.ISTAR(I).IDEC=DEC
    IF I<>1 THEN
      PRINT "V MAGNITUDE = ";
      IF X$<>"Y" THEN
            PRINT "("; STARG.ISTAR(I).IMAG; ")";
          ENDIF
          INPUT K$
          IF K$<>"" THEN
            STARG.ISTAR(I).IMAG=VAL(K$)
          ENDIF
        ELSE
          STARG.ISTAR(I).IMAG=0
```

```
                ENDIF
              NEXT I
              IF STARG.ISTAR(1).IRTA=0 THEN
                STARG.ISTAR(1).IRTA=(STARG.ISTAR(2).IRTA+STARG.ISTAR(
                  3).IRTA)/2
              ENDIF
              IF STARG.ISTAR(1).IDEC=0 THEN
                STARG.ISTAR(1).IDEC=(STARG.ISTAR(2).IDEC+STARG.ISTAR(
                  3).IDEC)/2
              ENDIF
              GOSUB 10
              INPUT "IS THIS OK Y/N ",A$
            UNTIL A$="Y"
            IF X$<>"Y" THEN
              SEEK #FILE,GROUP*SIZE(STARG)
            ENDIF
            PUT #FILE,STARG
          ENDLOOP
          CLOSE #FILE
          END
10        \(* SUBROUTINE TO DISPLAY CURRENT GROUP
          PRINT USING "'CURRENT CONTENTS OF GROUP ',I4<"; GROUP
          PRINT "NAME: "; STARG.NAMEIN;
          PRINT "      DIAPHRAGM = "; STARG.DIAPHRAGM
          PRINT "     NAME      RIGHT ASCEN.  DECLIN.    V-MAG."
          FOR I=0 TO 3
            RTA=12/PI*STARG.ISTAR(I).IRTA
            DEC=180/PI*STARG.ISTAR(I).IDEC
            R1=INT(RTA)
            R2=INT(MOD(RTA*60,60))
            R3=MOD(RTA*3600,60)
            D1=INT(ABS(DEC))
            D2=INT(MOD(ABS(DEC)*60,60))
            D3=INT(MOD(ABS(DEC)*3600,60))
            IF DEC<0 THEN
              DSIGN$="-"
            ELSE
              DSIGN$="+"
            ENDIF
            MAG=STARG.ISTAR(I).IMAG
            PRINT USING "S6<",STARNAM(I);
            PRINT USING "S10<,' '",STARG.ISTAR(I).INAM;
            PRINT USING "I3>,I3>,R5.1>,'    ',S1,I3>,I3>,I3>,R7.2>",R1
              ,R2,R3,DSIGN$,D1,D2,D3,MAG
          NEXT I
          RETURN
```

2. COEFFICENTS

```
0.36  0.0    -0.05
0.47 -0.036 -0.05
0.77 -0.036  0.01
KPRIME(VIS) KDOUBL(VIS) EPSILON(VIS)
KPRIME(BLU) KDOUBL(BLU) EPSILON(BLU)
KPRIME(ULT) KDOUBL(ULT) EPSILON(ULT)
```

3. DATREAD ▆▆▆▆▆▆▆▆

```
(* DATREAD (NO PARAMETERS)
(* PROGRAM TO DISPLAY CONTENTS OF RAW DATA FILES FROM THE AUTOMATED
(* PHOTOMETER.
BASE 0
(* ISTAR(): 0=CHECK , 1=SKY , 2=COMPARISON , 3=VARIABLE
(* COLOR(): 0=ULTRAVIOLET , 1=BLUE , 2=VISUAL
TYPE STARIN=INAM:STRING[10]; IHOUR,COLOR(3),ISECZ:REAL
DATA 0,1,2,3,2,3,2,3,2,1,0
TYPE GROUPIN=NAMEIN:STRING[10]; CORRECT:REAL; ISTAR(11):STARIN
DIM GIN:GROUPIN
DIM FILE,MAXFILE,INFILE:INTEGER \(* FILE NUMBERS
DIM T:REAL
DIM T3,T4,T5:INTEGER
DIM NAME:STRING[10]
DIM INNAME:STRING[16]
DIM STARTYPE:INTEGER
INFILE=4
LOOP
   INPUT "FILE PATH (/D_/MJD_____) OR Q = ",P$
EXITIF P$="Q" OR P$="q" THEN
ENDEXIT
   OPEN #INFILE,P$
   FILE=0
   MAXFILE=0
   WHILE NOT(EOF(#INFILE)) DO
     MAXFILE=MAXFILE+1
     GET #INFILE,GIN
   ENDWHILE
   PRINT MAXFILE; " FILES FOUND"
   LOOP
     INPUT " DISPLAY FORWARD OR BACKWARD (F,B,Q)",A$
     IF FILE<MAXFILE AND A$="F" OR A$="f" THEN FILE=FILE+1
     ENDIF
     IF FILE>1 AND A$="B" OR A$="b" THEN FILE=FILE-1
     ENDIF
   EXITIF A$="Q" OR A$="q" THEN
   ENDEXIT
     SEEK #INFILE,(FILE-1)*SIZE(GIN)
     GET #INFILE,GIN
     PRINT " FILE # "; FILE
     PRINT "GROUP = "; GIN.NAMEIN
     PRINT "HELIOCENTRIC CORRECITON = ";
     PRINT USING "R6.4>",GIN.CORRECT
     PRINT "TYPE         NAME            ULTRA    BLUE  VISUAL    SECZ   HH
                                                                   MM   SS"
     FOR ILINE=0 TO 10
       READ STARTYPE
       IF STARTYPE=0 THEN
         PRINT "CHECK      ";
       ENDIF
       IF STARTYPE=1 THEN
         PRINT "SKY        ";
       ENDIF
       IF STARTYPE=2 THEN
         PRINT "COMP       ";
       ENDIF
       IF STARTYPE=3 THEN
         PRINT "VARIABLE   ";
       ENDIF
```

```
        PRINT USING "S10<",GIN.ISTAR(ILINE).INAM;
        FOR FILTER=2 TO 0 STEP -1
          PRINT USING "R8.2>",GIN.ISTAR(ILINE).COLOR(FILTER);
        NEXT FILTER
        PRINT USING "R8.2>",GIN.ISTAR(ILINE).ISECZ;
        T=GIN.ISTAR(ILINE).IHOUR
        T3=INT(T)
        T=60*(T-T3)
        T4=INT(T)
        T=60*(T-T4)
        T5=INT(T)
        PRINT USING "3(I4>)",T3,T4,T5
      NEXT ILINE
      RESTORE
    ENDLOOP
ENDLOOP
```

4. DIGITAL

```
(* digital (paddle:integer)
(* bit 0 west, bit 1 east
(* bit 2 north,bit 3 south
(* bit 4 s0, bit 5 s1
(* bit 6 end-flag--shared with cloud sense
(* bit 7 not used--rain sence
(* returns integer betseen 1 and 127
PARAM paddle: INTEGER
temp1=PEEK($E011)
temp2=255-temp1
IF temp2>127 THEN
  paddle=temp2-128
ELSE
  paddle=temp2
ENDIF
END
```

5. HELIO

```
(* HELIO(RTA,DEC,MJUL,CORRECTION:REAL)
(* RTA= RIGHT ASCENSION IN RADIANS
(* DEC= DECLINATION IN RADIANS
(* MJUL= MODIFIED JULIAN DATE IN DECIMAL DAYS
(* CORRECTION = HELIOCENTRIC CORRECTION IN DECIMAL DAYS
PARAM RTA,DEC,MJUL,CORRECTION:REAL
DIM T,LOTS,RADIUS,COSDC:REAL
RAD
COSDC:=COS(DEC)
T:=(MJUL-15019.5)*2.73785079E-05
LOTS:=4.88162794+628.331951*T+5.279621E-06*SQ(T)
RADIUS:=.999720458/(1.+.016719*COS(LOTS+4.956105))
CORRECTION:=-.0057755*(RADIUS*(COS(LOTS)*COS(RTA)*COSDC+SIN
  (LOTS)*(.39794248*SIN(DC)+.91741037*SIN(RTA)*COSDC)))

END
```

6. HUNT ▃▃▃▃▃▃▃▃▃▃▃▃▃▃▃▃▃▃▃▃▃▃▃▃▃▃▃▃▃▃▃▃▃▃

```
(* HUNT(RADIUS,LOOPS,DURATION:INTEGER;THRESHOLD:REAL)
(* MODIFIED FOR USE WITH RAMP PROGRAM
(* THIS PROCEDURE WILL MOVE THE TELESCOPE IN A SQUARE SPIRAL CHECKING
(* for a star after move radius of diaphragm in steps
(* IF A STAR IS FOUND WHICH EXCEEDS THRESHOLD THEN THE LOOP IS ENDED.
(* IF NO STAR IS FOUND, THRESHOLD WILL REMAIN UNCHANGED.
BASE 0
PARAM RADIUS,LOOPS,duration:INTEGER
PARAM THRESHOLD:REAL
DIM I,Q,SIDE:INTEGER
TYPE ELEMENTS=LOW,MID,TOP,DSR,DSS,SNEW:BYTE
DIM X,Y:ELEMENTS
DIM COUNT:REAL
(* ENTER VALUES FOR RAMP AND SLEW HERE
X.DSR=200
X.DSS=20
GOSUB 200 \(* CHECK IF STAR ALREADY CENTERED
(* SET LENGTH OF MOVE (<32767 MAX)
X.LOW=MOD(RADIUS,256)
X.MID=RADIUS/256
X.TOP=0
(* BEGIN SQUARE SPIRAL SEARCH
(* LOOP PATTERN WEST,SOUTH,EAST,NORTH
SIDE=0
REPEAT
  SIDE=SIDE+1
  X.SNEW=1
  GOSUB 100
  X.SNEW=8
  GOSUB 100
  SIDE=SIDE+1
  X.SNEW=2
    GOSUB 100
    X.SNEW=4
    GOSUB 100
  UNTIL SIDE=LOOPS+LOOPS
  (* RETURN IN SOUTHEAST DIRECTION TO CENTER IF THRESHOLD NOT EXCEEDED
  X.LOW=MOD(RADIUS/2,256)
  X.MID=RADIUS/512
  X.SNEW=9
  GOSUB 100
  END  \(* END BECAUSE OF MAXIMUM # OF LOOPS
100   (* STEP THROUGH ONE SIDE
      FOR I=1 TO SIDE
        Y=X  \(* TRANSFER REQUIRED BECAUSE RAMP MODIFIES PARAMETERS
        RUN RAMP(Y)
        GOSUB 200
      NEXT I
      RETURN
200   (* TEST FOR STAR IN FIELD
      RUN MEAS(duration,COUNT)
      PRINT COUNT
      IF COUNT>THRESHOLD THEN  \(* IF STAR IS FOUND !
        THRESHOLD=(THRESHOLD+COUNT)/2
        END  \(* EXIT THE PROGRAM WITH THRESHOLD SET TO STAR READING
      ENDIF
      RETURN
      END
```

7. JOY4 ▄▄▄

```
(* joy4(oldrta,olddec:real)
(* entry into joy4 alows the telescope to be moved using the
(* paddle. a cumulative position in right ascension and declination
(* will be returned as real parameters..
(* small revision   by lws  on 9/7/84
DIM oldrta,olddec:REAL
DIM TEMPA,TEMPD:REAL
DIM movex,movey,mult:REAL
DIM paddle,flag:INTEGER
deg_rad=PI/180.
rad_deg=180./PI
hrs_rad=PI/12.
rad_hrs=12./PI
olddec=.0
oldrta=.0
INPUT "would you like to run the telescope to start position (y/n) ?
                                                                    "
  ,a$
IF a$="y" OR a$="Y" THEN
  RUN startscope(oldrta,olddec)
      TEMPA=oldrta
      TEMPD=olddec
    ENDIF
    REPEAT
      RUN shoco(oldrta,olddec)
      INPUT "set coordinates (y/n)",a$
      IF a$="Y" OR a$="y" THEN
        RUN manco(oldrta,olddec)
        TEMPA=oldrta
        TEMPD=olddec
      ENDIF
      LOOP
        POKE $E011,112
1       RUN digital(paddle)
        IF TEMPA<>oldrta OR TEMPD<>olddec THEN
          RUN shoco(oldrta,olddec)
          PRINT CHR$(11);
        ENDIF
        (* initialize to defaults
        movex=.0
        movey=.0
        mult=1
        (* check bit 6 for end flag
        IF paddle=64 THEN
          RUN SLEW2(oldrta,olddec)
          paddle=paddle-64
        ENDIF
        (* check bit 5 for slew speed
        IF paddle>=32 THEN
          mult=1000
          paddle=paddle-32
        ENDIF
        (* check bit 4 for set speed
        IF paddle>=16 THEN
          mult=50
          paddle=paddle-16
        ENDIF
        (* check bit 3 for south
        IF paddle>=8 THEN
```

```
             movey=-1.
             paddle=paddle-8
          ENDIF
          (* check bit 2 for north
          IF paddle>=4 THEN
             movey=1.
             paddle=paddle-4
          ENDIF
          (* check bit 1 for east
          IF paddle>=2 THEN
             movex=-1.
             paddle=paddle-2
          ENDIF
          (* check bit 0 for west
          IF paddle>=1 THEN
             movex=1.
             paddle=paddle-1
           ENDIF
           newrta=oldrta-movex*mult*hrs_rad/3600.
           newdec=olddec+movey*mult*deg_rad/240.
           TEMPA=oldrta
           TEMPD=olddec
           RUN travel(oldrta,olddec,newrta,newdec)
         ENDLOOP
         INPUT "Quit (y/n)",B$
       UNTIL B$="Y" OR B$="y"
       END
```

8. LOCK ▬▬▬▬▬▬▬▬▬▬▬▬▬▬▬▬▬▬▬▬▬▬▬▬▬▬▬

```
(* LOCK(RADIUS:INTEGER;THRESHOLD:REAL)
(* MODIFIED FOR USE WITH RAMP ( )
(* READS FOUR POSITIONS IN A DIAMOND PATTERN AND MAKES LOGICAL MOVE B

(* ON THE RESULT OF THE READINGS.
(* ENTER WITH BRIGHTNESS DECISION THRESHOLD WHICH WILL REMAIN THE SAM

(* IF NO STAR IS FOUND AND WILL BE SET TO THE READING OF THE STAR
(* BRIGHTNESS IF IT IS.
PARAM RADIUS:INTEGER
PARAM THRESHOLD:REAL
DIM SIDE,I,SUM,QRAD:INTEGER
DIM THRESH,COUNT:REAL
DIM TEST:BOOLEAN
TYPE ELEMENTS=LOW,MID,TOP,DSR,DSS,SNEW:BYTE
DIM X:ELEMENTS
(* SET RAMP AND SLEW DELAY
X.DSR=200
X.DSS=20
X.TOP=0 \(* NO MOVE ALLOWED > 32767 STEPS
QRAD=RADIUS/4
(* MAKE DIAMOND TEST : NE>R/2 , S>R , W>R , N>R , SE>R/2
THRESH:=THRESHOLD
SIDE:=QRAD*2 \X.SNEW:=6
100
SUM:=0
GOSUB 260
IF TEST THEN SUM:=SUM+1
ENDIF
SIDE:=QRAD*4 \X.SNEW:=8
```

```
        GOSUB 260
        IF TEST THEN SUM:=SUM+2
        ENDIF
        SIDE:=QRAD*4 \X.SNEW:=1
        GOSUB 260
        IF TEST THEN SUM:=SUM+4
        ENDIF
        SIDE:=QRAD*4 \X.SNEW:=4
        GOSUB 260
        IF TEST THEN SUM:=SUM+8
        ENDIF
        ON SUM+1 GOTO 250,110,120,130,140,240,150,160,170,180,240,190
        ,200,210,220,250
        END
110     SIDE:=QRAD*4 \X.SNEW:=6
        GOTO 230
120     SIDE:=QRAD*4 \X.SNEW:=10
        GOTO 230
130     SIDE:=QRAD*2 \X.SNEW:=2
        GOTO 230
140     SIDE:=QRAD*4 \X.SNEW:=9
        GOTO 230
150     SIDE:=QRAD*2 \X.SNEW:=8
        GOTO 230
160     SIDE:=QRAD \X.SNEW:=10
        GOTO 230
170     SIDE:=QRAD*4 \X.SNEW:=5
        GOTO 230
180     SIDE:=QRAD*2 \X.SNEW:=4
        GOTO 230
190     SIDE:=QRAD \X.SNEW:=6
        GOTO 230
200     SIDE:=QRAD*2 \X.SNEW:=1
        GOTO 230
210     SIDE:=QRAD \X.SNEW:=5
        GOTO 230
220     SIDE:=QRAD \X.SNEW:=9
        GOTO 230
230
        GUSUB 260
240
        SIDE:=QRAD*4 \X.SNEW:=2
        GOTO 100
250     SIDE:=QRAD*2 \X.SNEW:=10
        GOSUB 260
        IF COUNT>THRESHOLD THEN
          THRESHOLD:=COUNT
        ENDIF
        END
260
        X.LOW=MOD(SIDE,256)
        X.MID=INT(SIDE/256)
        RUN RAMP(X)
        FOR I=1 TO 2000 \(* DAMPEN VIBRATION
        NEXT I
        RUN MEAS(2,COUNT)
        IF COUNT>THRESH+THRESH THEN
          THRESH=COUNT*.7
        ENDIF
        IF COUNT>THRESH THEN
          TEST:=TRUE
```

```
ELSE
   TEST:=FALSE
ENDIF
RETURN
END
```

9. LUNAR

```
(* LUNAR (MJUL,RTA,DEC:REAL)
(* ROUGH APPROXIMATION OF R.A. AND DEC. OF THE MOON
(* MJUL= MODIFIED JULIAN DATE (INPUT)
(* RTA= RIGHT ASCENSION OF THE MOON IN RADIANS (OUTPUT)
(* DEC= DECLINATION OF THE MOON IN RADIANS (OUTPUT)
(* DERIVED FROM "CELESTIAL BASIC" (SYBEX)
PARAM MJUL,RTA,DEC:REAL
DIM P2,BD,LM:REAL
P2=6.28318531 \(* TWO PI
BD=MJUL-36934.5 \(* BASE DAY
LM=5.43090264+.229971506*BD \(* LONGITUDE OF THE MOON
RTA=MOD(LM+.109756775*SIN(LM-1.9443666E-03*BD-4.46356263),P2
 )
DEC=.403946*SIN(RTA)+.0898*SIN(RTA-3.11888592+9.24221652E-04
 *BD)
END
```

10. MAIN

```
(* MAIN - Main telescope control subroutine.
(* FOR FOE          10/27/84
(* Reads the list of stars to be measured from a file named STARLIST.
(* Writes the readings to a file named MJD(  ).
(*
(* Stars are always taken in groups of three and are read
(* in the following order : check,sky,comparison,variable,
(* comparison,variable,comparison,variable,
(* comparison,sky,check.
(* The check star should always be an easily locatable star as
(* it is acquired first.
(*
BASE 0
TYPE RISESET=RISE,SET:REAL; STATUS:INTEGER \(* RISE AND SET IN DECIMA
                                                                 L HOURS
TYPE STARIN=INAM:STRING[10]; IRTA,IDEC,IMAG:REAL
(* ISTAR(): 0=CHECK , 1=SKY , 2=COMPARISON , 3=VARIABLE
TYPE GROUPIN=NAMEIN:STRING[10]; PRIORITY:INTEGER; ISTAR(4):
 STARIN
TYPE STAROUT=ONAM:STRING[10]; OHOUR,READING(3),OSECZ:REAL
TYPE GROUPOUT=NAMEOUT:STRING[10]; CORRECT:REAL; OSTAR(11):STAROUT
DATA 0,1,2,3,2,3,2,3,2,1,0 \(* SEQUENCE FOR OUTPUT DATA
DIM TABLE(100):RISESET \(* INGROUP#,(RISETIME,SETTIME)
DIM GIN:GROUPIN
DIM GOUT:GROUPOUT
DIM GTEMP:GROUPOUT
DIM INFILE,OUTFILE:INTEGER \(* FILE NUMBERS
DIM MJUL,LMST,CORRECTION,PHOTONS,THRESHOLD:REAL
DIM RTA,DEC,OLDRTA,OLDDEC:REAL
DIM RTASUN,DECSUN,RTAMOON,DECMOON:REAL
DIM TWILIGHT:REAL
DIM ANGLE,MAXANGLE,COSZMAX:REAL
```

```
DIM TIM(6),TIMER:INTEGER
DIM DEPTH,GROUP,MAXGROUP,LASTGROUP:INTEGER
DIM skipgroup,paddle:INTEGER
DIM RADIUS(3),LOOPS,DURATION:INTEGER \(* FOR PROCEDURE HUNT AND LOCK
DIM OLINE,FILTER:INTEGER \(* FOR-NEXT COUNTERS
DIM VIS,BLU,ULT:INTEGER \(* FILTER NUMBERS
DIM NEWF,OLDF,NEWA,OLDA:INTEGER \(* FOR PROCEDURE APFILT
DIM ABORT:INTEGER
DIM VALCOUNT,VALTYPE,VALFILT,VALSTAR,VALLINE:INTEGER
DIM MEAN,MEANSQ,VALTEMP:REAL
DIM NAME:STRING[10]
DIM OUTNAME:STRING[16]
DIM TYPE$(4),FILT$(3):STRING[6]
DIM CLS:STRING[2]
DIM NIGHT,GROUPABORT:BOOLEAN
(* FIXED CONSTANTS
DEG_RAD=PI/180
RAD_DEG=180/PI
HRS_RAD=PI/12
RAD_HRS=12/PI
MIN_HRS=1/60.
SEC_HRS=1/3600.
VIS=0 \BLU=1 \ULT=2
TYPE$(0)="CHECK "
TYPE$(1)="SKY   "
TYPE$(2)="COMP  "
TYPE$(3)="VARI  "
FILT$(0)="VISUAL"
FILT$(1)="BLUE  "
FILT$(2)="ULTRA "
(* ADJUSTABLE CONSTANTS
(* RADIUS OF DIAPHRAGMS IN HALF STEPS
RADIUS(0)=25
RADIUS(1)=25
RADIUS(2)=25
LATITUDE=39.80075*DEG_RAD
SINLAT=SIN(LATITUDE)
COSLAT=COS(LATITUDE)
TWILIGHT=100
TIMER=100 \(* 10 SECOND INTEGRATION TIME
(* ADJUST PHTONMULT FOR PHOTMETER IN HUNT AND LOCK BAND
PHOTONMULT=2500
(* ADJUST EXTINCTION FOR BAND USED IN HUNT AND LOCK
EXTINCTION=.24
CLS=CHR$($1B)+"K" \(* CLEAR SCREEN
INFILE=3 \(* INPUT FILE TO READ IN GROUP INFORMATION
OUTFILE=4 \(* OUTPUT FILE TO STORE MEASUREMENTS
RUN TIME(TIM,MJUL,LMST)
PRINT "START OF JULIAN DAY ": INT(MJUL)+2400000.5
PRINT "OPENING INPUT FILE AND BUILDING LOOKUP TABLE"
OPEN #INFILE,"/DO/STARFILE":READ
(* LOAD TABLE WITH SIDEREAL TIME WHICH EACH GROUP WILL RISE
(* .    AND SET IN THE OBSERVABLE CONE ABOVE THE OBSERVATORY.
COSZMAX=COS(55*DEG_RAD) \(* SET MAXIMUM ZENITH ANGLE TO 45 DEGREES
MAXANGLE=15*DEG_RAD \(* SET MAXIMUM EAST OR WEST HR ANGLE (MOUNT LIMI

GROUP=0
OPEN #OUTFILE,"/P1":WRITE
WHILE NOT(EOF(#INFILE)) DO
  GET #INFILE,GIN
  RTA=GIN.ISTAR(1).IRTA
```

```
      DEC=GIN.ISTAR(1).IDEC
      ANGLE=ABS(ACS((COSZMAX-SINLAT*SIN(DEC))/(COSLAT*COS(DEC))
        ))
      IF ANGLE>MAXANGLE THEN  \(* LIMITED BY MOUNT ON PHX 10" TELESCOPE
        ANGLE=MAXANGLE
      ENDIF
      TABLE(GROUP).RISE=MOD((RTA-ANGLE)*RAD_HRS+24,24)
      TABLE(GROUP).SET=MOD((RTA+ANGLE)*RAD_HRS,24)
      TABLE(GROUP).STATUS=0
      PRINT USING "S10<,R8.3>,R8.3>",GIN.NAMEIN,TABLE(GROUP).RISE
        ,TABLE(GROUP).SET
      PRINT #OUTFILE,"GROUP # "; GROUP
      PRINT #OUTFILE,"RISE "; TABLE(GROUP).RISE,"SET "; TABLE(GROUP
        ).SET
      IF NOT(EOF(#INFILE)) THEN
        GROUP=GROUP+1
      ENDIF
    ENDWHILE
    CLOSE #OUTFILE
    MAXGROUP=GROUP
    (* OPEN OUTPUT FILE
    ON ERROR GOTO 20
    OUTNAME="/D1/MJD"+LEFT$(STR$(MJUL),5)
    PRINT "OPENING OUTPUT FILE "; OUTNAME
    OPEN #OUTFILE,OUTNAME  \(* TEST TO SEE IF FILE EXISTS (ERROR IF FALSE)
10
    CLOSE #OUTFILE
    GOTO 30

20
    CREATE #OUTFILE,OUTNAME  \(* MAKE NEW FILE IF NONE EXISTS
    GOTO 10
30
    ON ERROR  \(* RESET FOR NORMAL ERROR MESSAGES
    RUN STARTSCOPE(OLDRTA,OLDDEC)
    LASTGROUP=0
    DEPTH=0
    ABORT=0
    LOOP  \(* FOR EACH GROUP MEASURED
      (* CHECK LOOKUP TABLE FOR NEXT STAR TO RUN
      REPEAT  \(* UNTIL OBSERVABLE GROUP FOUND
        RUN TIME(TIM,MJUL,LMST)
        FIRSTSET=12.
        FOR TEMPGROUP=0 TO MAXGROUP
          IF TABLE(TEMPGROUP).STATUS<=DEPTH THEN
            TEMPRISE=LMST-TABLE(TEMPGROUP).RISE
            IF TEMPRISE<0 THEN
              TEMPRISE=TEMPRISE+24
            ENDIF
            TEMPSET=TABLE(TEMPGROUP).SET-LMST
            IF TEMPSET<0 THEN
              TEMPSET=TEMPSET+24
            ENDIF
            IF TEMPRISE<12 AND TEMPSET<12 AND TEMPSET<FIRSTSET THEN

              FIRSTSET=TEMPSET
              GROUP=TEMPGROUP
            ENDIF
          ENDIF
        NEXT TEMPGROUP
        IF FIRSTSET=12 THEN
          DEPTH=DEPTH+1  \(* IF NO UNMEASURED STARS START REPEATING
```

```
ENDIF
RUN SUNANGLE(ANGLE)
(* IF ANGLE>TWILIGHT THEN
NIGHT=TRUE
(* ELSE
(* NIGHT=FALSE
(* ENDIF
WHILE LASTGROUP<>GROUP DO
   SEEK #INFILE,GROUP*SIZE(GIN)
   GET #INFILE,GIN
   LASTGROUP=GROUP
ENDWHILE
RTA=GIN.ISTAR(1).IRTA \(* USE SKY POSITION FOR TEST
DEC=GIN.ISTAR(1).IDEC
RUN LUNAR(MJUL,RTAMOON,DECMOON)
RTAMOON=ABS(RTA-RTAMOON)
IF RTAMOON>PI THEN
   RTAMOON=RTAMOON-(PI+PI)
ENDIF
DECMOON=ABS(DEC-DECMOON)
      IF SQ(RTAMOON)+SQ(DECMOON)<.03 THEN  \(* LIMIT TO WITHIN 10 DEGRE
                                                         ES OF MOON
        PRINT "ATTEMPT TO MOVE NEAR MOON. DELETING GROUP "; GIN.NAMEIN
        TABLE(GROUP).STATUS=TABLE(GROUP).STATUS+1
        FIRSTSET=12.
      ENDIF
   UNTIL ABS(FIRSTSET-12.)>.00000001 OR NOT(NIGHT)
EXITIF NOT(NIGHT) THEN
   PRINT "NOT NIGHT - SOLAR ANGLE = "; ANGLE
ENDEXIT
   (* STAR CALCULATES TO BE OBSERVABLE SO MOVE TO FIRST STAR
   PRINT "MOVING TO GROUP "; TRIM$(GIN.NAMEIN); ", NUMBER = "
    ; GROUP
   PRINT " AT ";
   RUN SHOCO(RTA,DEC)
   RUN TRAVEL(OLDRTA,OLDDEC,RTA,DEC)
   (* FLIP STATEMENT REMOVED HERE AS NOT NEEDED DFM SYSTEM
   DURATION=1
   LOOPS=20 \(* 20 LOOPS FOR FIRST STAR OF A GROUP
   (* NEWF=0 STATMENT REMOVED HERE
   (* NEWA=GIN.DIAPHRAM*90 STATEMENT REMOVED HERE
   (* RUN APFILT(NEWF,OLDF,NEWA,OLDA) REMOVED HERE
   RUN THRESH(COUNT)
   PRINT "SKY BACKGROUND = ";  \ PRINT USING "R8.2<",COUNT
EXITIF COUNT>400 THEN  \(* INDICATES BRIGHT SKY OR CLOUDS
   PRINT "THRESHOLD COUNT > 400"
ENDEXIT
   (* MEASUREMENT SEQUENCE
   RESTORE
   PRINT "TYPE     THRESH    NAME        ULTRA    BLUE   VISUAL    SECZ
                                                   HH   MM   SS"
   FOR OLINE=0 TO 10
     READ STAR
     NAME=GIN.ISTAR(STAR).INAM
     RTA=GIN.ISTAR(STAR).IRTA
     DEC=GIN.ISTAR(STAR).IDEC
     RUN PRECESS(MJUL,RTA,DEC)
     RUN TRAVEL(OLDRTA,OLDDEC,RTA,DEC)
     RUN digital(paddle)
     IF paddle>63 THEN
       RUN options(skipgroup)
     ENDIF
```

```
EXITIF skipgroup=1 THEN
ENDEXIT
  PRINT TYPE$(STAR);
  IF STAR<>1 THEN
    (* IF NOT SKY MEASUREMENT THEN SET THRESHOLD
    (* BASED ON THRESHOLD COUNT AND STAR MAGNITUDE
    RUN ZENITH(RTA,DEC,LMST,COSZ)
    THRESH1=COUNT+PHOTONMULT*2.511886^-(GIN.ISTAR(STAR).IMAG
      +EXTINCTION/COSZ)
    PRINT USING "R8.2>",THRESH1;
    PRINT "   ";
    LOOP
      THRESH2=THRESH1
      RUN HUNT(RADIUS,LOOPS,DURATION,THRESH2)
      DURATION=10
      LOOPS=8 \(* EIGHT LOOPS PER STAR ONCE IN A GROUP
    EXITIF THRESH2=THRESH1 THEN
      PRINT "UNABLE TO FIND STAR ": NAME
      PRINT "ABORTING GROUP ": GIN.NAMEIN
      GROUPABORT=TRUE
      TABLE(GROUP).STATUS=TABLE(GROUP).STATUS+1
    ENDEXIT
      PRINT "HUNT COUNT = ":  \ PRINT USING "R8.2<",THRESH2
      THRESH2=THRESH1
      RUN LOCK(RADIUS.THRESH2)
    EXITIF THRESH2<>THRESH1 THEN
      PRINT "LOCK COUNT = ":  \ PRINT USING "R8.2<",THRESH2
      GROUPABORT=FALSE
      ABORT=0
    ENDEXIT
    ENDLOOP
  ELSE
    PRINT "             ";
  ENDIF
EXITIF GROUPABORT THEN
  ABORT=ABORT+1
  RUN STARTSCOPE(OLDRTA,OLDDEC)
  (* OLDF,OLDA REMOVED FROM STARTSCOPE CALL ABOVE
ENDEXIT
  (* MAKE MEASUREMENTS IN THREE COLORS
  PRINT USING "S10<",NAME;
  FOR FILTER=ULT TO VIS STEP -1
    RUN MEAS(TIMER,PHOTONS)
    GOUT.OSTAR(OLINE).READING(FILTER)=PHOTONS
    PRINT USING "R8.2>",PHOTONS;
  NEXT FILTER
  GOUT.OSTAR(OLINE).ONAM=GIN.ISTAR(STAR).INAM
  RUN TIME(TIM,MJUL,LMST)
  GOUT.OSTAR(OLINE).OHOUR=TIM(3)+TIM(4)*MIN_HRS+TIM(5)*SEC_HRS
  RUN ZENITH(RTA,DEC,LMST,COSZ)
  GOUT.OSTAR(OLINE).OSECZ=1/COSZ
  PRINT USING "R8.2>",1/COSZ;
  PRINT USING "3(I4>)",TIM(3),TIM(4),TIM(5)
  IF OLINE=5 THEN
    GOUT.NAMEOUT=GIN.NAMEIN
    RUN HELIO(RTA,DEC,MJUL,CORRECTION)
    GOUT.CORRECT=CORRECTION
  ENDIF
NEXT OLINE
(* RUN FLIP(1) REMOVED HERE
IF NOT(GROUPABORT) THEN
```

```
PRINT "GROUPS ON RECORD "; OUTNAME
OPEN #OUTFILE,OUTNAME
SEEK #OUTFILE,0
I=0
WHILE NOT(EOF(#OUTFILE)) DO
      GET #OUTFILE,GTEMP
      I=I+1
      PRINT USING "I3>,' ',S10<",I,GTEMP.NAMEOUT;
      IF MOD(I,4)=0 OR EOF(#OUTFILE) THEN
        PRINT
      ENDIF
    ENDWHILE
    PUT #OUTFILE,GOUT
    CLOSE #OUTFILE
    PRINT "GROUP "; GOUT.NAMEOUT; " ADDED TO FILE"
    DEPTH=0
    (* VALIDATE DATA
    MAXDEVI=0
    PRINT "STATISTICS ( RMS/MEAN )"
    PRINT "TYPE     VISUAL   BLUE    ULTRA"
    FOR VALSTAR=0 TO 3
      PRINT TYPE$(VALSTAR);
      FOR VALFILT=VIS TO ULT
        MEAN=0
        MEANSQ=0
        RESTORE
        VALCOUNT=0
        FOR VALLINE=0 TO 10
          READ VALTYPE
          IF VALSTAR=VALTYPE THEN
            VALCOUNT=VALCOUNT+1
            VALTEMP=GOUT.OSTAR(VALLINE).READING(VALFILT)
            MEAN=MEAN+VALTEMP
            MEANSQ=MEANSQ+VALTEMP*VALTEMP
          ENDIF
          NEXT VALLINE
          DEVI=SQRT(MEANSQ/VALCOUNT)/(MEAN/VALCOUNT)
          PRINT USING "R8.5>",DEVI;
          IF DEVI>MAXDEVI AND VALSTAR<>1 THEN
            MAXDEVI=DEVI
          ENDIF
        NEXT VALFILT
        PRINT
      NEXT VALSTAR
      TABLE(GROUP).STATUS=TABLE(GROUP).STATUS+1
      (* TABLE INC. FROM MIDDLE OF IF - ELSE *)
      IF MAXDEVI<1.001 THEN   \(* REMOVE FROM TABLE IF DATA OK
        PRINT " DATA VALID "
      ELSE
        PRINT "DATA NOT VALID. RMS/MEAN = ";
        PRINT USING "R8.5<",MAXDEVI
      ENDIF
  ENDIF
EXITIF ABORT>=4 THEN
  PRINT " FAILED FOUR TIMES TO FIND GROUP, NIGHTS RUN ABORTED"
ENDEXIT
ENDLOOP  \(* END MEASUREMENT LOOP
RUN STOPSCOPE
END
```

11. MANCO

```
(* MANCO(RTA,DEC:REAL)
(* GETS RIGHT ASCENSION AND DECLINATION IN RADIANS FROM
(* FREE FORM KEYBOARD INPUT.   ALLOWS ENTRY IN FORMAT
(* XX.XXX OR XX MM.MMM OR XX MM SS.SSS
PARAM RTA,DEC:REAL
DIM IN$,XX$(3):STRING[20]
DIM XX:REAL
DIM CHAR:STRING[1]
DIM I,N,SIGN:INTEGER
DIM DELIM:BOOLEAN
HRS_RAD=PI/12.
DEG_RAD=PI/180.
PRINT "RIGHT ASCENSION IN HOURS = ";
VALUE=RTA/HRS_RAD
GOSUB 10
RTA=HRS_RAD*VALUE
PRINT "  DECLINATION IN DEGREES = ";
VALUE=DEC/DEG_RAD
GOSUB 10
DEC=DEG_RAD*VALUE
END
10
INPUT IN$
IF IN$<>"" THEN
  FOR I=1 TO 3
    XX=.0
    XX$(I)="0"
  NEXT I
  N=1
  SIGN=1
  DELIM=TRUE
  FOR I=1 TO LEN(IN$)
    CHAR=MID$(IN$,I,1)
    IF CHAR<>" " AND CHAR<>"," AND CHAR<>"-" THEN
      XX$(N)=XX$(N)+CHAR
      DELIM=FALSE
    ELSE
      IF CHAR="-" THEN
        SIGN=-1
      ELSE
        IF NOT(DELIM) THEN
          DELIM=TRUE
          N=N+1
        ENDIF
      ENDIF
    ENDIF
  NEXT I
  VALUE=SIGN*(VAL(XX$(1))+VAL(XX$(2))/60.+VAL(XX$(3))/3600.
    )
ENDIF
RETURN
```

12. MEAS

```
(* MEAS(TIME:INTEGER,COUNT:REAL)
(* TIME IS THE INTEGRATION TIME IN TENTHS OF SECONDS,
(* AND COUNT IS ACTUAL COUNT NORMALIZED TO .1 SECONDS
PARAM TIME:INTEGER
```

```
      PARAM COUNT:REAL
      DIM TEMP:REAL
      DIM PORT,I,J:INTEGER
      PORT:=$E012
      COUNT:=.0
      FOR I:=1 TO TIME
         GOSUB 100
         COUNT:=COUNT+TEMP
      NEXT I
      COUNT:=COUNT/TIME
      END
100
      POKE PORT+1,0
      FOR J:=1 TO 510 \(* TRIM FOR 1/10 SECOND PER READING
      NEXT J
      TEMP:=.0
      FOR J:=1 TO 5
         TEMP:=TEMP*10+LAND(PEEK(PORT+1),$0F)
      NEXT J
      RETURN
```

13. MOVE ▄▄

```
(* MOVE(MOVEX,MOVEY:REAL)
(* MODIFIED FOR USE WITH RAMP ( )
(* DRIVES TELESCOPE BY NUMBER OF STEPS GIVEN BY MOVEX AND MOVEY
(* POSITIVE X IS WEST, POSITIVE Y IS NORTH
PARAM MOVEX,MOVEY:REAL
DIM X,Y,VECTOR:REAL
DIM I:INTEGER
TYPE ELEMENTS=LOW,MID,TOP,DSR,DSS,SNEW:BYTE
DIM K:ELEMENTS
(* SET SLEW AND RAMP SPEEDS
K.DSR=100
K.DSS=12
X:=INT(MOVEX+.5*SGN(MOVEX))
Y:=INT(MOVEY+.5*SGN(MOVEY))
REPEAT
   IF ABS(X)>.0 AND ABS(Y)>.0 THEN
      VECTOR:=(ABS(X)+ABS(Y)-ABS(ABS(X)-ABS(Y)))/2.
   ELSE
      IF ABS(X)>ABS(Y) THEN
         VECTOR:=ABS(X)
      ELSE
         VECTOR:=ABS(Y)
      ENDIF
   ENDIF
   K.SNEW:=0
   IF X>.0 THEN K.SNEW:=LOR(K.SNEW,1)
   ENDIF
   IF X<.0 THEN K.SNEW:=LOR(K.SNEW,2)
   ENDIF
   IF Y>.0 THEN K.SNEW:=LOR(K.SNEW,4)
   ENDIF
   IF Y<.0 THEN K.SNEW:=LOR(K.SNEW,8)
   ENDIF
   K.LOW=INT(MOD(VECTOR,256))
   K.MID=INT(MOD(VECTOR/256,256))
   K.TOP=INT(VECTOR/65536.)
   RUN RAMP(K)
```

```
   X:=X-SGN(X)*VECTOR
   Y:=Y-SGN(Y)*VECTOR
UNTIL ABS(X)=.0 AND ABS(Y)=.0
END
```

14. PRECESS

```
(* PRECESS(MJUL,RTA,DEC:REAL)
(* PRECESSES FROM EPOCH 2000.0 TO PRESENT EPOCH
(* MJUL= MODIFIED JULIAN DATE IN DECIMAL DAYS (INPUT)
(* RTA= RIGHT ASCENSION IN RADIANS (INPUT/OUTPUT)
(* DEC= DECLINATION IN RADIANS (INPUT/OUTPUT)
PARAM MJUL,RTA,DEC:REAL ·
DIM RDIF,DDIF,DAYS:REAL
DAYS:=51544.-MJUL
RDIF:=DAYS*(6.112E-07+2.668E-07*SIN(RTA)*TAN(DEC))
DDIF:=DAYS*2.655E-07*COS(RTA)
RTA=RTA-RDIF
DEC=DEC-DDIF
END
```

15. PTCLK

```
(* PTCLK(TIM(6):INTEGER) - Reads the Peripheral Technology
(* MC146818 clock chip.  Returns an array TIM which contains
(* 0=year,1=month,2=day,3=hour,4=minute,5=second
(* Executes in about .01 second.
BASE 0
PARAM TIM(6):INTEGER
DIM DAT,ADR:INTEGER
DAT=$E01C
ADR=$E01D
REPEAT
  POKE ADR,$00
  TIM(5)=PEEK(DAT)
  POKE ADR,$02
  TIM(4)=PEEK(DAT)
  POKE ADR,$04
  TIM(3)=PEEK(DAT)
  POKE ADR,$07
  TIM(2)=PEEK(DAT)
  POKE ADR,$08
  TIM(1)=PEEK(DAT)
  POKE ADR,$09
  TIM(0)=PEEK(DAT)
  POKE ADR,$00
UNTIL PEEK(DAT)=TIM(5)
IF TIM(0)>80 THEN
  TIM(0)=TIM(0)+1900
ELSE
  TIM(0)=TIM(0)+2000
ENDIF
END
```

16. QJOY2

```
(* procedure qjoy2 (replaces qjoy)
(* qjoy2 (no parameters)
(* main program for manual control
```

```
        (* runs digital, move2
        (* does not display coordinates
        (* west is plus movex
        (* north is plus movey
        DIM movex,movey,mult:REAL
        DIM paddle,flag:INTEGER
        POKE $E011,112
1       RUN digital(paddle)
        (* initialize to defaults
        movex=.0
        movey=.0
        mult=1
        (* check bit 6 for end flag
        IF paddle>=64 THEN
          PRINT "paddle>=64--end"
          END
        ENDIF
        (* check bit 5 for slew speed
        IF paddle>=32 THEN
          mult=10000
          paddle=paddle-32
        ENDIF
        (* check bit 4 for set speed
        IF paddle>=16 THEN
          mult=50
          paddle=paddle-16
        ENDIF
        (* check bit 3 for south
        IF paddle>=8 THEN
          movey=-1.
          paddle=paddle-8
        ENDIF
        (* check bit 2 for north
        IF paddle>=4 THEN
          movey=1.
          paddle=paddle-4
        ENDIF
        (* check bit 1 for east
        IF paddle>=2 THEN
          movex=-1.
          paddle=paddle-2
        ENDIF
        (* check bit 0 for west
        IF paddle>=1 THEN
          movex=1.
          paddle=paddle-1
        ENDIF
        (* calculate steps
        movex=movex*mult
        movey=movey*mult
        RUN move2(movex,movey)
        FOR i=1 TO 100
        NEXT i
        GOTO 1
        END
```

17. RAMP ▬▬▬▬▬

```
        * RAMP(ELEMENT(6):BYTE) - Called from Basic09 by STEPPER module.
        * There must be 6 bytes in the passed parameter which are:
        * ELEMENT(0):BYTE - LOW BYTE OF 24 BIT STEP COUNT
        * ELEMENT(1):BYTE - MID BYTE OF 24 BIT STEP COUNT
        * ELEMENT(2):BYTE - TOP BYTE OF 24 BIT STEP COUNT
        * ELEMENT(3):BYTE - DELAY FOR START OF RAMP
        * ELEMENT(4):BYTE - DELAY FOR SLEWING SPEED
        * ELEMENT(5):BYTE - DEFINES LEADS TO BE PULSED

        * Up to 4 motors may be operated simultaneously.
        * Bits 0 and 1 are for the first motor, 2 and 3 for second motor et
        * Directions to run should have the corresponding bit set to
        * one.  Pulsing both directions on one motor will produce
        * unpredictable results.
        *
        * Limit switches correspond to the output bits.  A limit switch
        * is normally high and must go low on reaching the limit.
        * Program will exit if all active limit switches have been
        * reached.  On a limit switch the limited motor stops immediately.
        * The residual count in ELEMENT 0,1,and 2 will indicate the
        * the remaining count if limit switches are encountered.
        * This can be used to detect a successful move.
        * The output port must consist of up/down counter inputs which
        * will respond to '0' level input.  Writing a '1' to the
        * output port should have no effect.
        * The drivers and limit switches share the write and read functions
        * of the same address, which in this case is set to $E010 which is
        * the low address of SWTPC port #4.
        *
                        NAM     RAMP
                        USE     /D0/DEFS/OS9DEFS

        *
        * OS-9 System Definition File Included
        *


                        opt     1
E010            PORT    EQU     $E010        ADDRESS OF MOTOR DRIVER AND LI
0021            TYPE    SET     SBRTN+OBJCT
0081            REVS    SET     REENT+1
0000 87CD0078          MOD     REND,RNAM,TYPE,REVS,RENT,0
000D 52414DD0  RNAM    FCS     /RAMP/
0011 EE64      RENT    LDU     4,S          GET ADDRESS OF ARRAY OF BYTES
0013 C600              LDB     #0           SET RAMP DELAY TO MAXIMUM
        * BEGINNING OF MAIN LOOP
0015 E143      MLOOP   CMPB    3,U          CHECK IF AT SLEWING SPEED
0017 2701              BEQ     TOPTEST
0019 5C                INCB                 DECREASE RAMP DELAY
001A 6D42      TOPTEST TST     2,U
001C 2610              BNE     DELAY1
001E 6D41              TST     1,U
0020 2612              BNE     DELAY2
0022 6DC4              TST     0,U          LESS THAN 256 STEPS LEFT
0024 274D              BEQ     EXIT         IF NO MORE STEPS LEFT
0026 E1C4              CMPB    0,U          IF RAMPCOUNT > REMAINING STEPS
0028 2310              BLS     RDEL         THEN DON'T BRANCH
002A E6C4              LDB     0,U          AND SET RAMPCOUNT = TO REMAINI
002C 200C              BRA     RDEL
        * EQUALIZATION DELAYS
```

```
002E 21FE       DELAY1   BRN     *         10 CLOCK CYCLES
0030 21FE                BRN     *
0032 12                  NOP
0033 12                  NOP
0034 8D3C       DELAY2   BSR     RETURN    18 CLOCK CYCLES
0036 21FE                BRN     *
0038 21FE                BRN     *
               * VARIABLE DELAY FOR RAMP
003A 1F98       RDEL     TFR     B,A
003C A143       RLOOP    CMPA    3,U
003E 2703                BEQ     SDEL
0040 4C                  INCA
0041 20F9                BRA     RLOOP
               * VARIABLE DELAY FOR SLEW
0043 8600       SDEL     LDA     #0        COUNT UP FOR SLEW DELAY
0045 A144       SLOOP    CMPA    4,U
0047 2703                BEQ     PULSE
0049 4C                  INCA
004A 20F9                BRA     SLOOP
               * GENERATE STEP PULSES
004C A645       PULSE    LDA     5,U
004E 43                  COMA              CONVERT TO NEGATIVE TRUE
W 004F BAE010            ORA     PORT      NO PULSE IF LIMIT SWITCH
0052 81FF                CMPA    #$FF      NO PULSE TO OUTPUT ?
0054 271D                BEQ     EXIT      QUIT IF ALL MOVES BLOCKED
W 0056 B7E010            STA     PORT      PULL SELECTED MOTOR COUNTERS L
0059 86FF                LDA     #$FF
W 005B B7E010            STA     PORT      RESET MOTOR COUNTERS HIGH
               * DECREMENT STEP COUNTER
005E A6C4                LDA     0,U
0060 8001                SUBA    #1
0062 A7C4                STA     0,U
0064 A641                LDA     1,U
0066 8200                SBCA    #0
0068 A741                STA     1,U
006A A642                LDA     2,U
006C 8200                SBCA    #0
006E A742                STA     2,U
0070 20A3                BRA     MLOOP
               * END OF MAINLOOP
0072 39        RETURN    RTS               FOR DELAY3
0073 5F        EXIT      CLRB              NO ERROR CODE
0074 39                  RTS               RETURN TO CALLING PROGRAM
0075 179D02              EMOD
0078          REND       EQU     *
                         END
```

18. SET

```
(* SET (NO PARAMETERS) - Sets the Peripheral Technology PT-69
(* clock chip.
BASE 0
DIM TIM(6):INTEGER
DATA "    YEAR"," MONTH","     DAY","    HOUR"," MINUTE"
DATA $09,$08,$07,$04,$02,$00
DAT=$E01C
ADR=$E01D
RUN PTCLK(TIM)
PRINT
FOR I=0 TO 4
```

```
  READ A$
  PRINT A$: " ";
  PRINT USING "I5>",TIM(I);
  INPUT " ",A$
  IF A$<>"" THEN
    TIM(I)=VAL(A$)
  ENDIF
NEXT I
TIM(5)=$00
TIM(0)=MOD(TIM(0),100)
(* SET DIVIDER RATIOS
POKE ADR,$0A
POKE DAT,$00
(* SET FORMAT REGISTER
POKE ADR,$0B
POKE DAT,$85
(* SET TIME-DATE REGISTERS
FOR I=0 TO 5
  READ REG
  POKE ADR,REG
  POKE DAT,TIM(I)
NEXT I
(* RELEASE SET BIT, 24HR FORMAT, NO DLST
POKE ADR,$0B
POKE DAT,$05
END
```

19. SHOCO

```
    (* SHOCO(RTA,DEC:REAL)
    (* DISPLAYS COORDINATES GIVEN RIGHT ASCENSION AND
    (* DECLINATION IN RADIANS.
    PARAM RTA,DEC:REAL
    DIM TEMP,RAD_SEC,RAD_ASEC:REAL
    DIM XX,MM,SS:INTEGER
    RAD_SEC=43200./PI
    RAD_ASEC=648000./PI
    TEMP=RTA*RAD_SEC+.5
    WHILE TEMP<0 DO
      TEMP=TEMP+86400.
    ENDWHILE
    GOSUB 10
    PRINT USING "'R.A. = ',3(I3>)",XX,MM,SS;

    TEMP=DEC*RAD_ASEC+.5
    IF TEMP>=0 THEN
      PRINT "  DEC. = ";
    ELSE
      PRINT "  DEC. = -";
    ENDIF
    GOSUB 10
    PRINT USING "3(I3>)",XX,MM,SS
    END
10
    TEMP=ABS(TEMP)
    XX=INT(TEMP/3600.)
    TEMP=TEMP-XX*3600.
    MM=INT(TEMP/60.)
    SS=INT(TEMP-MM*60.)
    RETURN
    END
```

20. SHOWTIM

```
(* SHOWTIME - DISPLAYS TIME IN YY/MM/DD HH:MM:SS FORMAT
BASE 0
DIM TIM(6):INTEGER
RUN PTCLK(TIM)
PRINT TIM(0); "/"; TIM(1); "/"; TIM(2); "   ";
PRINT TIM(3); ":"; TIM(4); ":"; TIM(5);
END
```

21. SLEW2

```
(* slew to coordinates     v 0.1
(* BY P. SCOTT HAWTHORN & LLOYD SLONAKER
DIM oldrta,olddec:REAL
DIM TIM(6):INTEGER
DIM MJUL,LMST:REAL
DIM I,J:INTEGER
DIM COUNT:REAL
DIM CHAR:STRING[1]
PARAM rta,dec:REAL
J=1
RUN shoco(rta,dec)

olddec=dec
oldrta=rta
PRINT ""
PRINT "MENU FOR SLEW TO STAR"
PRINT ""
PRINT "1.      b peg"
PRINT "2.      ALTAIR"
PRINT "3.      B CYG"
PRINT "4.      A CYG"
PRINT "5.      SOME OTHER STAR"
PRINT "6.      TAKE READING ON STAR"
PRINT ""
PRINT "ENTER YOUR CHOICE"
GET #0,CHAR
IF CHAR="1" THEN
  rta=23.06289
  dec=55.23667
ENDIF
IF CHAR="2" THEN
  rta=5.19576519
  dec=.154781616
ENDIF
IF CHAR="3" THEN
  rta=5.1083751
  dec=.488086173
ENDIF
IF CHAR="4" THEN
  rta=5.41676024
  dec=.790289933
ENDIF
IF CHAR="5" THEN
  PRINT " ENTER NEW COORDINATES (2000.0 EPOCH) "
  RUN manco(rta,dec)
ENDIF
IF CHAR="6" THEN
  INPUT "NAME OF THE STAR ? ",A$
```

```
      PRINT "OUTPUT GOES TO PRINTER AND SCREEN"
      OPEN #J,"/P1"
      PRINT #J,""
      PRINT #J,"READING FOR STAR--- "; A$
      FOR I=1 TO 3
        RUN MEAS(10,COUNT)
        PRINT "LOOP # "; I; " COUNT = "; COUNT
        PRINT #J,"LOOP # "; I; " COUNT = "; COUNT
      NEXT I
    END
  ENDIF
RUN TIME(TIM,MJUL,LMST)
RUN PRECESS(MJUL,rta,dec)
RUN travel(oldrta,olddec,rta,dec)
RUN shoco(rta,dec)
END
```

22. SOLAR

```
(* SOLAR (MJUL,RTA,DEC:REAL)
(* MJUL= MODIFIED JULIAN DATE (INPUT)
(* RTA= RIGHT ASCENSION OF THE SUN IN RADIANS (OUTPUT)
(* DEC= DECLINATION OF THE SUN IN RADIANS (OUTPUT)
(* CALCULATIONS BASED ON US. NAVAL OBS. ALMANAC FOR COMPUTERS
PARAM MJUL,RTA,DEC:REAL
RAD
DIM HALFPI,TWOPI,T,LOTS:REAL
HALFPI=1.57079633
TWOPI=6.28318531
(* T=JULIAN CENTURIES FROM 0 JANUARY 1900 12H ET.
T=(MJUL-15019.5)*2.73785079E-05
(* MAOS=MEAN ANOMALY OF SUN, TLOS=TRUE LONGITUDE OF SUN
MAOS=6.25658358+628.301946*T-2.61799388E-06*T*T
TLOS=MOD(MAOS+.03344051*SIN(MAOS)+.000349066*SIN(MAOS+MAOS)
 +4.93169,TWOPI)
RTA=ATN(.91746*TAN(TLOS))
IF TLOS>HALFPI THEN RTA=RTA+PI
  IF TLOS>HALFPI+PI THEN RTA=RTA+PI
  ENDIF
ENDIF
DEC=ASN(.39782*SIN(TLOS))
END
```

23. STARFILE

```
CURRENT CONTENTS OF GROUP   0
NAME: LAM AND        DIAPHRAGM = 1
        NAME        RIGHT ASCEN.   DECLIN.     V-MAG.
CHECK HD 222439   23 40 24.4   + 44 20  2    4.14
  SKY             23 41 47.8   + 46 26 21     .00
  COMP HD 223047  23 46  1.9   + 46 25 13    4.95
  VARI HD 222107  23 37 33.7   + 46 27 29    3.83

CURRENT CONTENTS OF GROUP   1
NAME: 39 CET        DIAPHRAGM = 1
        NAME        RIGHT ASCEN.   DECLIN.     V-MAG.
CHECK HD 7476      1 14 49.1   -  0 58 26    5.70
  SKY              1 14  9.8   -  2 22 32     .00
  COMP HD 7147     1 11 43.4   -  2 15  4    5.94
  VARI HD 7672     1 16 36.2   -  2 30  1    5.41
```

```
CURRENT CONTENTS OF GROUP  2
NAME: SIGMA GEM        DIAPHRAGM = 1
          NAME       RIGHT ASCEN.   DECLIN.    V-MAG.
CHECK HD 60522     7 35 55.3   + 26 53 45    4.06
   SKY             7 39 13.7   + 29 55 20     .00
  COMP HD 60318    7 35  8.7   + 30 57 40    5.33
  VARI HD 62044    7 43 18.7   + 28 53  1    4.28
```

24. STARTSCOPE

```
(* STARTSCOPE (OLDRTA.OLDDEC:REAL)
(* THIS PROCEDURE IS USED TO INITIALIZE THE TELESCOPE.
(* IT MAKES SURE THAT THE TELESCOPE IS EXACTLY AT THE HOME
(* POSITION. THEN STARTS THE SIDEREAL CLOCK WITH THE
(* R.A. AND DEC. OF THE HOME POSITION SET FOR THE STARTING
(* LOCAL MEAN SIDEREAL TIME. IT ALSO INSURES THAT THE
(* DIAPHRAM FILTERS ARE INITIALIZED.
PARAM OLDRTA.OLDDEC:REAL
(* DELETED FROM ABOVE STATMENT ": OLDF,OLDA:INTEGER"
DIM TIM(6):INTEGER; LMST,MJUL:REAL
(* DIM NEWF,NEWA:INTEGER DELETED HERE
(* DECLINATION OF HOME POSITION IN RADIANS
HOMEDEC=-.83761236
(* HOUR ANGLE OF HOME POSITION IN RADIANS
HOMERTA=5.4228765
HRS_RAD=PI/12
(* RESET DIAPHRAGM AND FILTER
(* THIS ENTIRE SECTION REMOVED--SEE ORIGINAL FOR OTHER SYSTEMS
(*
(* INITIALIZE $E011 PORT WITH SIDEREAL DRIVE OFF
POKE $E011,80
(*
(* INSURE TELESCOPE IS EXACTLY AT HOME POSITION
RUN MOVE(500.,500.)
RUN MOVE(-1000000.,-1000000.)
RUN MOVE(20.,20.)
RUN MOVE(-50.,-50.)
(* GET R.A. AND DEC. OF HOME POSITION AT THIS INSTANT
RUN TIME(TIM,MJUL,LMST)
OLDDEC=HOMEDEC
OLDRTA=LMST*HRS_RAD-HOMERTA
IF OLDRTA<0 THEN OLDRTA=OLDRTA+PI+PI
ENDIF
RUN SHOCO(OLDRTA.OLDDEC)
(*
(* CHANGES CONTRIOL PORT $E011 TO START SIDEREAL CLOCK
POKE $E011,112
END
```

25. STOPSCOPE

```
(* STOPSCOPE - NO PARAMETERS )
(* TURN OFF SIDEREAL RATE
POKE $E011,80
RUN MOVE(-1000000.,-1000000.) \(* MOVE TO SOUTHEAST LIMIT SWITCHES
END
```

26. SUNANGLE

```
(* SUNANGLE(ANGLE:REAL) - CALCULATES THE ZENITH ANGLE OF THE SUN
(* IN DEGREES
RAD_DEG=57.2957795
PARAM ANGLE:REAL
DIM TIM(6):INTEGER
DIM MJUL,LMST,RTA,DEC,COSZ:REAL
RUN TIME(TIM,MJUL,LMST)
RUN SOLAR(MJUL,RTA,DEC)
RUN ZENITH(RTA,DEC,LMST,COSZ)
ANGLE=RAD_DEG*ACS(COSZ)
END
```

27. THRESH

```
(* THRESH(COUNT:REAL)
(* MODIFIED FOR RAMP
(* MEASURES THE SKY IN FOUR LOCATIONS SURROUNDING THE CURRENT
(* POSITION, RETURNS A COUNT OF LOWEST READING FOUND
PARAM COUNT:REAL
DIM I,SIDE,LSIDE,MSIDE:INTEGER
DIM SPOT:REAL
TYPE ELEMENTS=LOW,MID,TOP,DSR,DSS,SNEW:BYTE
DIM X:ELEMENTS
SIDE=500 \(* LENGTH OF SIDE OF SEARCH PATTERN
LSIDE=MOD(SIDE,256)
MSIDE=INT(SIDE/256)
DATA 4,10,9,5,2
X.TOP=0
X.DSR=125 \(* SET RAMP DELAY
X.DSS=15 \(* set Slew Delay
COUNT=1.0E+30
FOR I=1 TO 5
   X.LOW=LSIDE
   X.MID=MSIDE
   READ X.SNEW
   RUN RAMP(X)
   RUN MEAS(10,SPOT)
   IF SPOT<COUNT THEN
     COUNT=SPOT
   ENDIF
NEXT I
END
```

28. TILDARK

```
(* TILDARK (NO PARAMETERS)
(* WAIT UNTIL DARK AND THEN RUN MAIN PROGRAM
REPEAT
   RUN SUNANGLE(ANGLE)
   ANGLE=INT(ANGLE*100)/100
   WHILE ANGLE<>OLDANGLE DO
     PRINT USING "'SUN ANGLE = ',R6.2>",ANGLE;
     PRINT CHR$(13);
     OLDANGLE=ANGLE
   ENDWHILE
UNTIL ANGLE>100
RUN MAIN
END
```

29. TIME ▬▬▬▬▬▬▬▬▬▬▬▬▬▬▬▬▬▬▬▬▬▬▬▬▬▬▬▬▬▬▬▬

```
(* TIME(MJUL,LMST:REAL)
(* OUTPUT:
(* MJUL= MODIFIED JULIAN DATE (J.D.- 2400000.5) IN DECIMAL DAYS
(* LMST= LOCAL MEAN SIDEREAL TIME IN HOURS
BASE 0
PARAM TIM(6):INTEGER: MJUL,LMST:REAL
DIM MJUL0,FRACT,LONG:REAL
LONG:=5.598174 \(* LOCAL LONGITUDE IN HOURS
RUN PTCLK(TIM)
MJUL0:=-678987.+367.*TIM(0)+INT(INT(275*TIM(1)/9)+TIM(2)-INT
 (7*(TIM(0)+INT((TIM(1)+9)/12))/4))
FRACT=(TIM(3)*3600.+TIM(4)*60.+TIM(5))/86400.
MJUL:=MJUL0+FRACT
LMST:=MOD(6.67170278+.065709823*(MJUL0-33282.)+1.00273791*(
 FRACT*24)-LONG,24)
END
```

30. TRANSFORM ▬▬▬▬▬▬▬▬▬▬▬▬▬▬▬▬▬▬▬▬▬▬▬▬▬▬

```
(* READS DATA FROM A DISK AND INFORMATION ON A PARTICULAR
(* STAR GROUP.  THE DATA IS REDUCED AND STORED ON ANOTHER DISK.
(* THE FILE NAME OF THE OUTPUT WILL BE THE HD NUMBER OF THE VARIABLE
(* CALLS A FILE NAME COEFFICIENTS TO GET CURRENT VALUES FOR
(* TRANSFORMATION AND EXTINCTION.
(* THIS PROGRAM DOES NOT ADJUST THE COEFFICIENTS.
BASE 0
TYPE STARIN=INAM:STRING[10]: IHOUR.READING(3),ISECZ:REAL

TYPE GROUPIN=NAMEIN:STRING[10]; CORRECT:REAL; ISTAR(11):STARIN
DIM GIN:GROUPIN
(* STAROUT(0,1,2,3)= VARI,VARI,VARI,CHECK
TYPE STAROUT=OHH,OMM,OSS:BYTE; DIFFMAG(3):REAL
TYPE GROUPOUT=HELIOJD:STRING[12]; PHASE,OMEAN(3),OSIGMA(3):
 REAL; OSTAR(4):STAROUT
DIM GOUT:GROUPOUT
DIM DATFILE,TEMPFILE,INFILE,OUTFILE:INTEGER \(* FILE NUMBERS
DIM STAR,DIR_COUNT:INTEGER
DIM MJDAY,SECZ,X1,X2,X3,AIRMASS:REAL
DIM T:REAL
DIM NAME(4):STRING[10]
DIM DI$,DO$:STRING[3]
DIM X$:STRING[64]
DIM DATNAME,OUTNAME:STRING[12]
DIM QUARTER:STRING[4]
DIM FILENAME(30):STRING[8]
DIM VIS,BLU,ULT:INTEGER
DIM CHECK,SKY,COMP,VARI:INTEGER
(* INFILE 0 1 2 3 4 5 6 7 8 9 10
(* STAR    0   1 2 3 4 5 6 7   8
(* TYPE    0 1 2 3 2 3 2 3 2 1 0
(* CHECK-COMP = (0+8)/2 - (1+3+5+7)/4
(* VARI-COMP(1) = 2 - (1+3)/2
(* VARI-COMP(2) = 4 - (3+5)/2
(* VARI-COMP(3) = 6 - (5-7)/2
DIM INSTMAG(11,3):REAL
DIM STDMAG(11,3):REAL
DIM KPRIME(3):REAL
DIM KDOUBL(3):REAL
```

```
DIM EPSILON(3):REAL
DIM AVGSKY(3):REAL
DIM VARI_COMP(3,3):REAL
DIM CHECK_COMP(3):REAL
DIM MEAN(3)
DIM SIGMA(3)
DIM FILT$(3):STRING[6]
DIM TYPE$(4):STRING[6]
DIM GROUP_OK:BOOLEAN
DATFILE=3 \(* HEADER DATA PATH
INFILE=4 \(* DATA INPUT PATH
OUTFILE=5 \(* OUTPUT DATA PATH
TEMPFILE=6 \(* MISC FILE PATH
CHCK=0 \SKY=1 \COMP=2 \VARI=3
VIS=0 \BLU=1 \ULT=2
FILT$(0)="VISUAL"
FILT$(1)="BLUE  "
FILT$(2)="ULTRA "
TYPE$(0)="CHECK "
TYPE$(1)="SKY   "
TYPE$(2)="COMP  "
TYPE$(3)="VARI  "
DATA 0,1,2,3,2,3,2,3,2,1,0
(* PROMPT FOR INPUT AND OUTPUT DRIVE
INPUT "Source drive number (0,1,2) ",DI$
INPUT "Destination drive    (0,1,2) ",DO$
DI$="/D"+DI$ \DO$="/D"+DO$
(*
(* GET NAME OF STAR TO BE RECORDED ON THIS PASS
(*
INPUT "Variable common name = ",N$
OUTNAME=""
FOR I=1 TO 10
  A$=MID$(N$,I,1)
  IF A$<>" " AND A$<>"" THEN
    OUTNAME=OUTNAME+A$
  ELSE
    OUTNAME=OUTNAME+"_"
  ENDIF
NEXT I
INPUT "NAME OF QUARTER i.e. 1Q84 ",QUARTER
(*
(* OPEN GENERAL DATA FILE
(*
ON ERROR GOTO 100
OPEN #DATFILE,DO$+"/d_"+OUTNAME:READ
FOR STAR=CHCK TO VARI
  READ #DATFILE,NAME(STAR)
  PRINT TYPE$(STAR),NAME(STAR)
NEXT STAR
PRINT "FILTER","K-PRIME","K-DOUBLE","EPSILON"
FOR FILTER=VIS TO ULT
  READ #DATFILE,KPRIME(FILTER),KDOUBL(FILTER),EPSILON(FILTER
    )
  PRINT FILT$(FILTER),KPRIME(FILTER),KDOUBL(FILTER),EPSILON
    (FILTER)
NEXT FILTER
READ #DATFILE,MJEPOCH,PERIOD
PRINT "EPOCH (MJD) "; MJEPOCH,"PERIOD (DAYS) "; PERIOD
READ #DATFILE,MJDAY
GOTO 110
```

```
100
      CREATE #DATFILE,DO$+"/d_"+OUTNAME:UPDATE
      PRINT "No data file exists on the output disk."
      INPUT "Enter reduction coeff. from /DO/CMDS/COEFFICIENTS (Y/N) "
      ,Q$
      IF Q$="Y" OR Q$="y" THEN
        OPEN #TEMPFILE,"/DO/CMDS/COEFFICIENTS":READ
        FOR FILTER=VIS TO ULT
          READ #TEMPFILE,KPRIME(FILTER),KDOUBL(FILTER),EPSILON(FILTER
          )
        NEXT FILTER
        CLOSE #TEMPFILE
      ELSE
        FOR FILTER=VIS TO ULT
          PRINT FILT$(FILTER);
          INPUT " KPRIME,KDOUBL,EPSILON ",KPRIME(FILTER),KDOUBL(FILTER
          ),EPSILON(FILTER)
        NEXT FILTER
      ENDIF
      INPUT "Epoch in MJD ",MJEPOCH
      INPUT "Period in days ",PERIOD
      FOR STAR=CHCK TO VARI
        NAME(STAR)=""
      NEXT STAR
      MJDAY=.0
110
      ON ERROR
      INPUT "Data OK (Y/N) ",Q$
      IF Q$="N" OR Q$="n" THEN
        CLOSE #DATFILE
        SHELL "DEL "+DO$+"/d_"+OUTNAME
        GOTO 100
      ENDIF
      CLOSE #DATFILE
      ON ERROR GOTO 120
      OPEN #OUTFILE,DO$+"/f_"+OUTNAME+QUARTER:WRITE
      GOTO 130
120
      CREATE #OUTFILE,DO$+"/f_"+OUTNAME+QUARTER:WRITE
130
      ON ERROR
      (*
      (* PROMPT FOR EACH DATA DISK
      (*
      LOOP
        PRINT "Last date reduced = MJD"; MJDAY
        PRINT "Insert data disk into drive "; DI$
        INPUT "Type 'Y' when ready ",Q$
      EXITIF Q$<>"Y" AND Q$<>"y" THEN
      ENDEXIT
        (*
        (* BUILD TABLE OF INPUT DATA FILE NAMES
        (*
        SHELL "DIR E "+DI$+" >"+DO$+"/INDIR"
        OPEN #TEMPFILE,DO$+"/INDIR":READ
        DIR_COUNT=0
        WHILE NOT(EOF(#TEMPFILE)) DO
          READ #TEMPFILE,X$
          IF MID$(X$,49,3)="MJD" THEN
            DIR_COUNT=DIR_COUNT+1
            FILENAME(DIR_COUNT)=MID$(X$,49,8)
```

```
            ENDIF
          ENDWHILE
          CLOSE #TEMPFILE
          SHELL "DEL "+DO$+"/INDIR"
          FOR FCOUNT=1 TO DIR_COUNT
            MJDAY=VAL(MID$(FILENAME(FCOUNT),4,5))
            OPEN #INFILE,DI$+"/"+FILENAME(FCOUNT):READ
            PRINT "FILE "; FILENAME(FCOUNT)
            WHILE NOT(EOF(#INFILE)) DO
              GET #INFILE,GIN
    IF GIN.NAMEIN=N$ THEN
      IF NAME(0)="" THEN
        FOR STAR=CHCK TO VARI
          NAME(STAR)=GIN.ISTAR(STAR).INAM
        NEXT STAR
      ENDIF
      (*
      (* REDUCTION TO INSTRUMENTAL MAGNITUDES
      (*
      (* DETERMINE AVERAGE SKY BRIGHTNESS
      FOR FILTER=VIS TO ULT
        AVGSKY(FILTER)=(GIN.ISTAR(1).READING(FILTER)+GIN.ISTAR
          (9).READING(FILTER))/2
      NEXT FILTER
      FOR STAR=0 TO 10
        IF STAR<>1 AND STAR<>9 THEN
          FOR FILTER=VIS TO ULT
            IF GIN.ISTAR(STAR).READING(FILTER)>AVGSKY(FILTER
              ) THEN
              INSTMAG(STAR,FILTER)=-2.5*LOG10(GIN.ISTAR(STAR
                ).READING(FILTER)-AVGSKY(FILTER))
            ELSE
              INSTMAG(STAR,FILTER)=1000
            ENDIF
          NEXT FILTER
          (* BEMPORADS CORRECTION FOR AIRMASS
          X0=GIN.ISTAR(STAR).ISECZ
          X1=X0-1
          X2=X1*X1
          X3=X2*X1
          AIRMASS=X0-.0018167*X1-.002875*X2-.0008083*X3
          (* STANDARD MAGNITUDE CALCULATION
          BLU_VIS=(INSTMAG(STAR,BLU)-INSTMAG(STAR,VIS))*(
            1-(KDOUBL(BLU)-KDOUBL(VIS))*AIRMASS)-(KPRIME
            (BLU)-KPRIME(VIS))*AIRMASS
          ULT_BLU=(INSTMAG(STAR,ULT)-INSTMAG(STAR,BLU))*(
            1-(KDOUBL(ULT)-KDOUBL(BLU))*AIRMASS)-(KPRIME
            (ULT)-KPRIME(BLU))*AIRMASS
          STDMAG(STAR,VIS)=INSTMAG(STAR,VIS)-KPRIME(VIS)*
            AIRMASS+EPSILON(VIS)*BLU_VIS
          STDMAG(STAR,BLU)=INSTMAG(STAR,BLU)-KPRIME(BLU)*
            AIRMASS-KDOUBL(BLU)*AIRMASS*BLU_VIS+EPSILON
            (BLU)*BLU_VIS
          STDMAG(STAR,ULT)=INSTMAG(STAR,ULT)-KPRIME(ULT)*
            AIRMASS-KDOUBL(ULT)*AIRMASS*ULT_BLU+EPSILON
            (ULT)*ULT_BLU
        ENDIF
      NEXT STAR
      FOR FILTER=VIS TO ULT
        MEAN(FILTER)=0
      NEXT FILTER
```

```
FOR STAR=0 TO 2
  FOR FILTER=VIS TO ULT
    VARI_COMP(STAR,FILTER)=STDMAG(STAR*2+3,FILTER)-
      (STDMAG(STAR*2+2,FILTER)+STDMAG(STAR*2+4,
      FILTER))/2
    MEAN(FILTER)=VARI_COMP(STAR,FILTER)/3+MEAN(FILTER
      )
  NEXT FILTER
NEXT STAR
FOR FILTER=VIS TO ULT
  CHECK_COMP(FILTER)=(STDMAG(0,FILTER)+STDMAG(10,FILTER
    ))/2-(STDMAG(2,FILTER)+STDMAG(4,FILTER)+STDMAG
    (6,FILTER)+STDMAG(8,FILTER))/4
NEXT FILTER
(*
(* CALCULATE MEAN ERROR
(*
GROUP_OK=TRUE
FOR FILTER=VIS TO ULT
  SUMSQ=0
  FOR STAR=0 TO 2
    X_XBAR=VARI_COMP(STAR,FILTER)-MEAN(FILTER)
    SUMSQ=SUMSQ+X_XBAR*X_XBAR
  NEXT STAR
  SIGMA(FILTER)=SQRT(SUMSQ/6)
  IF SIGMA(FILTER)>.02 THEN
    GROUP_OK=FALSE
  ENDIF
NEXT FILTER
IF GROUP_OK THEN (* DONT PRINT BAD DATA
  (*
  (* DEVELOP HELIOCENTRIC CORRECTED JULIAN DATE (TO BIG FOR R
                                                   EAL #)
  (*
  JDFRACT=.5+GIN.ISTAR(5).IHOUR/24+GIN.CORRECT
  JDINT=2400000.+MJDAY+INT(JDFRACT)
  JDFRACT=JDFRACT-INT(JDFRACT)+1 \(* FRACTIONAL PART OF JD +

  GOUT.HELIOJD=MID$(STR$(JDINT),1,8)+MID$(STR$(JDFRACT
    ),3,4)
  PRINT "        "; GOUT.HELIOJD
  (* DETERMINE PHASE
  IF PERIOD>0 THEN
    GOUT.PHASE=MOD((VAL(MID$(GOUT.HELIOJD,3,10))-.5
    -MJEPOCH)/PERIOD,1)
  ELSE
    GOUT.PHASE=.0
  ENDIF
  (* SET UP OUTPUT FOR MEAN AND SIGMA
  FOR FILTER=VIS TO ULT
    GOUT.OMEAN(FILTER)=MEAN(FILTER)
    GOUT.OSIGMA(FILTER)=SIGMA(FILTER)
  NEXT FILTER
  (* FOR VARI-COMP
  FOR STAR=0 TO 2
    FOR FILTER=VIS TO ULT
      GOUT.OSTAR(STAR).DIFFMAG(FILTER)=VARI_COMP(STAR
        ,FILTER)
    NEXT FILTER
    TTIME=GIN.ISTAR(STAR*2+3).IHOUR
    GOUT.OSTAR(STAR).OHH=INT(TTIME)
```

```
                    TTIME=(TTIME-GOUT.OSTAR(STAR).OHH)*60
                    GOUT.OSTAR(STAR).OMM=INT(TTIME)
                    TTIME=(TTIME-GOUT.OSTAR(STAR).OMM)*60
                    GOUT.OSTAR(STAR).OSS=INT(TTIME)
                  NEXT STAR
                  (* FOR CHECK-COMP
                  FOR FILTER=VIS TO ULT
                    GOUT.OSTAR(3).DIFFMAG(FILTER)=CHECK_COMP(FILTER
                      )
                  NEXT FILTER
                  TTIME=(GIN.ISTAR(0).IHOUR+GIN.ISTAR(10).IHOUR)/2
                  GOUT.OSTAR(3).OHH=INT(TTIME)
                  TTIME=(TTIME-GOUT.OSTAR(3).OHH)*60
                  GOUT.OSTAR(3).OMM=INT(TTIME)
                  TTIME=(TTIME-GOUT.OSTAR(3).OMM)*60
                  GOUT.OSTAR(3).OSS=INT(TTIME)
                  PUT #OUTFILE,GOUT
                ENDIF
              ENDIF
            ENDWHILE
            CLOSE #INFILE
         NEXT FCOUNT
ENDLOOP
CLOSE #OUTFILE
DELETE DO$+"/d_"+OUTNAME
CREATE #DATFILE,DO$+"/d_"+OUTNAME:WRITE
FOR STAR=CHCK TO VARI
  WRITE #DATFILE,NAME(STAR)
NEXT STAR
FOR FILTER=VIS TO ULT
  WRITE #DATFILE,KPRIME(FILTER),KDOUBL(FILTER),EPSILON(FILTER
    )
NEXT FILTER
WRITE #DATFILE,MJEPOCH,PERIOD
WRITE #DATFILE,MJDAY
CLOSE #DATFILE
END
```

31. TRAVEL ▬▬▬▬▬▬▬▬▬▬▬▬▬▬▬▬▬▬▬▬▬▬

```
(* TRAVEL(OLDRTA,OLDDEC,NEWRTA,NEWDEC:REAL)
(* CONVERT R.A. AND DEC. TO HALFSTEPS FOR STEPPER
(* DRIVER AND MAKE MOVE.  UPDATE CURRENT POSITION.
PARAM OLDRTA,OLDDEC,NEWRTA,NEWDEC:REAL
DIM MOVEX,MOVEY:REAL
MOVEX=NEWRTA-OLDRTA
IF MOVEX>PI THEN MOVEX=MOVEX-(PI+PI)
ENDIF
IF MOVEX<=-(PI) THEN MOVEX=MOVEX+(PI+PI)
ENDIF
MOVEY=NEWDEC-OLDDEC
IF MOVEY>PI THEN MOVEY=MOVEY-(PI+PI)
ENDIF
IF MOVEY<=-(PI) THEN MOVEY=MOVEY+(PI+PI)
ENDIF
(* REM EXACT NUMBER OF STEPS/RADIAN MAY NOT BE SAME
(* FOR X AND Y AXIS.  NOTE THAT INCREASING R.A. GIVES
(* NEGATIVE STEPS.
(* CONVERSION OF RADIANS TO HALF-STEPS (DEPENDS ON DRIVE RATIO)
```

```
MOVEX=MOVEX*-218139.
MOVEY=MOVEY*218139.
RUN MOVE(MOVEX,MOVEY)
OLDRTA=NEWRTA
OLDDEC=NEWDEC
END
```

32. ZENITH

```
(* ZENITH(RTA,DEC,LMST,COSZ,:REAL)
(* RTA= RIGHT ASCENSION OF OBJECT IN RADIANS (INPUT)
(* DEC= DECLINATION OF OBJECT IN RADIANS (INPUT)
(* LMST= LOCAL MEAN SIDEREAL TIME IN HOURS (INPUT)
(* COSZ= COSINE OF THE ZENITH ANGLE (OUTPUT)
PARAM RTA,DEC,LMST,COSZ:REAL
LATITUDE=39.80075*PI/180
SLAT:=SIN(LATITUDE)
CLAT:=COS(LATITUDE)
COSZ:=SLAT*SIN(DEC)+CLAT*COS(DEC)*COS(.2618*LMST-RTA)
END
```

```
]LIST

10    REM
20    REM    =====================================================
30    REM
40    REM          (S)imple (T)elescope (O)perating (P)rogram
50    REM
60    REM    =====================================================
70    REM
80    REM
90    REM
100   GOSUB 6610: REM         ----> LOAD SCREEN STRINGS & VARIABLES
120   REM
130   REM    =====================================================
140   REM       This is setup module, with lat, long, alt and temp !!!
150   REM    =====================================================
160   REM
180   PRINT  CHR$ (12)
200   PRINT  CHR$ (30) CHR$ (32 + 0) CHR$ (32 + 1)M0$
220   PRINT  CHR$ (30) CHR$ (32 + 0) CHR$ (32 + 23)M0$: REM   SCREEN
240   PRINT  CHR$ (30) CHR$ (32 + 0) CHR$ (32 + 1)M1$
250   PRINT  CHR$ (30) CHR$ (32 + 70) CHR$ (32 + 1)P4$
260   PRINT  CHR$ (30) CHR$ (32 + 0) CHR$ (32 + 2)M0$
280   PRINT  CHR$ (30) CHR$ (32 + 15) CHR$ (32 + 3)L1$: REM    TITLE
282   PRINT  CHR$ (30) CHR$ (32 + 15) CHR$ (32 + 5)BP$;: INPUT AN$
284   IF AN$ = "Y" THEN  PRINT  CHR$ (12): GOTO 1105
300   FOR ZZ = 1 TO 200: NEXT ZZ
320   PRINT  CHR$ (30) CHR$ (32 + 20) CHR$ (32 + 8)L0$
340   FOR ZZ = 1 TO 200: NEXT ZZ
360   PRINT  CHR$ (30) CHR$ (32 + 0) CHR$ (32 + 10)L3$
380   PRINT  CHR$ (30) CHR$ (32 + 9) CHR$ (32 + 10)"";: INPUT LA
400   PRINT  CHR$ (30) CHR$ (32 + 13) CHR$ (32 + 10)L4$
420   PRINT  CHR$ (30) CHR$ (32 + 18) CHR$ (32 + 10)"";: INPUT LB
440   PRINT  CHR$ (30) CHR$ (32 + 22) CHR$ (32 + 10)L5$
460   PRINT  CHR$ (30) CHR$ (32 + 26) CHR$ (32 + 10)"";: INPUT LC
480   PRINT  CHR$ (30) CHR$ (32 + 30) CHR$ (32 + 10)L9$
500   PRINT  CHR$ (30) CHR$ (32 + 40) CHR$ (32 + 10)L6$
520   PRINT  CHR$ (30) CHR$ (32 + 50) CHR$ (32 + 10)"";: INPUT LD
540   PRINT  CHR$ (30) CHR$ (32 + 55) CHR$ (32 + 10)L4$
560   PRINT  CHR$ (30) CHR$ (32 + 60) CHR$ (32 + 10)"";: INPUT LE
580   PRINT  CHR$ (30) CHR$ (32 + 64) CHR$ (32 + 10)L5$
600   PRINT  CHR$ (30) CHR$ (32 + 68) CHR$ (32 + 10)"";: INPUT LF
620   PRINT  CHR$ (30) CHR$ (32 + 72) CHR$ (32 + 10)L9$
630   REM    ------------- WE JUST GOT LAT. AND LONG. ----------------
640   FOR ZZ = 1 TO 200: NEXT ZZ
660   PRINT  CHR$ (30) CHR$ (32 + 28) CHR$ (32 + 12)L2$
680   FOR ZZ = 1 TO 200: NEXT ZZ
700   PRINT  CHR$ (30) CHR$ (32 + 28) CHR$ (32 + 14)L7$
720   PRINT  CHR$ (30) CHR$ (32 + 37) CHR$ (32 + 14)"";: INPUT FT
725   REM         ---------- WE JUST GOT ALTITUDE IN FEET ! ------------
740   PRINT  CHR$ (30) CHR$ (32 + 44) CHR$ (32 + 14)L8$
742   FOR ZZ = 1 TO 200: NEXT ZZ
744   PRINT  CHR$ (30) CHR$ (32 + 20) CHR$ (32 + 17)TP$;: INPUT TP
750   REM    ------------- WE JUST GOT TEMP. IN FAHRENHEIT -----------
760   PRINT  CHR$ (30) CHR$ (32 + 20) CHR$ (32 + 20)LA$
780   PRINT  CHR$ (30) CHR$ (33 + 58) CHR$ (32 + 20)"";: INPUT A$
800   IF A$ = "N" THEN  GOTO 180
```

```
860   FOR ZZ = 1 TO 200: NEXT ZZ
880   L1 = LA + (LB / 60) + (LC / 3600): REM  LOCATION  LAT. IN DECIMAL
900   LO = LD + (LE / 60) + (LF / 3600): REM     "      LON. IN DECIMAL
920   REM
940   REM   ===============================================================
1020  REM                       BEGIN THE  MAIN PROGRAM HERE
1060  REM   ===============================================================
1100  REM
1105  PRINT  CHR$ (12): REM                      FORMFEED
1110  CALL 49152 + (256 * 7): REM                GET THE CLOCK GOIN'
1115  PRINT  CHR$ (26) CHR$ (51)
1120  PRINT  CHR$ (30) CHR$ (32 + 0) CHR$ (32 + 1)M1$
1125  PRINT  CHR$ (26) CHR$ (50)
1130  REM
1135  PRINT  CHR$ (07): PRINT  CHR$ (07): PRINT  CHR$ (07)
1140  PRINT  CHR$ (30) CHR$ (32 + 0) CHR$ (32 + 0)M0$
1145  PRINT  CHR$ (30) CHR$ (32 + 0) CHR$ (32 + 8)M0$
1150  PRINT  CHR$ (30) CHR$ (32 + 0) CHR$ (32 + 19)M0$
1155  PRINT  CHR$ (30) CHR$ (32 + 0) CHR$ (32 + 21)M0$
1160  PRINT  CHR$ (30) CHR$ (32 + 0) CHR$ (32 + 3)ST$
1165  PRINT  CHR$ (30) CHR$ (32 + 25) CHR$ (32 + 3)TI$
1170  PRINT  CHR$ (30) CHR$ (32 + 0) CHR$ (32 + 5)AZ$
1175  PRINT  CHR$ (30) CHR$ (32 + 0) CHR$ (32 + 7)AL$
1180  PRINT  CHR$ (30) CHR$ (32 + 40) CHR$ (32 + 3)HX$
1185  PRINT  CHR$ (30) CHR$ (32 + 40) CHR$ (32 + 5)RF$
1190  PRINT  CHR$ (30) CHR$ (32 + 40) CHR$ (32 + 7)RG$
1195  PRINT  CHR$ (30) CHR$ (32 + 40) CHR$ (32 + 1)FC$
1200  PRINT  CHR$ (30) CHR$ (32 + 64) CHR$ (32 + 1) PDL (0)
1205  FOR ZZ = 1 TO 200: NEXT ZZ
1210  REM
1300  GOSUB 9500
1708  REM   ===============================================================
1710  REM                     INITIAL POSTION DATA INPUT
1714  REM   ===============================================================
1718  REM
1720  M1 = 49360:M2 = 49344
1740  PRINT  CHR$ (30) CHR$ (32 + 6) CHR$ (32 + 11)M5$;: INPUT DD$
1760  DK = 1
1780  IF  LEFT$ (DD$,1) = "-" OR  LEFT$ (DD$,1) = "+" THEN 1820
1800  GOTO 1740
1820  DD =  ABS ( VAL (DD$))
1840  IF DD < 0 OR DD > 90 THEN 1740
1860  IF  LEFT$ (DD$,1) = "-" THEN DK =  - 1
1880  PRINT  CHR$ (30) CHR$ (32 + 6) CHR$ (32 + 12)M6$;: INPUT DM$
1890  IF DM$ = "" THEN 1880
1900  DM =  ABS ( VAL (DM$)): IF DM < 0 OR DM > 59 THEN 1880
1920  PRINT  CHR$ (30) CHR$ (32 + 6) CHR$ (32 + 13)M7$;: INPUT DC$
1930  IF DC$ = "" THEN 1920
1940  DC =  ABS ( VAL (DC$)): IF DC < 0 OR DC > 59 THEN 1920
1960  PRINT  CHR$ (30) CHR$ (32 + 6) CHR$ (32 + 15)N4$;: INPUT RH$
1970  IF RH$ = "" THEN 1960
1980  RH =  ABS ( VAL (RH$)): IF RH < 0 OR RH > 23 THEN 1960
2000  PRINT  CHR$ (30) CHR$ (32 + 6) CHR$ (32 + 16)N5$;: INPUT RM$
2010  IF RM$ = "" THEN 2000
2020  RM =  ABS ( VAL (RM$)): IF RM < 0 OR RM > 59 THEN 2000
2040  PRINT  CHR$ (30) CHR$ (32 + 6) CHR$ (32 + 17)N6$;: INPUT RS$
2050  IF RS$ = "" THEN 2040
2060  RS =  ABS ( VAL (RS$)): IF RS < 0 OR RS > 59 THEN 2040
2080  REM
2100  REM       >>>>>>>>>>>> GOSUB HOUR ANGLE  <<<<<<<<<<<<<<<<<
2120  GOSUB 8060
2140  REM       >>>>>>>>>>>> GOSUB ALTITUDE <<<<<<<<<<<<<<<<<<<<
2160  GOSUB 8500
2180  REM       >>>>>>>>>>>>> GOSUB REFRACTION <<<<<<<<<<<<<<<<<<<<
2185  GOSUB 9085
2190  REM
2222  REM
```

```
2230  REM   ==========================================================
2232  REM                 SMALL COMMAND LINE SWITCHING
2234  REM   ==========================================================
2236  WAIT M1 + 8,128: WAIT M2 + 8,128: REM     MOTOR WAIT
2238  PRINT  CHR$ (30) CHR$ (32 + 1) CHR$ (32 + 20)O1$;: INPUT C$
2240  IF C$ = "I" THEN   PRINT  CHR$ (30) CHR$ (32 + 0) CHR$ (32 + 20)E1$
      : GOTO 940
2244  IF C$ = "E" THEN   GOTO 7660
2248  IF C$ = "C" THEN   PRINT  CHR$ (30) CHR$ (32 + 0) CHR$ (32 + 20)E1$
      :: GOTO 2294
2249  IF C$ = "R" THEN   PRINT  CHR$ (30) CHR$ (32 + 0) CHR$ (32 + 20)E1$
      : GOSUB 9500: GOTO 2238
2250  IF C$ = "T" THEN   PRINT  CHR$ (30) CHR$ (32 + 0) CHR$ (32 + 20)E1$
      :: PRINT  CHR$ (30) CHR$ (32 + 25) CHR$ (32 + 3)TI$;: GOTO 2238
2255  IF C$ = "F" THEN   PRINT  CHR$ (30) CHR$ (32 + 0) CHR$ (32 + 20)E1$

2257  PRINT  CHR$ (30) CHR$ (32 + 64) CHR$ (32 + 1) PDL (0)
2259  GOTO 2238
2260  REM
2288  REM   ==========================================================
2290  REM             BEGIN DATA INPUT OF "GOTO" POSITION
2292  REM   ==========================================================
2293  REM
2294  PRINT  CHR$ (30) CHR$ (32 + 40) CHR$ (32 + 11)M9$;: INPUT ED$
2296  DS = 1
2298  IF  LEFT$ (ED$,1) = "-" OR  LEFT$ (ED$,1) = "+" THEN 2320
2300  GOTO 2294
2320  ED =  ABS ( VAL (ED$))
2340  IF  LEFT$ (ED$,1) = "-" THEN DS =  - 1
2360  PRINT  CHR$ (30) CHR$ (32 + 40) CHR$ (32 + 12)N1$;: INPUT EM$
2370  IF EM$ = "" THEN 2360
2380  EM =  ABS ( VAL (EM$)): IF EM < 0 OR EM > 59 THEN 2360
2400  PRINT  CHR$ (30) CHR$ (32 + 40) CHR$ (32 + 13)N2$;: INPUT ES$
2410  IF ES$ = "" THEN 2400
2420  ES =  ABS ( VAL (ES$)): IF ES < 0 OR ES > 59 THEN 2400
2440  PRINT  CHR$ (30) CHR$ (32 + 40) CHR$ (32 + 15)N7$;: INPUT SH$
2450  IF SH$ = "" THEN 2440
2460  SH =  ABS ( VAL (SH$)): IF SH < 0 OR SH > 23 THEN 2440
2480  PRINT  CHR$ (30) CHR$ (32 + 40) CHR$ (32 + 16)N8$;: INPUT SM$
2490  IF SM$ = "" THEN 2480
2500  SM =  ABS ( VAL (SM$)): IF SM < 0 OR SM > 59 THEN 2480
2520  PRINT  CHR$ (30) CHR$ (32 + 40) CHR$ (32 + 17)N9$;: INPUT SS$
2530  IF SS$ = "" THEN 2520
2540  SS =  ABS ( VAL (SS$)): IF SS < 0 OR SS > 59 THEN 2520
2580  REM
2600  REM   ==========================================================
2640  REM                 MAIN  COMMAND LINE SWITCHING
2680  REM   ==========================================================
2700  REM
2720  PRINT  CHR$ (30) CHR$ (32 + 0) CHR$ (32 + 20)O2$;: INPUT C$
2740  IF C$ = "A" THEN   PRINT  CHR$ (30) CHR$ (32 + 0) CHR$ (32 + 20)E1$
      : GOTO 2200
2760  IF C$ = "E" THEN 7660: REM                THE STOP LINE !!!
2780  IF C$ = "I" THEN   PRINT  CHR$ (30) CHR$ (32 + 0) CHR$ (32 + 20)E1$
      : GOTO 940
2800  IF C$ = "P" THEN   PRINT  CHR$ (30) CHR$ (32 + 0) CHR$ (32 + 20)E1$
      : GOTO 3040: REM    THIS IS THE PRECESS ROUTINE
2820  IF C$ = "R" THEN   PRINT  CHR$ (30) CHR$ (32 + 0) CHR$ (32 + 20)E1$
      :: GOTO 4290: REM     RUN ROUTINE ONLY!
2840  GOTO 2720
2860  REM
2880  REM   ==========================================================
2920  REM                 THE COORDINATE PRECESSER
2960  REM   ==========================================================
2980  REM
3000  REM                             (RIGOROUS)
3020  REM
3040  P = 3.14159265
3060  RD = 180 / P:DR = 1 / RD:RZ = RD * 3600
```

```
3080   PRINT   CHR$ (30) CHR$ (32 + 0) CHR$ (32 + 20)E1$
3100   PRINT   CHR$ (30) CHR$ (32 + 0) CHR$ (32 + 20)Y0$;: INPUT Y0
3120   PRINT   CHR$ (30) CHR$ (32 + 0) CHR$ (32 + 20)E1$
3140 H = SH:M = SM:S = SS
3160 A7 = (SH + SM / 60 + SS / 3600) * 15 * DR
3180 D$ = ED$:M = EM:S = ES
3200 DS = 1: IF  LEFT$ (D$,1) = "-" THEN DS =  - 1
3220 ED =  ABS ( VAL (D$))
3240  FOR ZZ = 1 TO 200: NEXT ZZ
3260 D7 = DS * (ED + EM / 60 + ES / 3600) * DR
3280   PRINT   CHR$ (30) CHR$ (32 + 0) CHR$ (32 + 20)Y$;:  INPUT Y
3300   PRINT   CHR$ (30) CHR$ (32 + 0) CHR$ (32 + 20)E1$
3320   PRINT   CHR$ (30) CHR$ (32 + 0) CHR$ (32 + 20)Y1$
3340  FOR ZZ = 1 TO 200: NEXT ZZ
3360   PRINT   CHR$ (30) CHR$ (32 + 0) CHR$ (32 + 20)E1$
3380 A0 = A7 + (Y - Y0) * M4 * 15 / RZ
3400 D0 = D7 + (Y - Y0) * M5 / RZ
3420 T0 = (Y0 - 1900) / 100
3440 T = (Y - Y0) / 100:T2 = T * T:T3 = T2 * T
3460 Z0 = (2304.25 + 1.396 * T0) * T + 0.302 * T2 + 0.018 * T3
3480 Z = Z0 + 0.791 * T2 + 0.001 * T3
3500 T4 = (2004.682 - 0.853 * T0) * T - 0.426 * T2 - 0.042 * T3
3520 Z0 = Z0 / RZ:Z = Z / RZ:T4 = T4 / RZ
3540 AM =  COS (D0) *  SIN (A0 + Z0)
3560 BM =  COS (T4) *  COS (D0) *  COS (A0 + Z0) -  SIN (T4) *  SIN (D0)

3580 CM =  SIN (T4) *  COS (D0) *  COS (A0 + Z0) +  COS (T4) *  SIN (D0)

3600 AZ =  ATN (AM / BM): IF BM < 0 THEN AZ = AZ + P
3620 D9 =  ATN (CM /  SQR (AM * AM + BM * BM))
3640 A9 = (AZ + Z):A = A9 * RD / 15: REM   R.A. IN HOURS HERE
3660 D = D9 * RD: REM                   DEC. IN DEGREES HERE
3680  IF A >  = 0 AND A < 24 THEN 3760
3700  IF A < 0 THEN A = A + 24
3720  IF A >  = 24 THEN A = A - 24
3740  GOTO 3680
3760 A1 =  INT (A):A2 = (A - A1) * 60:A3 = (A2 -  INT (A2)) * 60:A2 =  INT
      (A2)
3780 A3 =  INT (1000 * A3 + .5) / 1000
3800 S4 =  SGN (D):D =  ABS (D):DS$ = "+": IF S4 < 0 THEN DS$ = "-"
3820 D1 =  INT (D):D1$ =  STR$ (D1)
3840 D1$ = DS$ + D1$
3860 D2 = (D - D1) * 60:D3 = (D2 -  INT (D2)) * 60:D2 =  INT (D2)
3880 D3 =  INT (100 * D3 + .5) / 100
3900  REM
3920 SH = A1:SM = A2:SS =  INT (A3): REM         R.A. NOS. BACK !!!!
3940 ED$ = D1$:ED =  ABS ( VAL (ED$))
3960 EM = D2:ES =  INT (D3): REM                  DEC NOS. BACK !!!!
3980  REM
4000  REM
4120  REM
4270  REM  =======================================================
4275  REM               Initial calculation of position
4280  REM  =======================================================
4285  REM
4290 X =  INT (DD * 1200 + (DM * 1200 / 60) + (DC * 1200 / 3600))
4293  REM         INITIAL DEC POSITION IN MOTOR PULSE
4295 Y =  INT (ED * 1200 + (EM * 1200 / 60) + (ES * 1200 / 3600))
4297  REM        NEW DEC POSITION IN MOTOR PULSES
4300 A =  INT (RH * 20000) + (RM * 333.33) + (RS * 5.5555)
4305  REM        INITIAL RA POSITION IN MOTOR PULSES
4310 B =  INT (SH * 20000) + (SM * 333.33) + (SS * 5.5555)
4315  REM        NEW RA POSITION IN MOTOR PULSES
4320  REM  =======================================================
4360  REM  =                 THE R.A. ROUTINE                   =
4400  REM  =======================================================
4420  REM
4440 M2 = 49344: REM          (THE  R.A.  MOTOR IS SLOT 5)
4460  REM
```

```
4480   REM
4500   IF A = B THEN  GOTO 5140: REM        NO MOVE AT ALL    <------
4520   IF A < B THEN 4640
4540   IF A - B < 160000 THEN 4600
4560   DIR = 1:Q = (320000 - A) + B: REM              EQUATION # 4
4580   GOTO 4800
4600   DIR = 0:Q = A - B: REM                         EQUATION # 2
4620   GOTO 4800
4640   IF B - A > 160000 THEN 4700
4660   DIR = 1:Q = B - A: REM                         EQUATION # 1
4680   GOTO 4800
4700   DIR = 0:Q = (160000 - B) + A: REM              EQUATION # 3
4750   REM
4800   Q =  ABS (Q): IF Q > 65535 THEN  PRINT  CHR$ (30) CHR$ (32 + 10) CHR$
       (32 + 20)M2$: GOTO 4804
4802   GOTO 4818
4804   FOR ZZ = 1 TO 600: NEXT ZZ
4808   PRINT  CHR$ (30) CHR$ (32 + 0) CHR$ (32 + 20)E1$
4810   GOTO 2200: REM        ------- ABHORTING MOVE ---------
4818   B2 =  INT (Q / 256):B1 = Q - 256 * B2
4820   POKE M2,125: POKE M2 + 1,0: REM              STANDARD PULSE INTERVAL
4840   POKE M2 + 2,B1: POKE M2 + 3,B2: REM          LOAD PULSE COUNTS
4860   POKE M2 + 5 + DIR,1: REM                     LOAD THE DIRECTION
4880   POKE M2 + 4,1: REM                           >>>> GO <<<<
4900   IF Q = < 2000 THEN  GOTO 5140
4920   FOR QQ = 1 TO 100 STEP 2
4940   FOR ZZ = 1 TO 5: NEXT ZZ
4960   POKE M2,200 - QQ: NEXT QQ
4980   REM
5000   REM
5020   REM    =================================================================
5060   REM    =                 THE DEC SUBROUTINES                       =
5100   REM    =================================================================
5120   REM
5140   IF  LEFT$ (DD$,1) = "-" THEN NE = 0: REM        DEC INITIAL POSITI
       ON
5160   IF  LEFT$ (DD$,1) = "+" THEN NE = 1: REM        DEC INITIAL POSITI
       ON #2
5180   IF  LEFT$ (ED$,1) = "-" THEN NA = 0: REM        DEC GO TO   POSITI
       ON
5200   IF  LEFT$ (ED$,1) = "+" THEN NA = 1: REM        DEC GO TO   POSITI
       ON#2
5220   REM
5240   IF NE + NA = 2 THEN  GOTO 5320: REM        TYPE 1 MOVE HERE
5260   IF NE + NA = 1 THEN  GOTO 5400: REM        TYPE 2 MOVE HERE
5280   IF NE + NA = 0 THEN  GOTO 5460: REM        TYPE 3 MOVE HERE
5300   REM
5320   IF X > Y THEN DR = 0:Z = X - Y: GOTO 5540
5340   REM        .........SOUTH MOVE
5360   DR = 1:Z = Y - X: GOTO 5540
5380   REM        ..........NORTH MOVE
5400   IF  LEFT$ (DD$,1) = "+" THEN DR = 0:Z = X + Y: GOTO 5540
5420   DR = 1:Z = X + Y: GOTO 5540
5440   REM
5460   IF X < Y THEN DR = 0:Z = Y - X: GOTO 5540
5480   REM        .........NORTH MOVE
5500   DR = 1:Z = X - Y: GOTO 5540
5520   REM        .........SOUTH MOVE
5540   IF Z > 65535 THEN  PRINT  CHR$ (30) CHR$ (32 + 10) CHR$ (32 + 20)M
       4$: GOTO 5542
5541   GOTO 5560
5542   FOR ZZ = 1 TO 600: NEXT ZZ
5546   PRINT  CHR$ (30) CHR$ (32 + 0) CHR$ (32 + 20)E1$
5548   GOTO 2200: REM        ---------- ABHORTING MOVE ------------
5560   Y2 =  INT (Z / 256):Y1 = Z - 256 * Y2
5580   M1 = 49360: REM                          SLOT 5
5600   POKE M1,200: POKE M1 + 1,0: REM          STD INTERVAL
5620   POKE M1 + 2,Y1: POKE M1 + 3,Y2: REM      LOADING THE PULSE COUNT
5640   POKE M1 + 5 + DR,1: REM                  LOAD THE DIRECTION
```

```
5660  POKE M1 + 4,1: REM                        >>>>>  GO  <<<<<
5680  IF Z =  < 2000 THEN  GOTO 5860
5700  FOR QQ = 1 TO 100 STEP 2
5720  FOR ZZ = 1 TO 10: NEXT ZZ: REM          STD RAMP UP TO SPEED
5740  POKE M1,200 - QQ: NEXT QQ
5750  REM
5760  REM                  ======= SCREEN UPDATING TIME ! ========
5780  REM
5800  REM      Blank out old DESTINATION side nos., and then......
5820  REM      put "NEW POSITION" stuff in the  CURRENT POS. instead
5840  REM
5860  PRINT   CHR$ (30) CHR$ (32 + 6) CHR$ (32 + 11)M5$;: PRINT E2$
5880  PRINT   CHR$ (30) CHR$ (32 + 6) CHR$ (32 + 11)M5$;: PRINT ED$
5900  PRINT   CHR$ (30) CHR$ (32 + 6) CHR$ (32 + 12)M6$;: PRINT E2$
5920  PRINT   CHR$ (30) CHR$ (32 + 6) CHR$ (32 + 12)M6$;: PRINT EM
5940  PRINT   CHR$ (30) CHR$ (32 + 6) CHR$ (32 + 13)M7$;: PRINT E2$
5960  PRINT   CHR$ (30) CHR$ (32 + 6) CHR$ (32 + 13)M7$;: PRINT ES
5980  PRINT   CHR$ (30) CHR$ (32 + 6) CHR$ (32 + 15)N4$;: PRINT E2$
6000  PRINT   CHR$ (30) CHR$ (32 + 6) CHR$ (32 + 15)N4$;: PRINT SH
6020  PRINT   CHR$ (30) CHR$ (32 + 6) CHR$ (32 + 16)N5$;: PRINT E2$
6040  PRINT   CHR$ (30) CHR$ (32 + 6) CHR$ (32 + 16)N5$;: PRINT SM
6060  PRINT   CHR$ (30) CHR$ (32 + 6) CHR$ (32 + 17)N6$;: PRINT E2$
6080  PRINT   CHR$ (30) CHR$ (32 + 6) CHR$ (32 + 17)N6$;: PRINT SS
6100  PRINT   CHR$ (30) CHR$ (32 + 40) CHR$ (32 + 11)M9$;: PRINT E2$
6120  PRINT   CHR$ (30) CHR$ (32 + 40) CHR$ (32 + 12)N1$;: PRINT E2$
6140  PRINT   CHR$ (30) CHR$ (32 + 40) CHR$ (32 + 13)N2$;: PRINT E2$
6160  PRINT   CHR$ (30) CHR$ (32 + 40) CHR$ (32 + 12)N3$;: PRINT E2$
6180  PRINT   CHR$ (30) CHR$ (32 + 40) CHR$ (32 + 15)N7$;: PRINT E2$
6200  PRINT   CHR$ (30) CHR$ (32 + 40) CHR$ (32 + 16)N8$;: PRINT E2$
6220  PRINT   CHR$ (30) CHR$ (32 + 40) CHR$ (32 + 17)N9$;: PRINT E2$
6240  DD = ED:DM = EM:DC = ES::RH = SH:RM = SM:RS = SS:X = Y:A = B:DK = D
      S:NE = NA:DD$ = ED$
6260  PRINT   CHR$ (30) CHR$ (32 + 25) CHR$ (32 + 3)TI$
6280  PRINT   CHR$ (30) CHR$ (32 + 64) CHR$ (32 + 1)"       "
6300  PRINT   CHR$ (30) CHR$ (32 + 64) CHR$ (32 + 1) PDL (0)
6320  GOSUB 8060
6340  GOSUB 8500
6350  GOSUB 9085
6360  REM
6380  GOTO 2236: REM            -----> FIRST DATA INPUT POINT <-----
6400  REM                       -----> OF THE "GOTO" POSITION <-----
6420  REM
6440  REM
6460  REM
6500  REM  ==========================================================
6520  REM  =                THE  SCREEN  STRINGS                     =
6540  REM  ==========================================================
6580  REM
6600  REM
6605  RO$ = "Do you want REFRACTION in effect or not <Y or N>"
6610  P4$ = "P4"
6615  RO$ = "Do you want REFRACTION in effect or not <Y or N>"
6620  M0$ = "=================================================================
      ====================="
6640  02$ = "<R>un  <P>recess  <I>nit  <A>bhort  <E>xit  -- YOUR CHOICE -
      -> "
6660  03$ = "
                            "
6680  M1$ = " SIMPLE TELESCOPE OPERATING PROGRAM "
6700  L1$ = "SETUP MODULE -- Please answer the following questions:"
6710  BP$ = "Do you wish to BYPASS this routine < Y or N > "
6720  L0$ = "---)  What is the location of PIV ? I need ....
6740  L2$ = ".. and .. I'll also need your"
6760  LA$ = "Is this  stuff correct or not <Y or N> ?"
6780  L3$ = "LATITUDE    DEG."
6800  L4$ = "DEG.      MIN"
6820  L5$ = "MIN.     SEC"
6840  L9$ = "SEC"
```

```
6860 L6$ = "LONGITUDE        DEG."
6880 L7$ = "ALTITUDE        FEET"
6900 L8$ = "FEET"
6910 TP$ = "What is the current temperature (F.) "
6920 M2$ = "Move is too far in RA for the  counters ... repeat please."
6940 M3$ = "&"
6960 M4$ = "Move is too far in Dec for the  counters ... repeat please."

6980 M5$ = "CURRENT DECLINATION DEGREES "
7000 M6$ = "                    MINUTES "
7020 M7$ = "                    SECONDS "
7040 M9$ = "THE NEW DECLINATION DEGREES "
7060 N1$ = "                    MINUTES "
7080 N2$ = "                    SECONDS "
7100 N4$ = " CURRENT R. ASCENSION HOURS "
7120 N5$ = "                    MINUTES "
7140 N6$ = "                    SECONDS "
7160 N7$ = " THE NEW R. ASCENSION HOURS "
7180 N8$ = "                    MINUTES "
7200 N9$ = "                    SECONDS "
7220 O1$ = "<I>nit <E>xit <C>ontinue <F>ocus <R>efract <T>ime -- WHICH -
     ->"
7240 E1$ = "
                   "
7260 E2$ = "     "
7280 TI$ = "19:00:26"
7300 ST$ = "SIDEREAL TIME IS -----> "
7320 FC$ = "FOCAL POSITION -------> "
7340 EX$ = ">> PROGRAM IS TERMINATING << "
7360 Y0$ = "        WHAT YEAR ( XXXX.X ) WERE THOSE I JUST GOT "
7380 Y$ = "         WHAT YEAR ( XXXX.X ) IS THIS ONE NOW "
7400 Y1$ = "                        .....THANKS BUD......"
7420 AZ$ = "CURRENT AZIMUTH ------> "
7440 AL$ = "CURRENT ALTITUDE -----> "
7460 HX$ = "CURRENT H. ANGLE -----> "
7480 RF$ = "REFR. IN ALPHA -------> "
7500 RG$ = "REFR. IN DELTA -------> "
7520 RETURN
7540 REM  =========================================================
7580 REM                 THE EXIT ROUTINE
7620 REM  =========================================================
7640 REM
7660 POKE 49281 + (16 * 7),47
7680 CALL 49406 + (256 * 7)
7700 PRINT  CHR$ (30) CHR$ (32 + 0) CHR$ (32 + 20)E1$
7720 PRINT  CHR$ (30) CHR$ (32 + 20) CHR$ (32 + 20)EX$
7740 FOR ZZ = 1 TO 300: NEXT ZZ
7760 PRINT  CHR$ (30) CHR$ (32 + 0) CHR$ (32 + 20)E1$
7780 SPEED= 200
7800 PRINT  CHR$ (30) CHR$ (30 + 20) CHR$ (32 + 20)"..........SO, GET B
     USY, YOU GEEK ................!"
7820 FOR ZZ = 1 TO 500: NEXT ZZ
7840 PRINT  CHR$ (07)
7860 REM
7880 SPEED= 255
7900 END
7920 REM  =========================================================
7960 REM                 HOUR ANGLE ROUTINE
8000 REM  =========================================================
8020 REM              HA = ST - RA
8040 REM
8060 X1$ =  RIGHT$ (TI$,2): REM     SIDEREAL TIME, SECONDS
8080 X2$ =  MID$ (TI$,4,2): REM              "       MINUTES
8100 X3$ =  LEFT$ (TI$,2): REM               "       HOURS
8120 X3 =  VAL (X3$):X2 =  VAL (X2$):X1 =  VAL (X1$)
8140 XQ = X1 + (X2 * 60) + (X3 * 3600): REM     S.T. IN SECONDS
8160 XR = RS + (RM * 60) + (PH * 3600): REM   CUR. POS. IN SECONDS
8180 HA = XQ - XR: IF XQ > XR THEN  GOTO 8240
8200  IF HA > 43200 THEN HA = 86400 - HA
```

```
8220   IF XQ < XR THEN   GOTO 8280
8240   IF HA < 43200 THEN HQ$ = " W": GOTO 8320
8260  HA = 86400 - XQ + XR:HQ$ = " E": GOTO 8320
8280   IF   ABS (HA) < 43200 THEN HQ$ = " E": GOTO 8320
8300   IF   ABS (HA) > 43200 THEN HA = 86400 - XR + XQ:HQ$ = " W"
8320  HA =   ABS (HA):H1 =   INT (HA / 3600):H2 =   INT ((HA - H1 * 3600) /
         60):H3 = HA - (3600 * H1 + 60 * H2)
8340  H1$ =   STR$ (H1):H2$ =   STR$ (H2):H3$ =   STR$ (H3)
8360   IF   LEN (H2$) = 1 THEN H2$ = "0" + H2$
8380   IF   LEN (H3$) = 1 THEN H3$ = "0" + H3$
8400   REM
8420   REM                                 RECONSTRUCT STRING
8440  HA$ = H1$ + ":" + H2$ + ":" +   LEFT$ (H3$,4) + " " + HQ$
8460   PRINT   CHR$ (30) CHR$ (32 + 64) CHR$ (32 + 3)HA$
8480   RETURN
8500   REM   ======================================================
8540   REM                          ALT / AZ CONVERSION
8580   REM   ======================================================
8600  P = 3.141596
8620  L2 = 57.295775131: REM                  DEGREES TO RADIANS
8640  R1 = RH + (RM / 60) + (RS / 3600): REM  DECIMAL RIGHT ASC.
8660  F1 = DD + (DM / 60) + (DS / 3600): REM   DECIMAL DECLINATION
8680  F2 = (XQ - XR) / 3600 * 15
8700  L = L1 / L2
8720  H = F2 / L2
8740  D = F1 * DK / L2
8760  A = ( SIN (D) *  SIN (L)) + ( COS (D) *  COS (L) *  COS (H))
8780  AA =  ATN (A /  SQR ( - A * A + 1))
8800  AQ = AA * L2: REM  -----> AQ IS ALTITUDE IN DEGREES
8820  AZ = ( SIN (D) - ( SIN (L) *  SIN (AA))) / ( COS (L) *  COS (AA))
8840   IF AZ ^ 2 = 1 THEN 8900
8860  AF =  -  ATN (AZ /  SQR ( - AZ * AZ + 1)) + 1.5708
8880   GOTO 8920
8900  AF = 0
8920  AR = AF * L2
8940   IF   SIN (H) > 0 THEN AR = 360 - AR
8960   PRINT   CHR$ (30) CHR$ (32 + 25) CHR$ (32 + 7)"                         "
8980   PRINT   CHR$ (30) CHR$ (32 + 25) CHR$ (32 + 7)AQ
9000   PRINT   CHR$ (30) CHR$ (32 + 25) CHR$ (32 + 5)"                         "
9020   PRINT   CHR$ (30) CHR$ (32 + 25) CHR$ (32 + 5)AR
9040   REM
9050   RETURN
9060   REM   ======================================================
9065   REM                          REFRACTION ROUTINE
9070   REM   ======================================================
9080   REM
9085   IF RN$ = "N" THEN   PRINT   CHR$ (30) CHR$ (32 + 64) CHR$ (32 + 5)"T
       URNED OFF NOW!": PRINT   CHR$ (30) CHR$ (32 + 64) CHR$ (32 + 7)"TURN
       ED OFF NOW!": RETURN
9100   REM   ------> NOTE THAT CZ (THE REFRACTION COEFF.) IS IN RADIANS
9120   REM
9140  CZ = .142159382 / (460 + TP) *   EXP ( - FT / 27171)
9160  ZD = P / 2 - AA
9180  CE = ( SIN (L) -  SIN (D) *  COS (ZD)) / ( COS (D) *  SIN (ZD))
9200  SE =   SQR (1 - CE * CE)
9220   IF HQ$ = " E" THEN SE =  - SE
9226   IF ZD * L2 > 85 THEN 9234
9230  R = CZ * ( TAN ((90 - AQ) / L2) - 8.4622625E - 04 *   TAN ((90 - AQ)
        / L2) ^ 3)
9232   GOTO 9240
9234  R = CZ * (14.9873066) * ( EXP ( - 29.91341939 * AQ) +   EXP ( - 5.98
       2683878 * AQ))
9240  ZA = ZD - R
9260  OD = R * CE: REM     THIS QTY IS IN RADIANS
9280  OF = R * SE /   COS (D + OD): REM    THIS QTY IN RADIANS
9290   PRINT   CHR$ (30) CHR$ (32 + 64) CHR$ (32 + 5)"                      "
9300   PRINT   CHR$ (30) CHR$ (32 + 64) CHR$ (32 + 5)OF * L2 * 60
9310   PRINT   CHR$ (30) CHR$ (32 + 64) CHR$ (32 + 7)"                      "
9320   PRINT   CHR$ (30) CHR$ (32 + 64) CHR$ (32 + 7)OD * L2 * 60
```

```
9340   RETURN
9360   REM   ============================================================
9380   REM                        REFRACTION SWITCH ROUTINE
9400   REM   ============================================================
9450   REM
9500   PRINT   CHR$ (30) CHR$ (32 + 10) CHR$ (32 + 20)RO$;: INPUT RN$
9505   IF RN$ = "Y" OR RN$ = "N" THEN 9520
9510   PRINT   CHR$ (30) CHR$ (32 + 1) CHR$ (32 + 20)E1$: GOTO 9500
9520   PRINT   CHR$ (30) CHR$ (32 + 1) CHR$ (32 + 20)E1$
9530   RETURN
9600   REM   ============================================================
```

Appendix N—CALIBRATING ENCODERS USING KALMAN FILTERING

A. Weighted Least Squares

This appendix describes a method of obtaining small gains in pointing accuracy for large expenditures of effort. These gains are made by observing a large number of stars all over the sky, then using modern mathematical techniques to estimate the values of several error constants all at once. Only those with advanced mathematical skills and the need for very high pointing accuracy should continue reading. The rest of us are better advised to use special observations to measure each error source independently of the others, as discussed in Chapter 7.

To obtain estimates of the mechanical error constants identified in Chapter 14, a large number of stars can be used in conjunction with a digital filter designed to operate when there is an overabundance of data. There should be at least 5-10 times as many data points as there are parameters to be estimated. To provide a means of converting raw encoder readings to useful coordinates, stars of known position are aligned on the optical axis of the telescope, and the corresponding raw encoder readings are recorded. When high accuracy pointing is a requirement, the goal is to extract the most information possible about the error constants, that is, to obtain optimal estimates of the values of the constants.

One method of obtaining an optimal estimate of the error constants is to use a weighted least squares filter. Gelb (1974, p.23) states the well-known theory of weighted least squares in the form used below. The notation is derived from Dunham (1983), and uses the convention that small letters denote differences between vectors, while names in capital letters denote the main vectors themselves. Another excellent source for digital filtering techniques is Wertz (1978).

To perform a weighted least squares estimate, the following information is needed:

1. The initial condition of the system. The system is described by the state vector X, whose m elements are the parameters being estimated and which uniquely determine the system. For the WMO alt-az mount, the azimuth parameters that are to be estimated are the following: A1 (zero offset), a and b (the azimuth tilt angles), p (the angle between the axes - 90°), and C_{ew} (the east-west collimation error). The zenith distance parameters are the following: Z1 (zero offset), a, b, p, C_{ns} (the north-south collimation error), and F(90°) (the flexure constant). The filter requires initial rough guesses of these values.

2. A set of n measurements Y_i of one or more variables related to the system parameters by a set of observation equations. The measurements are the raw encoder readings from the azimuth and zenith distance axes. The observation equations are used to produce a matrix of observations of the state parameters $G(X_i)$. This generalization permits the estimation of parameters which are not observed directly.

3. An estimate of the errors in the measurements, given in the matrix R. The inverse of this matrix, R^{-1}, is the matrix of weights used to weight the effect of each measurement on the estimate that is made of the system parameters. It is assumed that the measurements are subject to Gaussian noise statistics.

Assuming there exists a set of "true" values of the state vector $[X]$, the filter is designed to estimate a correction vector x given by

$$x = [X] - X$$

which, when added to the original guess X, gives a "best estimate" of $[X]$.

To use this method, first a set of differences $y_i = Y_i - G(X_i)$ is defined, which contains the observed values (O) minus the computed values (C), using the initial estimate X. In a linear system, there exists a matrix H of the observation partial derivatives

$$H_i = \frac{dG}{dX_i}$$

(where d/dX_i is the partial derivative with respect to X_i) such that $y_i = H_i x_i$. The H_i are evaluated using the initial estimate of the state of the system X. Next, the covariance matrix

$$P = (H^T R^{-1} H)^{-1}$$

and the estimate of the correction vector

$$x = P H^T R^{-1} y$$

are computed. The best estimate of the state of the system is then

$$X' = X + x$$

This filter works on measurements taken together as a batch. The filter produces the same numerical estimates, regardless of how the measurements are ordered in the measurements vector Y (which is usually the order in which the measurements were taken). If the initial errors are unknown, the filter can be run once with R = I, the unit matrix, then run again with a new initial state using X (run 2) = X' (run 1) and using the O-C's as the elements of R.

To apply this approach to the problem of estimating the error constants, the following corrections to azimuth and zenith distance are defined, using the equations given above:

A1 = azimuth encoder zero offset constant
A2 = azimuth tilt correction to azimuth

$$= \arccos \left\{ \frac{\sin Z''}{\sin Z} (\cos A'' \cos b - \sin A'' \sin b) \right\} - A''$$

where sin Z = sin [arccos (sin Z" cos A" a sin b + sin Z" sin A" a cos b + cos Z")]
and (A",Z") are the coordinates in the system that is tilted through angle **a** and
rotated through angle **b**

A3 = non-perpendicular axes error
= p cot Z
A4 = collimation error
= C_{ew} csc Z
Z1 = zenith distance encoder zero offset constant
Z2 = azimuth tilt correction to zenith distance
= arccos (sin Z" cos A" a sin b + sin Z" sin A" a cos b + cos Z") - Z"
Z3 = non-perpendicular axes error
= p ΔA
Z4 = collimation error
= C_{ns}
Z5 = tube flexure
= F(90°) sin Z

The servo lag error is a constant multiplied by the motor error command step,
which is not directly a function of raw encoder readings. To keep the filter linear, it
is ignored, and the servo time constant is obtained by direct measurement using a
step function input.

The observation equations relating the observed quantities (raw encoder
readings) to the computed quantities (the (A,Z) coordinates of the telescope) are as
follows:

$$A = K_a E_a + A1 + A2 + A3 + A4$$

$$Z = K_z E_z + Z1 + Z2 + Z3 + Z4 + Z5$$

where A is the azimuth coordinate and Z is the zenith distance coordinate (obtained
to high accuracy from a catalog, then reduced to apparent place) of the star that is
centered in the field of the telescope, E_a is the raw azimuth encoder reading, and
E_z is the raw zenith distance encoder reading. The conversion factors K_a or K_z (in
arc seconds per encoder count) are also constants, which are determined by
autocollimating off the telescope axes and counting the encoder ticks while each
axis is swung through precisely 360° back into autocollimation. The Y_i are the E_a
and E_z.

The O-C equations are of the form

$$y_a = K_a E_a + A1 + A2 + A3 + A4 - A$$

$$y_z = K_z E_z + Z1 + Z2 + Z3 + Z4 + Z5 - Z$$

The state parameters for the two axes overlap, in that **a, b,** and p are related to
both axes, so that measurements from both encoders are coupled to each other
through these state parameters. Therefore, the O-C's are placed in a 2 x n matrix y,

and the partials matrix H contains the partial derivatives of the observation equations for both axes with respect to the eight state parameters A1, Z1, a, b, p, C_{ew}, C_{ns}, and F(90°). Much of this H matrix consists of zeroes, which act as "place holders". The other matrices are as described above.

This method, although it provides optimal estimates for the correction constants, has three main drawbacks:

1. It is computationally inefficient, since matrix inversions need to be performed. As the number m of state parameters grows, the size of the covariance matrix P to be inverted grows by m^2, which increases the number of operations needed to perform the inversion by a factor of m^4.

2. The measurements must all be taken ahead of time, then processed as a batch. There is no way to perform the calculations while pointing the telescope for the next error constant measurement. Thus there is no way to monitor the progression of the O-C matrix y as an increasing number of processed measurements diminishes the effects of the Gaussian measurement noise. Later, during observing operations, there is no way to take advantage of setting or guiding inputs made by the telescope operator to adjust the correction constants in real-time to improve pointing accuracy over the course of the first few hours of an observing run.

3. There is no computationally efficient way to extract more information from the observations, such as obtaining better estimates of the observer's longitude and latitude, since the resulting constraint equations are non-linear.

B. Kalman Filter

The Kalman, or sequential, filter is mathematically identical to the weighted least squares filter, in that the same inputs produce a numerically identical optimal estimate (Gelb, 1974, p.105). The Kalman filter is computationally more efficient, because it does not require the inversion of a matrix. Instead, the same basic information in the least squares filter has been re-arranged so that only a scalar is inverted.

The nature of the sequential filter algorithm requires initial best guesses of the state X, the error in the state x, and the covariance matrix P. Rather than treating all observations at once as a batch, the sequential filter processes each observation as it is received. To illustrate the procedure as an iterative process, the initial guesses will be treated as the outputs of the i-1th step, and are labelled X_{i-1}, x_{i-1}, and P'_{i-1}.

When an observation Y_i is received by the sequential filter, the O-C

$$y_i = Y_i - G(X_{i-1})$$

and the partials matrix

$$H_i = \frac{dG_i}{dX_{i-1}}$$

are computed as before. Next, the Kalman gain

$$K_i = P_i H_i^T [H_i P_i H_i^T + R_i]^{-1}$$

is computed. R_i is the variance in the observation at the ith step. Since the quantity to be inverted in the brackets is a scalar, no matrix inversions need be done. The covariance matrix is computed using K_i and the old covariance matrix P_i

$$P'_i = [I - K_i H_i] P_i$$

where I is the identity matrix, then the new state correction is computed

$$x'_i = x_{i-1} + K_i (y_i - H_i x_{i-1})$$

The process is repeated using the most recent values of P' and x' for the new P and x, respectively, for each observation until all observations have been processed. Then the estimated state is found X' = x' + X from the initial estimate of the state and the most recent state correction.

The Kalman filter has several advantages over the weighted least squares filter.

1. The results may be monitored while the measurements are being taken, since the filter is applied to each observation as it becomes available, not after all measurements have been completed.

2. It is computationally more efficient, since no matrices need to be inverted.

3. Using linearizing techniques (e.g., Gelb, 1974, Chapter 6), estimates of other parameters previously taken as known quantities can be made, including longitude, latitude, sidereal clock error, and the servo time constant. Although these parameters could be estimated using similar linearizing techniques in the weighted least squares filter, the resulting increase in the number of elements in the covariance matrix would have caused its inversion to take much longer.

The Kalman filter has two drawbacks. The first is that if the initial estimates of the state parameters and their errors are themselves too much in error, rather than converging on the true values, the filter will diverge to a very bad estimate of the state. This means that relatively good estimates of the error constants must be made initially. This can be done most easily by making special calibrating observations to measure the constants first, then the Kalman filter can be used to refine these values.

The second problem is that if the estimate of the state deviation gets very good, the covariance matrix elements become very small and the resulting Kalman gain K becomes small, so that further measurements are ignored. This can be remedied by adding a process noise matrix Q to the covariance matrix P before computing the Kalman gain K. This represents an attempt to account for unmodelled errors in the system, and keeps the filter open enough to allow new measurements to affect the estimate of the state parameters. Note that if a process noise matrix is used, the numerical result is no longer identical to that given by the weighted least squares filter.

One may not always have a very good initial estimate of the state parameters, the measurement errors, or the state deviation, and one may have no idea what to use for the process noise values. Kalman filters require a good deal more tuning to implement than the weighted least squares filter, but often the extra effort is rewarded by the advantages of the Kalman filter approach.

C. Extended Kalman Filter

The extended Kalman filter is quite similar to the Kalman filter, with the difference being that the state vector is updated with every observation, rather than at the end, after all observations have been processed. This allows use of the estimated state quantities in real-time. For example, in a portable telescope system for which the observer's longitude and latitude may not be known to high accuracy, one could include them in the state vector and use the pointing residuals to obtain better estimates of the observer's location, which, in turn, give better pointing accuracy the next time a motor speed command is computed.

Again, initial best guesses of the state X and the covariance matrix P are needed, but since no state error matrix x is used in this filter, it is not needed. For the ith step of the filter, the initial X and P are treated as X_{i-1} and P'_{i-1}.

As before, when an observation Y_i is received by the extended sequential filter, the O-C

$$y_i = Y_i - G (X_{i-1})$$

and the partials matrix

$$H_i = \frac{dG_i}{dPX_{i-1}}$$

are computed. Also, as before, any process noise Q is added to the covariance matrix

$$P_i = P'_{i-1} + Q$$

and the Kalman gain

$$K_i = P_i H_i^T [H_i P_i H_i^T + R_i]^{-1}$$

is computed. Again, R is the variance in the observation. The covariance matrix is updated,

$$P'_i = [I - K_i H_i] P_i$$

and now the state vector itself is updated (rather than the state error vector),

$$X_i = X_{i-1} + K_i y_i$$

The process is repeated for the next observation using X_i and P'_i.

For a dynamic system, such as an artificial satellite (e.g., Landsat 4), which contains an on-board computer to estimate its location by observing navigation satellites (e.g., Navstar GPS), the state vector contains Landsat's X, Y, and Z position vectors, and the observables are the arrival times and transmittal times of the GPS transmissions, and the GPS satellite positions. The GPS positions and message transmit times are part of the GPS message received by the GPS user. In this system, the state itself is dynamic, and would have to be "propagated" to the next observation time by integrating the equations of motion between the last and the next observation, and updating the X and P arrays to the new integrated position before computing the y_i in the first step. We have eliminated this complication by estimating the error coefficients of the telescope, which are static, rather than the telescope position itself (in A,Z), which is dynamic. This allows one to monitor the error coefficients during an evening to see how they behave at different telescope pointing angles.

The extended Kalman filter suffers from the same drawbacks as the Kalman filter, namely, the filter can diverge if a good initial value of the state vector is not known, and tuning the filter parameters, such as the process noise Q, is difficult. However, the extended Kalman filter offers the same advantage as the Kalman filter of providing an optimal estimate of the state parameters with a minimum of computation overhead.

One advantage of the extended filter is the accessability of the state parameters in real-time, as mentioned earlier. Whenever the telescope is commanded to point to a new location, the hand paddle inputs to do final centering or to re-center the star during tracking can be used as observation inputs into the extended filter. The result will be better estimates of the error coefficients, so that fewer and fewer hand paddle corrections will be needed as the evening progresses.

In developing a computerized project, the Kalman filter method of calibrating encoders should receive no attention until the system is functioning well. After that stage is reached, a Kalman filter can be installed to improve the estimates of the error coefficients.

GLOSSARY OF TERMS

A	azimuth
AAT	Anglo-Australian Telescope
α	right ascension
AC	alternating current
A/D	analog to digital
alt-az	altitude-azimuth telescope mount
APT	automatic photoelectric telescope
bus	a digital signal pathway consisting of several parallel lines used to move data between different circuit boards which are connected to the pathway
CCW	counter-clockwise
chip	integrated circuit
CMOS	a technology for manufacturing integrated circuits using Complementary Metal Oxide and doped Silicon to achieve very low power drains and high noise immunity
CPU	Central Processing Unit (of a computer)
CW	clockwise
D	days
δ	declination
D/A	digital to analog
DC	direct current
DEC	registered trademark of the Digital Equipment Corporation
Dec	declination
DMA	direct memory access
D/S	digital to synchro
E	eccentric anomaly of the Earth in its orbit
e	eccentricity of the Earth's orbit
ε	the mean obliquity of the ecliptic
ECL	a standard family of very high speed logic devices using a non-saturating Emitter Coupled Logic technology
EMI	electromagnetic interference
EPROM	Eraseable (usually with ultraviolet light) Programmable Read Only Memory
GPS	Global Positioning System
h	hour angle
I/O	input/output
IR	infra-red
KB	kilobytes (1024 bytes)
Kb	kilobits (1000 bits)

Kbaud   kilobits per second along a single communications line
L_m geometric mean longitude of the Sun
LRC   inductance-resistance-capacitance
lvdt   linear variable differential transformer
L_t true longitude of the Sun
M   months, or mean anomaly of the Sun
m   minutes of time
'   minutes of arc
MB   megabytes (millions of bytes)
MOS   a technology for manufacturing integrated circuits using Metal Oxide
           and doped Silicon to achieve very high component densities and
           relatively moderate power drains (used to make memories and
           CPUs)
mV   millivolts (thousanths of a volt)
nV   nanovolts (billionths of a volt)
PC   personal computer
pc   printed circuit
pixel   picture element (a.k.a. pel or point)
PMT   photomultiplier tube
pot   potentiometer
PROM   programmable read only memory
PWM   pulse width modulation, in which information is conveyed by varying the
           width of each pulse in a constant stream of pulses
RA   right ascension
RAM   random access memory (the main read/write memory used by the CPU)
RFI   radio frequency interference
RMS   root mean square
ROM   read only memory
RSS   root sum square
s, S   seconds of time
"   seconds of arc
S/D   synchro to digital
STD bus   a standard bus developed by Pro-Log and Mostek
TTL   a standard family of logic devices using a bipolar Transistor Transistor
           Logic technology
μV   microvolts (millionths of a volt)
WMO   Winer Mobile Observatory

BIBLIOGRAPHY

Abdel-Gawad, M. K., **Analysis for Redesign of 150-inch Stellar Telescope Serrurier Truss Structure**, Kitt Peak National Observatory Engineering Department Technical Report No. 9. (Tucson:1969)

Abt, H.A., 1980, "The Cost-Effectiveness in Terms of Publications and Citations of Various Optical Telescopes at the Kitt Peak National Observatory", **PASP, 547**, 249.

A'Hearn, M.F., 1984a, private communication.

A'Hearn, M.F., 1984b, "When Not to Automate", talk delivered to the "Microcomputers in Astronomy II" Symposium, Fairborn, Ohio, July, 1984.

Almanac for Computers, published by the Nautical Almanac Office, United States Naval Observatory. (Washington:1979)

Arfken, G., **Mathematical Methods for Physicists,** Academic Press, Inc. (New York:1970)

Basili, V., 1980, private communication.

Beale, J.S., Gietzen, J. W., and Read, P. D., 1975, "Computer Control of the INT", in Huguenin and McCord (1975), 352.

Beaumont, G., and Wolfe, J., 1975, "Canada-France-Hawaii Telescope Computer and Control Systems", in Huguenin and McCord (1975), 275.

Berry, R.L., 1983, "RE MEM BER THE YUMAN FAC TOR", in Genet (1983), 9.

Booch, G., **Software Engineering with Ada,** Benjamin/Cummings Publishing Company. (Menlo Park:1983)

Bothwell, G.W., 1975, "Development of the Computer System for the 3.9 Metre Anglo-Australian Telescope", in Huguenin and McCord (1975), 310.

Bourlon, P., and Vin, A., 1975, "Automation of a 40cm Telescope", in Huguenin and McCord (1975), 68.

Boyce, P., 1975, in Huguenin and McCord (1975), 39.

Boyd, L.J. and Genet, R.M. in Wolpert and Genet (1983).

Boyd, L.J., Genet, R.M., and Hall, D.S., 1984a, **IAPPP Communications, 15,** 20.

Boyd, L.J., Genet, R.M., and Hall, D.S., 1984b, in Wolpert and Genet (1984).

Boyd, L.J., Genet, R.M., and Hall, D.S., 1984c, in Genet and Genet (1984).

Burgess, **Celestial Basic,** Sybex. (Berkeley:1982)

Burke, B. F., 1975, "Computer Control of Aperture Synthesis Interferometers", in Huguenin and McCord (1975), 378.

Colgate, S., 1975, in Huguenin and McCord (1975), 88.

Cox, R.E., and Sinnott, R. W., 1977, "Gleanings for ATM's", **Sky and Telescope, 54,** 4, 330.

Dijkstra, E. W., **A Discipline of Programming,** Prentice-Hall, Inc. (New Jersey:1976)

Dunham, D.W., 1984, private communication.

Dunham, J.B., 1983, private communication.

Eisele, J.A. and Shannon, P. E. V., "Atmospheric Refraction Corrections for Optical Sightings of Astronomical Objects", NRL Memorandum Report 3058, Naval Research Laboratory. (Washington:1975)

Electronic Products, April, 1982.

Fridenberg, J.T., Westphal, J. A., and Kristian, J., 1975,   "Microprocessors: A New Alternative for Automatic Telescope Control", in Huguenin and McCord (1975), 218.

Garfinkel, B., 1967, "Astronomical Refraction in a Polytropic Atmosphere", **AJ, 72,** 235.

Gelb, A., **Applied Optimal Estimation,** The M.I.T. Press. (Cambridge:1974)

Genet, R.M., 1979, **Telescope Making,5,** 28.

Genet, R.M., 1980, **IAPPP Communications,3,** 12.

Genet, R.M., Sauer, D.J., Kissell, K.E., and Roberts, G.C., 1982a, **IAPPP Communications,9,** 43.

Genet, R.M., **Real-Time Control With the TRS-80,** H.W. Sams, Inc. (Indianapolis:1982b).

Genet, R.M., ed., **Microcomputers in Astronomy,** Fairborn Observatory. (Fairborn:1983)

Genet, R.M., Boyd, L.J., and Sauer, D.J., 1984, "Interfacing for Real-Time Control--An Astronomical Example", **Byte,** April, 1984.

Genet, R.M. and Genet, K.A., **Microcomputers in Astronomy II,** Fairborn Observatory. (Fairborn:1984)

Ghedini, S., **Software for Photoelectric Astronomy,** Willmann-Bell, Inc. (Richmond:1982)

Giovane, F., Wood, F.B., Oliver, J. P., and Chen, K.-Y., 1983, "Automatic South Pole Telescope", in Genet (1983), 86.

Goldberg, L., 1983, private communication.

Goldberg, **Automatic Controls: Principles of Systems Dynamics,** Allyn and Bacon, Inc. (Boston:1964)

Gurnette, B.L. and Woolley, R.v.d.R., **Explanatory Supplement to The Astronomical Ephemeris and The American Ephemeris and Nautical Almanac,** Her Majesty's Stationery Office. (London:1961)

Hall, D.S., and Genet, R.M., **Photoelectric Photometry of Variable Stars,** International Amateur-Professional Photoelectric Photometry. (Fairborn, 1982)

Halliday, D. and Resnick, R., **Physics,** John Wiley and Sons, Inc. (New York:1966)

Harwood, J.V., 1975, "Institute for Astronomy 88-inch Telescope", in Huguenin and McCord (1975), 245.

Hill, S.J., 1975, in Huguenin and McCord (1975), 60.

Hirschfeld, A., and Sinnott, R.W., **Sky Catalog 2000.0,** Sky Publishing Corporation. (Cambridge, 1982)

Honeycutt, R.K., Kephart, J.E., and Hendon, A.A., 1977, **Sky and Telescope,56,** 495.

Huguenin, M.K. and McCord, T.B., 1975, Eds., **Telescope Automation: Proceedings of a Conference Held 29,30 April, 1 May, 1975 at Massachusetts Institute of Technology, Cambridge, Massachusetts and Sponsored by the National Science Foundation.** (Copies are available from the Remote Sensing Laboratory, Bldg. 24, Room 422, Massachusetts Institute of Technology, 77 Massachusetts Avenue, Cambridge, MA 02139.)

Ingalls, R., 1975, in Huguenin and McCord (1975), 24.

Jaworski, A., 1983, private communication.

Jelly, J.V., 1980, **Q. Jl. R. Astr. Soc., 21,** 14-31.

Kaplan, G.H., ed., **The IAU Resolution on Astronomical Constants, Time Scales, and the Fundamental Reference Frame,** United States Naval Observatory Circular No. 163, December 10, 1981.

Kibrick, R., 1984, in Genet and Genet (1984).

Klim, K. and Ziebell, M., 1985, "Coude Auxiliary Telescope: A low cost telescope at European Southern Observatory", **Optical Engineering, 24,** 2, 363.

Linnell, A.P., and Hill, S.J., 1975, "The MSU Computer Assisted Photometric System", in Huguenin and McCord (1975), 60.

Mansfield, A.W., 1984, private communication.

McCord, T.B., Paavola, S.H., and Snellen, G.H., 1975, "The MIT Automated Optical Telescope", in Huguenin and McCord (1975), 160.

McNall, J.F., Miedaner, T.L., and Code, A.D., 1968, **AJ,73,** 756.

Meeks, M. L., 1975, in Huguenin and McCord (1975), 23.

Meeus, J., **Astronomical Formulae for Calculators,** Second Edition, Willmann-Bell, Inc. (Richmond:1982)

Melsheimer, F., 1983, private communication.

Melsheimer, F., 1984, private communication.

Miller, R. W., **Servomechanisms: Devices and Fundamentals,** Reston Publishing Co., Inc. (Virginia:1977)

Moore, C.H., Rather, E.D., and Conklin, E.K., 1975, "Modular Software for On-Line Telescope Control", in Huguenin and McCord (1975), 1.

Moore, E., Merillat, P., Colgate, S., and Carlson, R., "Software for a Digitally Controlled Telescope", in Huguenin and McCord (1975), 95.

Opal, C.B., 1980, "Rotary Dial Mechanism for Digitally Tuned Receivers", **Ham Radio,** July, 14.

Orlov, B.A., **Refraction Tables of Pulkovo Observatory.** Fourth edition (Pulkovo:1956)

Otis, M.G., 1983, **Sky and Telescope, 65,** 551.

Paavola, S.H., 1975, in Huguenin and McCord (1975), 160.

Paffrath, L., 1975, "Data Acquisition and Telescope Control Systems at Kitt Peak National Observatory", in Huguenin and McCord (1975), 392.

Racine, R., 1975, "Rampant Conservatism in Telescope Controls", in Huguenin and McCord (1975), 34.

Radick, R.R., Hartman, L., Mihalas, D., Worden, S. P., Africano, J. L., Klimke, A., and Tyson, E. T., 1982, **PASP, 94,** 934.

Rafert, J.B., and Cone, G., 1983, "Stepper Motor Controlled Telescope", in Genet (1983), 50.

Rather, E., 1975, in Huguenin and McCord (1975), 390.

Rodin, B., 1982, private communication.

Schnurr, R.G., and A'Hearn, M.F., 1983, "Apple-Based Occultation Photometer", in Genet (1983), 130.

Seidelmann, P.K., and Kaplan, G.H., 1982, **Sky and Telescope,** 64, 409.

Skillman, D.R., 1981, **Sky and Telescope,** 61, 71.

Smart, W.M., **Textbook on Spherical Astronomy,** (Sixth Edition, reprinted), Cambridge University Press. (Cambridge:1979)

Spaulding, C.P., **How to Use Shaft Encoders,** DATEX Corporation. (Monrovia:1965)

Stephenson, T.P., 1975, "The Multiple Mirror Telescope Mount Control System", in Huguenin and McCord (1975), 365.

Stoll, M., and Jenkner, H., 1983, "Vienna 150-cm Telescope Control", in Genet (1983), 33.

Tausworthe, R.C., **Standardized Development of Computer Software,**   Prentice-Hall, Inc. (New Jersey:1977)

Taylor, J.H., 1975, "Automated Pulsar Observations at the Arecibo Observatory and the Five College Radio Astronomy Observatory", in Huguenin and McCord (1975), 28.

Titus, J., Titus, C., and Larson, J., **The STD Bus,** H.W. Sams, Inc. (Indianapolis:1982)

Tomer, A.J., and Bernstein, R., 1983, "Computer-Operated Portable Telescope", in Genet (1983), 74.

Tomer, A.J., 1984, "Towards Developing a Digital Telescope", manuscript circulated privately.

Van der Lans, J., and Lorenson, S., 1975, "The ESO Telescope Control Systems", in Huguenin and McCord (1975), 253.

Wallace, P.T., 1975, "Programming the Control Computer of the Anglo-Australian 3.9 Metre Telescope", in Huguenin and McCord (1975), 284.

Wellnitz, D., 1983, private communication.

Wertz, J.R., ed., **Spacecraft Attitude Determination and Control,** D. Reidel Publishing Company. (Dordrecht:1978)

White, N., 1984, private communication.

Wolpert, R.C. and Genet, R.M., eds., **Advances in Photoelectric Photometry Volume I,** Fairborn Observatory. (Fairborn, 1983)

Wolpert, R.C. and Genet, R.M., eds., **Advances in Photoelectric Photometry Volume II,** Fairborn Observatory. (Fairborn, 1984)

Woolard, E.W. and Clemence, G.M., **Spherical Astronomy,** Academic Press. (New York:1966)

Worden, S. P., Schneeberger, T.J., Kuhn, J.R., and Africano, J.L., 1981,   **Ap. J., 244,** 520.

INDEX